ROWAN UNIVERSITY
CAMPBELL LIBRARY
201 MULLICA HILL RD.
GLASSBORO, NJ 08028-1701

Design of Biomedical Devices and Systems

Second Edition

Design of Biomedical Devices and Systems

Second Edition

Paul H. King
Richard C. Fries

CRC Press is an imprint of the
Taylor & Francis Group, an **informa** business

CRC Press
Taylor & Francis Group
6000 Broken Sound Parkway NW, Suite 300
Boca Raton, FL 33487-2742

© 2009 by Taylor & Francis Group, LLC
CRC Press is an imprint of Taylor & Francis Group, an Informa business

No claim to original U.S. Government works
Printed in the United States of America on acid-free paper
10 9 8 7 6 5 4 3 2 1

International Standard Book Number-13: 978-1-4200-6179-6 (Hardcover)

This book contains information obtained from authentic and highly regarded sources. Reasonable efforts have been made to publish reliable data and information, but the author and publisher cannot assume responsibility for the validity of all materials or the consequences of their use. The authors and publishers have attempted to trace the copyright holders of all material reproduced in this publication and apologize to copyright holders if permission to publish in this form has not been obtained. If any copyright material has not been acknowledged please write and let us know so we may rectify in any future reprint.

Except as permitted under U.S. Copyright Law, no part of this book may be reprinted, reproduced, transmitted, or utilized in any form by any electronic, mechanical, or other means, now known or hereafter invented, including photocopying, microfilming, and recording, or in any information storage or retrieval system, without written permission from the publishers.

For permission to photocopy or use material electronically from this work, please access www.copyright.com (http://www.copyright.com/) or contact the Copyright Clearance Center, Inc. (CCC), 222 Rosewood Drive, Danvers, MA 01923, 978-750-8400. CCC is a not-for-profit organization that provides licenses and registration for a variety of users. For organizations that have been granted a photocopy license by the CCC, a separate system of payment has been arranged.

Trademark Notice: Product or corporate names may be trademarks or registered trademarks, and are used only for identification and explanation without intent to infringe.

Library of Congress Cataloging-in-Publication Data

King, Paul H., 1941-
   Design of biomedical devices and systems / Paul King and Richard C. Fries. -- 2nd ed.
      p. ; cm.
  Includes bibliographical references and index.
  ISBN 978-1-4200-6179-6 (alk. paper)
   1. Biomedical engineering. 2. Medical instruments and apparatus--Design and construction. 3.
Biomedical materials. I. Fries, Richard C. II. Title.
   [DNLM: 1. Biomedical Engineering--standards. 2. Biocompatible Materials. 3. Biotechnology. 4.
Equipment Design--standards. 5. Software Design. QT 36 K54d 009]

  R856.K53 2009
  610.284--dc22
                                                       2008027124

Visit the Taylor & Francis Web site at
http://www.taylorandfrancis.com

and the CRC Press Web site at
http://www.crcpress.com

3 3001 00981 6918

Dedication

To Sue

my wife and best friend

Paul H. King

To my wife June

whose friendship, support, and love

make me whole

Richard C. Fries

Contents

Preface

The design and functional complexity of medical devices and systems have increased during the past 50 years, evolving from the use of a metronome circuit for the initial cardiac pacemaker to functions that include medical bookkeeping, electrocardiogram analysis, delivery of anesthesia, laser surgery, magnetic resonance imaging, and intravenous delivery systems that adjust dosages based on patient feedback. As device functionality becomes more intricate, concerns arise regarding efficacy, safety, and reliability. Both the user and the patient want the device to operate as specified, perform in a safe manner, and continue to perform over a long period without failure. To be successful, the designer of medical devices must ensure that all devices meet these requirements.

Medical device design is a complex process that requires careful integration of diverse disciplines, technical activities, standards, regulatory requirements, and administrative project controls. The need for systematic approaches to product development and maintenance is necessary to ensure a safe and effective device for the user and the patient, an economical and competitive success for the manufacturer, and a reliable, cost-effective investment for the user.

This book is generally aimed at senior bioengineering students who are in the formative stages of deciding what to do for a senior design project and who need to consider what the factors are that may or may not impact their project now or in the future if brought to a useful conclusion. Portions of the book may be used in lower level classes, such as sections on brainstorming and elementary idea generation techniques. Portions of the book may also be used in early graduate level classes if students have had little exposure to the Food and Drug Administration (FDA) and CE mark information. The book is meant to be comprehensive enough that students working on a variety of topics (from databases to process analysis and device improvement) may have adequate information to begin a fairly comprehensive project. Additionally, it is aimed at design engineers new to the medical device industry who have not had access to such a comprehensive book or course in their background. This book should prove to be an excellent resource for these individuals as they enter the workforce.

The emphasis of the book is on the practical, hands-on approach to device design. The layout of the book follows the typical device design process. The mathematics included here is that which is necessary to conduct everyday tasks. Equations, where needed, are merely given, not derived. It is assumed the reader has a basic knowledge of statistics. References are given at the end of each chapter for those wishing to delve more deeply into the mathematics of the subject.

The first three chapters are a general introduction to the subject. Chapter 1 opens with a general overview of the process and definition of design. Chapter 2 is an outline of some fundamental ideas, generation techniques, and design, decision, and comparison tools with a brief introduction to the process of inventive problem solving. The use of quality function deployment (QFD) diagrams is also introduced as a comparison tool at this stage. Fundamental to successful design processes is the generation of a good design team, and the management thereof. Chapter 3 introduces the need for documentation techniques and requirements, and the use of databases in this endeavor. Reporting techniques for the student, through industry, are briefly covered in a discussion on posters, oral presentations, and progress reports.

Essential to a good design is a correct and customer-driven product definition. Chapter 4 summarizes the product definition process, and reiterates and concludes on the use of QFD in this process. Product documentation, record keeping, and levels of effort mandated by quality regulations and medical device regulations are reviewed in Chapter 5.

Chapter 6 delves into the product development process and gives an overview of product requirements, design and development planning, and requirements documentation. Specifically addressed are design inputs, design outputs, formal design reviews, verification, validation, and design transfer. Chapter 7 discusses hardware development methods and tools. Examples of discussion topics are design for six sigma, redundancy, component derating, safety margin, load protection, environmental protection, product misuse, and TRIZ. Chapter 8 discusses similar topics from the point of view of software development. Topics addressed include software planning, the software model, design levels, software architecture, language choice, software risk analysis, coding, and tools.

Chapters 9 and 10 comprise a good introduction to human factors issues and industrial design. Several of the techniques used to guard against human-caused errors are reviewed, as are techniques to increase usability. Workstation design and human expectations are also discussed, as are the methods used to test them in use.

Biomaterials and materials selection are the themes of Chapter 11, covering the various FDA (and some international) tests and test methods used for materials that may be used. Tests for toxicity, hemocompatibility, irritation, reactivity, and sensitization are summarized.

Chapter 12 covers some safety topics that are not dealt with anywhere else in this book, specifically addressing safety as a component of the design process and one of the several structured approaches to the consideration of safety in a design. One medical disaster is used as an exemplar.

Once the design is completed, it must be tested to prove whether it meets its requirements. The subject of testing is summarized in Chapter 13. Types of tests, parsing test requirements, establishing a test protocol, and defining failure are all discussed in detail. The chapter also includes the methodology for determining test sample size and test length.

Once the testing is completed, the test data must be analyzed to determine the success or failure in testing. Chapter 14 explains the mathematical basis of analyzing test data. Metrics that are covered here include failure rate, reliability, mean time between failures, confidence level, confidence limits, and minimum life. There is also a discussion of graphical analysis of data, including Pareto charts.

Chapter 15 discusses the legal ramifications of medical device development and failure. Topics include negligence, liability, breach of warranty, failure to warn of dangers, accident reconstruction, and forensics.

Chapter 16 discusses the impact of regulations and standards on medical device development. It reviews the FDA, both its history and the methods required to obtain clearance to market medical devices. Classification of medical devices and the related requirements are reviewed. Also included are the requirements of institutional review boards for human subject tests. Chapter 17 discusses regulations outside of the United States as well as within the United States and the rest of the world's standards. Included in the topics are the Medical Device Directives, CE marking, and a list of U.S. and international standards organizations that have an effect on medical device design.

Good design will likely generate intellectual property; Chapter 18 gives a summary of the protection of intellectual property via patents, copyrights, trademarks, and trade secrets.

The next two chapters cover manufacturing, quality control, and miscellaneous issues. Chapter 19 covers manufacturing processes and how quality control issues continue during this phase of the design process and how it must be addressed. Chapter 20 covers miscellaneous issues in medical device design, including learning from failure, design for assembly, design for the environment, Poka-Yoke, and product life issues. Chapter 21 covers liability issues that remain after the final users put the device in use, and some of the safety issues that arise. Investigation of medical device accidents is also reviewed, as is investigation of traffic accidents.

Chapter 22 is a brief synopsis of professional issues that must be considered by the biomedical professional. Specifically, membership in professional societies, licensure, and professional ethics are discussed. Forensics and consulting are also briefly covered.

Chapter 23 is a resource chapter; nine different design case studies are reviewed. This material may be read as one, or used in conjunction with earlier chapters as examples.

Design in biomedical engineering and bioengineering (discussed in Chapter 24) is a moving target; this is an interesting and demanding field in terms of breadth and depth. The text concludes with a chapter that briefly captures some snapshots of "hot" design areas right now and in the near future.

Design of Biomedical Engineering Devices and Systems is the joint effort of two licensed engineers, one with more than 40 years of teaching and research experience in teaching biomedical engineering, and the other 18 years as a sole instructor of a senior design course, and more than 30 years of experience as a reliability engineer in the biomedical device industry. The book is a result of the class notes and class experiences of Paul H. King, the industrial experiences of Richard C. Fries, and the first edition of this textbook by both.

Richard C. Fries
Paul H. King

Acknowledgments

We are deeply indebted to many people for their encouragement, help, and constructive criticism in the preparation of this book.

We would like to thank the students of BME 272 at Vanderbilt University for their review and constructive criticism of the first edition of this book. Their input made this edition a better text. We also thank those of you who have adopted and reviewed the prior edition.

Mostly, we would like to thank our wives, Sue and June, who constantly encouraged us and who sacrificed much quality time with us during the preparation of this book.

1 Introduction to Biomedical Engineering Design

Give a man a fish; you have fed him for today.
Teach a man to fish and you have fed him for a lifetime.

Anonymous

This chapter is designed to cover the design of biomedical engineering devices and systems. It is intended as a reference to guide thoughts and actions, by using prior experiences, classroom instruction, and otherwise, to the problem of designing something relevant to this field.

What is relevant to this field? Biomedical engineering can be very broad in scope, dependent on interests and circumstances. Biomedical engineers are expected to have some familiarity with medical devices, their design, their regulation, and use. They are further expected to consider safety aspects of the devices, and should consider the potential misuse of a device. Designers may be expected to involve themselves in the improvement of a process, such as the tending of patients in a hypertension clinic. They may get involved in biotechnology to manufacture products derived from mammalian cells; they may wind up in the manufacturing of implant devices for the treatment of diabetes. They may design a specialized brace for a single individual or design a medical device to be used by thousands of patients. It is vital to understand the many meanings of the term design, and have some experience at problem solving using the design principles outlined in this chapter.

1.1 WHAT IS DESIGN?

It is useful to discuss design from two viewpoints this early in this chapter, first by discussing what it is not, then by discussing what it is and its various forms.

Design is not research, which may be defined as a careful investigation or study, especially of a scholarly or scientific nature.* A design task may require research to accomplish a task, but it typically involves the integration of knowledge, rather than the generation of knowledge. Research may be done into the process of design and as such is sponsored by such groups as the National Science Foundation (NSF) (see http://www.eng.nsf.gov/dmii/index.htm for the design, manufacturing, and industrial innovation research division).

On the other hand, design is not craftsmanship. Designers are not nor should they be viewed as a craftsman. This work will involve brains and skills, not just skills.

Design as an action verb is to

- Conceive, invent
- Formulate a plan for; devise
- Have as a goal or purpose; intend*

Design work thus does not necessarily involve the manufacture of a physical device; it can be a plan or process, or a study to determine the same. Naturally, it can range from this level to the complete specification of a device and its manufacture.

---

* Microsoft Encarta, 1999.

Design as a noun (thing) incorporates the following:

- Drawing or sketch; especially a detailed plan for construction or manufacture
- Purposeful arrangement of parts or details
- Art or practice of making designs
- Ornamental pattern
- Plan or project
- Reasoned purpose; intent
- Often a secretive plot or scheme (Latin)*

Each of these terms has validity in the types of work that will be discussed in this chapter. Even the ornamental pattern qualifies as it is a product of the intellect, is therefore an invention, and may qualify as patentable intellectual property. How does a secretive plot qualify? Perhaps under the category of trade secret, for example, the recipe for the manufacture of Coca-Cola.

1.2   WHAT IS THE THRUST OF THIS CHAPTER?

This chapter is aimed at introducing one to the application of design processes to a wide category of design problems in biomedical engineering. It is anticipated that the user of this chapter will be involved during the reading of this chapter in one or more design projects or exercises. It likely will best be used in parallel with some early design exercises, and then referred to occasionally as a major design project is pursued. It is meant to be a part of the learning triad of hear or see or do, but not all.

The chapter also attempts to place the various steps in the design process in a logical order, typically that followed in engineering best practices for conducting and completing a design project. The process is generic and flexible, so that processes may be included or not, depending on the project.

1.3   WHAT MIGHT BE DESIGNED?

A partial listing of senior level design projects follows:

Biomedical devices

- Modified patient brace for an individual
- Patient (or pet) tracking device
- Development of a hand exerciser
- Improved safety warning system for an intensive care unit
- Improved patient monitoring for premature infants
- Development of a voice training system for patients with Parkinson's disease
- Development of a surgical tool for use in spina bifida surgery
- Development of an adjustable tray for a spinal cord-injured patient
- Modification of a riding mower for use by a paraplegic
- Development of a laser spot size measurement system

Biomedical systems

- Improved patient record-keeping system
- Revised and improved vaccine database system
- Comprehensive pain clinic data collection/billing system

---

* Microsoft Encarta, 1999.

- Development of a prostate cancer screening test
- Development of a skin disease database
- Development of a research ward database system
- Improved feeding apparatus for cystic fibrosis patients
- Development of a device for laparoscopic band pressure regulation
- Biofeedback system for wheelchair propulsion systems
- Development of a cauterizing biopsy catheter
- Development of a drug eluting stent

Biomedical processes

- Study of patient flow in an emergency room
- Improved patient communication in a breast cancer clinic
- Determination of clinic space and facility needs
- Development of a system to measure foot impressions and transmit same
- Optimization of T cell trapping in a microfluidic device

Note the key words improve, develop, revise, or study. Also note the key words device, process, and system. Each of these terms will see major elaboration in the ensuing chapters. Design will on occasion involve invention, but generally will involve an application (extension) of existing technology. In addition, as will be noted in the solution of design problems, there will be no exact answer but instead, there will be best attempts given constraints involving timing, financing, etc. of project work.

1.4 ESSENTIALS OF DESIGN

A well-written newspaper article quickly answers the following questions: Who? What? Where? When? Why? How? The process of design typically begins with such a listing, with the how portion being the major part of the endeavor. The most important part is the "what" section. If this section of the overall task is done well, one will not need to backtrack and rework a design (normally). The first five steps are required for proper task clarification, the final term, unless specified, typically is the end result of the design task. Figure 1.1 is the generic design process in a flowchart form.

If one properly defines the problem (i.e., understand the who/what/why/when/where part) then one can hope for a tracing, directly to the solution evaluation section. If the solution is wrong, or if the problem definition is wrong, one will have to backtrack and rework the overall solution. There will be other rearrangements of this basic structure as the tasks of documentation, standards, testing, codes, trademarks, patents, etc. are added. The most vital part of the design work will be to figure out what it is that one is asked to do.

Part of this problem will hopefully be minimized by prior educational experiences, which should have included medical nomenclature, some systems physiology, and medical instrumentation. If one is working with a nonengineer on your design project, some new communication skills may be needed. This will be especially true in dealing with most physician collaborators, who commonly go from diagnosis to treatment (or therapy) on a generally nonmodifiable patient, while the designer is charged with the modification of a device or process.

1.5 BIOMEDICAL ENGINEERING DESIGN IN AN INDUSTRIAL CONTEXT

Figure 1.2 is a concept map* describing, in a hierarchical fashion, the overall elements of the biomedical engineering design process in the context of society. It is a consensus document as to the

---

* Tutorial and software available from Institute of Human and Machine Cognition, Pensacola, Florida.

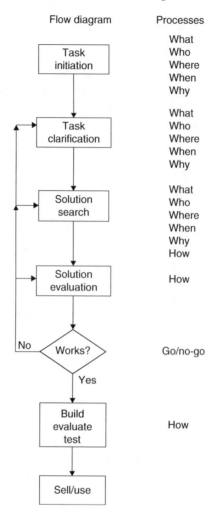

FIGURE 1.1 Generalized flowchart for the design process.

generic elements that must be considered in the overall design process. To understand this system, one normally would read from the top down and from left to right.

Biomedical engineering design projects generally involve opportunities, subject to certain constraints. A company will not typically pursue a project unless a market analysis has been done, there exists a potentially desired return on investment, and the product is a fit with industry needs (and intellectual property rights may be retained). (In contrast, for a student design process in a generic academic setting, projects are generally proposed to students based upon the potential advisor [medical faculty, engineering faculty, industry advisor] needs, tempered by what might be expected from a student design team, etc.) For industry, there are several overriding societal concerns involved in the design of devices and products; these involve multiple regulatory requirements (licensing, waste disposal, liability issues, etc.), regulatory agencies (the Food and Drug Administration [FDA], etc.), and bioethical constraints (animal care rules, Human Subjects Committee approvals, etc.). For student projects, Human Subjects Committee approvals are sometimes necessary, a knowledge of FDA rules is useful, a project may brush with the group Persons for the Ethical Treatment of Animals (PETA), etc.

Design projects typically arise from studies of medical and clinical problems (such as device complaints), a literature review, the scientific project needs of clinical and academic investigators,

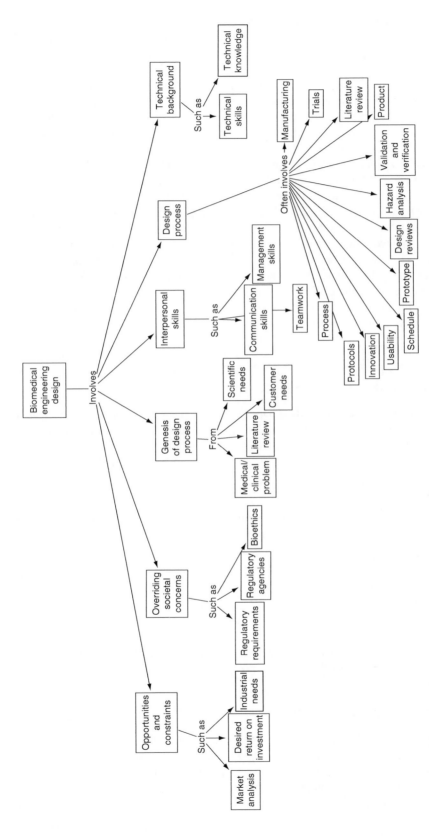

FIGURE 1.2 Generic biomedical engineering design process.

and occasionally from the needs of individual customers. Design projects seldom involve a single designer, and thus issues of interpersonal skills, such as communication skills, management skills (especially time management), and teamwork skills come to the forefront of the design process.

The design process itself may involve developing a process (as opposed to a device.) The industry design process will often involve development of manufacturing methods and testing of devices in trials according to specific protocols. Devices and processes must be tested for usability. Specific schedules must be developed to keep a competitive edge. Periodic design reviews are mandated to maintain schedules, and to determine validity and verification of the design. Hazard analysis should be done as a product is being developed, rather than later when liability becomes an issue. If possible, a prototype of the device or process should be developed and used for testing purposes before final manufacture of a product.

Design teams typically require technical skills and knowledge from a variety of disciplines; this is normally accomplished by the generation of multidisciplinary design teams. Depending on the particulars of a problem, the engineering team should include a required mix of electrical, mechanical, biomedical, computer, and chemical engineers. Persons with these backgrounds, as well as with an engineering management background, might form the core of the engineering product development team. Manufacturing and industrial management engineers may or may not be included in the prototyping or concept development team, but will be needed on the manufacturing part of the overall development.

1.6 GENERIC STEPS IN THE DESIGN AND DEVELOPMENT OF PRODUCTS AND PROCESSES

The National Academy of Science Publication, *Design in the New Millennium* has a useful figure outlining the generic development of products and processes (Figure 2.1, p. 12.) The essential nine steps indicated on their diagram are listed below (somewhat paraphrased):

1. Overview of requirements or strategy
2. Product specification
3. Concept development
4. Preliminary design
5. Refinement and verification of detailed designs
6. Prototype development
7. Preparation for production/manufacture
8. Production, testing, certification, rollout
9. Operation, maintenance, disposal

The above nine steps are generic to most engineering fields. This chapter will cover, for the biomedical engineering field, all but production and roll out in step 8 (leaving that material to the manufacturing engineers). Disposal (step 9) also will not be covered.

1.7 HOW IS THIS TEXT STRUCTURED?

This remainder of this text will approach the process of design as generically as possible; the special constraints relevant to biomedical engineering will be added as necessary. Such constraints include, but are not limited to, the FDA and its device classification and licensing rules, the Medical Device Directives (MDD; European), Clinical Trials issues, and others. Chapter 2 will outline some fundamental idea generation techniques. It will further introduce some design decision and comparison tools, followed with a brief introduction to the process of inventive problem solving. The use of quality function deployment (QFD) diagrams is also introduced at this stage as a comparison tool.

Chapter 3 covers fundamentals of successful design processes in the generation of a good design team, and the management thereof. This chapter discusses design team evaluation and peer review techniques. It further segues into the need for documentation techniques and requirements, and the use of databases in this endeavor. Reporting techniques are briefly covered in discussions on posters, oral presentations, and progress reports. Expectations and assessments of design teams are briefly covered.

Fundamental to a good design is correct and customer-driven product definition. Chapter 4 summarizes the product definition process, and reiterates and concludes on the use of QFD in this process. The FDA definition of a medical device is documented here also. Chapter 5 reviews product documentation, record keeping, and levels of effort mandated by quality regulations and medical device regulations.

Chapter 6 gives an overview of the product development process and several models for this process. Validation and verification are well covered.

Chapter 7 surveys several important hardware development methods and tools, such as component selection, design for six sigma, load protection, and safety margins.

Chapter 8 includes an overview of software design techniques that ensue from the earlier product specification tasks. The choice of language and techniques for programming are extensively discussed.

Chapters 9 and 10 introduce human factors issues and industrial design. Several of the techniques used to guard against human-caused errors are reviewed, as are techniques to increase usability. Workstation design and human expectations are also discussed, as are the methods used to test these in use.

Biomaterials and materials selection are the theme of Chapter 11, with heavy coverage of the various FDA (and some international) tests and test methods used for materials that may come into contact with users. Tests for toxicity, hemocompatibility, irritation, reactivity, and sensitization are summarized.

Chapter 12 covers some safety topics not elsewhere dealt with in the chapter, specifically addressing safety as a component of the design process and one of the several structured approaches to the consideration of safety in a design. Several legal cases are used to introduce the reader to the ramifications of bad designs.

Chapter 13 summarizes testing of samples. Types of tests and considerations to determine mean-time-to-failure are introduced. This is a good introduction to the concept of reliability testing. This concept is then extended in Chapter 14, with an introduction to calculation of reliability, mean time between failures, confidence levels, and graphical analysis of data. Chapter 15 extends this information with overviews of quality control and improvement and a formal introduction to reliability and the possible outcome of its converse, liability. Medical device errors as well as errors by medical personnel are discussed.

Chapter 16 reviews the FDA and the methods it requires one to use to obtain clearance to market medical devices. Classification of medical devices and the related requirements are reviewed. Also included are the requirements for institutional review boards for human subject tests. Chapter 17 reviews the major relevant standards and regulations involved in United States and international device regulations.

Good designs will likely generate intellectual property; Chapter 18 summarizes protection of intellectual property via patents, copyrights, trademarks, and trade secrets.

Chapter 19 continues the theme of product testing and validation, and total system testing, with special reference to good manufacturing practices as mandated by the FDA. Chapter 20 covers miscellaneous issues, from design for failure to design for the environment.

Chapter 21 introduces various product issues, such as product safety and legal issues and accident reconstruction.

Chapter 22 gives a brief synopsis of professional issues that must be considered by the biomedical professional. Specifically, membership in professional societies, licensure, and professional ethics are discussed. Forensics and consulting are also briefly covered.

Chapter 23 introduces nine design examples, which may provide insight into methods of design approaches not generally found in the literature.

Chapter 24 briefly discusses some of the trends observed in design work in biomedical engineering. An overview of the current U.S. government structure for funding of design and research projects is introduced. The chapter concludes with a discussion of new areas of design and research that may be developed soon.

1.8   REAL PURPOSE OF THIS CHAPTER

The real purpose of this chapter is to guide one in the tackling of a real-world design task relating to biomedical engineering. It is meant to be supportive of the newly involved engineer in the medical device market. It is meant to prepare bio and biomedical engineering students for development of senior design projects (primarily) and for careers in the medical device development industry. An ultimate goal is to prepare one for a career in design as advertised in the following Web-based example advertisement (2000):

R&D Engineer
This position offers excellent growth opportunities for the highly motivated individual. Seeking engineer to assist our product development team in developing and testing proprietary medical device concepts. Candidate must be hands-on and able to work independently. Working knowledge of ISO and FDA requirements preferred. Principle responsibilities will include product design, testing, and analysis. Qualified candidate should have 1–3 years' experience with medical device company and BS in mechanical engineering, materials engineering, or biomedical engineering. Additional responsibilities may include animal testing, clinical evaluation, patent/literature searches, and support of ongoing development projects as required. Experience with biomaterials and mechanical design a plus.*

This in fact was an advertisement that a class of 2000 student applied for, and continues with as of this writing (2008).

EXERCISES

1. Often, design projects are generated by persons concerned about improving the welfare of persons close to them (patients, family, friends). Think about your acquaintances and develop a design project definition. Be sure to detail the who/what/where/why/when specifics as much as is necessary.
2. Orthopedic physician has proposed that you study the effect of electrical stimulation on the healing rate of a bone fracture. Write this request up—briefly—as a research project. Rewrite this as a design project. Discuss the differences in the approaches.
3. You have done some form of design project in your personal life (device, plan of action, project, college choice, etc.). Briefly describe, for your instructor, your favorite project and any lessons learnt from it.

---

* Job opening listing Kensey Nash Corporation, Exton PA, May 2000. See http://www.kenseynash.com/ for current information.

4. This is a good time to look at your background to determine what areas you will be qualified to work in. Convey to your instructor the following information: are you familiar with html usage? FrontPage or the equivalent? VB or VBA use? Excel? Flowcharting? Access? PowerPoint? Microsoft project? Pert diagrams? Web survey form generation? What are your special skills and interests? What professional experience have you had to date? What area are you most interested in for a design project?

5. Who/what/where/why/how/when construct is often used in newspaper writing. From your Sunday newspaper, extract one short news story and one obituary and analyze it for the above content. Turn in both articles and your commentary to the instructor.

2 Fundamental Design Tools

Men have become the tools of their tools.

Henry David Thoreau

There are a series of design tools in current use that will be valuable in the design process as discussed in the remainder of this chapter. Some of these tools will be covered on an introductory level in this chapter. As design is in truth an information gathering and processing activity, these tools will reflect this process. Some of the tools involve interaction with humans, some with computer programs, and some with physical devices. This chapter will cover solution search methodologies and function structure abstraction, including flowcharting techniques.

2.1 BRAINSTORMING AND IDEA GENERATION TECHNIQUES

Without knowledge of idea generation techniques a designer may rely too heavily on prior knowledge or on making minor adjustments to a device or process that could be dramatically overhauled. Some of the more common idea generation techniques include brainstorming, Method 635, the Delphi Method, and synetics. There are variations on the themes as discussed below, and computer support and training tools exist for these and many other methods.

2.1.1 BRAINSTORMING

A typical brainstorming group will consist of 5–15 individuals generally chosen by the person who will be the discussion leader. The individuals should consist of the design team searching for a solution, and involve an additional mix of lay or other people who might be able to contribute due to their backgrounds. In a university environment, brainstorming teams consisting of several engineering students and two or three arts and science students are often much more effective than teams of just engineering students. The additional viewpoints are useful, as is the extra brainpower.

In preparation for the session, the leader should set a reasonable duration for the meeting (20–40 min) and stress that there will be no hierarchy and no criticism of any ideas presented. At the outset of the meeting, the leader should state or restate the problem to be solved and reiterate the rules for the meeting. The brainstorming discussion then begins, with the leader primarily trying to maintain flow of information from single individual, rather than having multiple people trying to talk at once. The leader may not lead the discussion, but may, during periods of long silence, suggest elaboration or expansion of earlier suggestions. All ideas are to be heard and posted, even the ridiculous ones, as they may in fact lead to a novel solution. No derogatory or dismissive comments are allowed. Someone, the leader or a designated secretary, should take the minutes on the meeting. One suggested method is to write each idea on a post-it note for later reclassification.

At the end of the session, the group can evaluate, rank, and if desired classify all ideas. The rank ordering and evaluation of the list can then become an agenda for the design team and its efforts. If the ideas are also classified, a design tool such as concept mapping can be used to categorize ideas and guide further work. Such a mapping for a brainstorming session on grades in a design course is illustrated in Figure 2.1. Note that the ridiculous ideas remain at this stage.

Brainstorming is generally useful in conditions when new ideas are needed and when a design group is deadlocked on methods to use in a design process. A drawback is that it relies on the

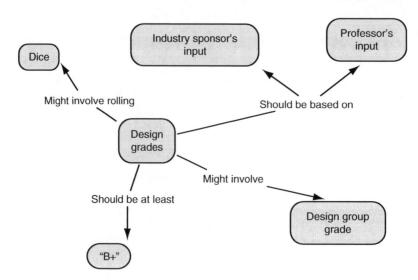

FIGURE 2.1   Concept map for a brainstorming session on grade sources in a design course.

abilities and backgrounds of the practitioners and on their willingness to speak up and give suggestions, even in areas outside of their competence.

2.1.2   Method 635

If shyness is a problem, or there is a problem getting people together, Method 635 is ideal. Once again, a problem is presented in sufficient detail to specify what needs to be addressed, but without presenting a solution. Each of the six individuals writes down three ideas for solution of the problem, typically on a large sheet of paper or tablet. At a preselected time these tablets are exchanged, each individual may now elaborate on the new ideas received or may generate three new ideas, or may simply relate the three original ideas (not preferred). Five exchanges take place such that no person gets back their original tablet and each of the six tablets holds as many as 18 ideas for problem solution.

This method is useful in that it is very systematic and can generate a large number of solutions if each individual has a modicum of originality and can build on other's ideas. It has a drawback in that each person works in isolation and thus the group synergy is missing.

2.1.3   Delphi Method

This method is a polling process and can be valuable in both design processes and in situations where the future of a process is a subject of concern. The process typically consists of three steps. First, a series of suggested starting points for a design problem (or the future or a process) are generated by the process leader or by a panel of experts. These suggestions are regrouped and returned to the panelists, who are asked to add to the list. The resultant response list is then recollected and reordered, and the panel is asked to evaluate and comment on the reordered and expanded list.

An example of the use of this method might be to predict the future of the process of anesthesia during major surgery. Your preliminary letter to the chairs of anesthesiology at major medical centers might request a simple listing of key events in place now, and some suggested future modifications of these processes. Your next letter would list the several responses that you obtained and might ask for both a timetable and a list of additional comments. Your third round would ask for advice on the practicality of each of the predictions, and the necessity for related events to occur to make an event happen.

This method relies on the willingness of experts to respond to a questionnaire in a timely fashion. With most surveys running at a maximum of 30% return you will need strongly motivated individuals on your panel.

2.1.4 SYNETICS

Synetics is a term that describes a way of looking at analogies of terms relating to a problem as described. The process, briefly described, involves the following steps. First, the problem to be solved is presented to a small group. The group discusses the problem and the environment of the problem to the point that they are familiar with the problem and are able to articulate it. They then operate by looking at analogies and forcing relationships.

An example of the use of this methodology might be looking at ways to close a defect in the heart, for example in the case of atrial septal defect. The problem is that there is a hole in the heart between the two atria, this hole needs to be closed by something that is small when introduced into the patient but large enough to cover and close the hole when deployed or opened. The key to this problem solution is to look at the key words in the last sentence: cover, close, open, small, and large. These key terms should enable you to make a number of analogies regarding devices that do one or more of these things; the most obvious would be to make the analogy to an umbrella. Another could be to compare this with patching a tire on your car, thus a patch would be needed.

There are variations on this theme that can be used, such as by randomly picking words from a dictionary and trying to apply them to the problem solution (forcing a fit). This method is somewhat better than brainstorming, if the problem to be solved has an analogy and your group is willing to explore seemingly bizarre trains of thought.

2.1.5 OTHER METHODS

One interesting variation on brainstorming is a fairly structured approach proposed by Dr. Edward de Bono that utilizes "six thinking caps." During a meeting on problem solving the facilitator (blue hat) guides the conversations of others who wear white hats while gathering information, or red hats while expressing emotions, black hats while expressing caution, yellow hats while being enthusiastic, or green hats while being creative. The objective of the hats is to allow persons to express feelings representative of the hats exclusively, rather than a combination of unstructured feelings as might be the case in a brainstorming situation. Changing hats allows one to "change gears" without "exposing" oneself.

2.2 CONVENTIONAL SOLUTION SEARCHES

Section 2.1 discussed techniques that required the use of other live humans to begin a solution search. A far more vast supply of information exists in Web-based and print literature, especially patent databases. Nature holds many examples of solutions to specific problems as plants and animals evolved to solve their own niche problems. Existing solutions and analogies should also be pursued.

2.2.1 WEB-BASED AND PRINT LITERATURE

Several Web-based search engines exist; these include AltaVista, Direct Hit, Excite, Yahoo, Google, HotBot, Infoseek, Lycos, Northern Light, Web Crawler, DogPile, etc. Some services, such as Go Express Search spawn as many as 11 other search engines in an attempt to quickly search the Web for requested information. Good services allow one to sort by relevance and to search using Boolean operators such that you can, for example, search only for "Patent Ductus Arteriosis" and "repair" to limit the amount of information that needs to be searched through.

Many libraries have computerized systems so that one can search, for example, book titles for a specific term, such as electrocardiogram. It is useful to access Amazon.Com (www.amazon.com) on the Web and use their search engine to find recent books that they stock. The search term "electrocardiogram" gave 12 hits in 2001 while in 2007 the same term gave 8958 hits. The more generic term "design" gave 32,000 hits in 2001, while in 2007 there were 600,674.

Another excellent source for information is the U.S. patent and trademark site (www.uspto.gov), which allows one to search granted or pending patents in a period of time for key words. For example, the search term embolus searched for in the title of granted patents in the time period of 1996–2007 yielded 1285 "hits." The site also lists other patent office sites, such as the Japanese patent site, some of which allow for similar patent key word searches.

Many trade magazines exist for product design and several are specific for medical product design. Some magazines (e.g., *Medical Design Online* and *Medical Industry Today*) send daily e-mails regarding work in the field, several maintain Web sites that allow search functions for devices and products.

2.2.2 SOLUTIONS IN NATURE AND ANALOGIES

Many design problems may have been solved in nature and may be transferred to design problems at hand. For example, the motion and flexibility of a worm should be studied if one is to look at improved catheter designs. The Eiffel Tower design was said to be mimicry of how bones support weight. The Amazon Web site lists 10,055 books on biomimetics (2007), several of which may apply depending on the problem at hand. The Web site www.nature.com/nature has a search function that allows a search on such terms as biomimicry. One article, "Lifes' lessons in design" is a standout example of design and biomimicry.

2.3 FUNCTION ANALYSIS

Many design problems will involve the use of flowcharting tools to assist in understanding the processes under study to improve or modify them. Properly done, these flowcharts will assist in the analysis of delays, patient irritations, added costs, and the like. Several levels of analyses may be of use, from simple process charts, to fairly complicated combinations of signal, material, and information flows. Overall, the process of flowcharting can be an excellent communication tool.

2.3.1 SIMPLE PROCESS CHARTS

Process charts can be extremely simplistic and tell an unequivocal story. Figure 2.2 details the process of applying a band-aid, one can almost imagine going through the process oneself while reading through the description. The shapes used in the diagram are fairly common, with many flowcharting systems typically using circles or ellipses for processes, arrows for designation of transport, diamonds for storage, squares for inspections, and "D" for delays or wait states. The example below is an extremely short diagram, but is illustrative of ways to display this data and potentially to make use of it. An extensive diagramming of a system could allow one to ask the questions: Where can I get rid of delays? How is the transport of several parts to be facilitated to speed up this operation? How many operations are my workers being asked to do per minute? Why am I seeing carpal tunnel syndrome? Expansion of this diagram, with addenda on each line indicating distance traveled during an operation, for example, may facilitate studies for the improvement of the entire process (minimize distance traveled, minimize delays, etc.). Most flowcharting programs will additionally have the ability to annotate, with arrows, decision points that allow for retracing of steps backward, if, for instance, there is a failed inspection.

2.3.2 CLINIC FLOWCHARTS

Figure 2.3 represents the path of a patient in a hypertension clinic. The rectangles represent operations such as sign-in and physician interaction. The large "D"s represent delays in the system.

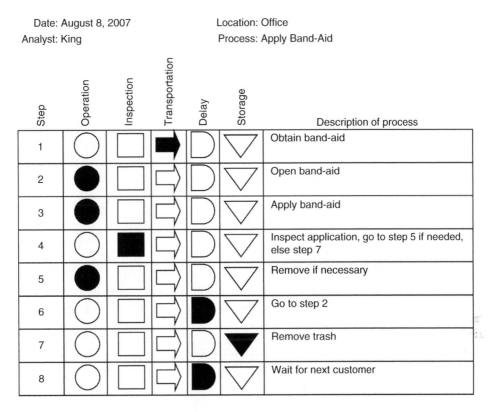

Date: August 8, 2007   Location: Office
Analyst: King   Process: Apply Band-Aid

FIGURE 2.2   Process diagram for applying a band-aid.

Note the nature of the path through the clinic and the emphasis on delays. Figure 2.3 emphasizes the fact that this system is linear, that each event must be completed before the next beginning. From a patient point of view, this can be extremely frustrating, as the delay times accumulate. Overall waiting time for a patient is from 12 to 70 min for this process which involves only 8 min of interaction with professionals, 5 min with the physician. As the physician is the only one in charge of emptying rooms, he/she can be blamed for this waste of time. A wait time of over 60 min (once!) for the first room caused one of the authors to switch physicians!

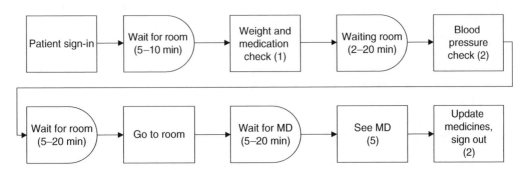

FIGURE 2.3   Blood pressure (hypertension) clinic flowchart. Numbers indicate length of delay or interaction time in minutes.

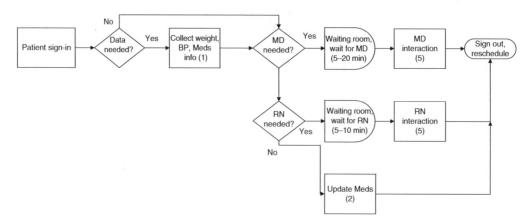

FIGURE 2.4   A faster hypertension clinic flowchart.

2.3.3   FLOWCHARTS WITH DECISION POINTS

To speed up a clinic process, at least for some patients, ancillary personnel can be used in the patient care taking and advising process. Figure 2.4 demonstrates a modification of the above flowchart wherein a nurse or other medical practitioner can take the weight, pulse rate, medication listing and blood pressure data (if needed), and screen the patient to see if the patient needs to see the physician or can be seen by an RN or other health care worker for any update on medications or other matters. The diamond in the flowchart is a decision point; the branches signify a yes or no branch condition.

This revision of the hypertension clinic allows three branching points, which on the average will allow for faster patient clinic visits by decreasing wait times via the use of ancillary personnel. Overall wait time has been reduced from 22–70 min to 5–20 for the MD path, 5–10 for the RN path, and zero for the ancillary personnel path in this model. The physician time is utilized for "needy patients" rather than the entire clinic scheduled patient load. This can lead to improved clinic utilization if planned wisely, as many patients will be status quo and will not need a physician interaction. This model is analogous to well-run dentist offices where dental hygienists take care of the majority of patients' minor needs while the dentist takes care of the remainder.

Presented above has been some very rudimentary flowcharting information, the minimum necessary to understand some of the essentials needed for a starting analysis of clinic and other process analysis as a prelude to redesign of the process. Many other variations and embellishments on the above material are part of normal flowcharting programs, such as the use of color to identify particular paths, the use of additional annotation to indicate complaints about sections of the process, and the use of additional symbols specific to a number of other processes, such as design of databases, process diagrams for the food industry, etc. Some common variations for systems involving electro-mechanical devices involve the simultaneous overlay of material flows, information flows, and energy flows. Many commercial products have a wide variety of template patterns available for use. Two of the more commonly used commercial packages are Microsoft's Visio and Micrografx FlowCharter, a few other programs are available on the Web for free but most have a limited time trial associated with them. Software packages exist that emulate processes such as patient clinic visits such that variables such as the effect of the number of examination rooms or number of clinic staff may be varied and the effect of this on patient throughput estimated (e.g., High Performance Systems program Stella).

2.4   ELEMENTARY DECISION-MAKING TECHNIQUES

In Chapter 4 the need to define well the parameters involved in the design problem statement will be very strongly emphasized. Terms such as demand and wish will be a part of the design analysis when design choices are being considered. Alternate terms involve objectives, quality, and function.

	Choice Number			
Demand Number	1	2	3	4
1	+	+	?	+
2	+	−	−	+
3	+	+	−	?
Summary	Go	No go	No go	Recheck

FIGURE 2.5 A simplified design selection process diagram.

The purpose of this section is to introduce some elementary concepts in the decision-making processes that may be selected in a solution search.

2.4.1 SELECTION CHART

A selection chart that might be used in a design process is shown schematically in Figure 2.5. The essence of the design chart is that design demands are listed vertically and concepts or design choices being considered to perform these demands are listed horizontally. If a choice will not work, a "−" is entered in the intersection, if the opposite, a "+" is used. If a design choice is uncertain, a "?" symbol is employed. The final scoring with this simplistic chart simply asks the question— Does this column (choice) meet all criteria? If not, the design choice is rejected. If there are only + and ? symbols, the particular choice will need to be investigated further.

As an example, consider the design decision for a proper writing implement for a grade school class in a damp environment. Figure 2.6 might be an example of a product selection matrix. This example is simplistic and is meant only as an example. The process can work very nicely if there are few choices and most are go-no go in nature. A drawback is that there are no "shades of gray" or partial solutions allowed, and, as will be seen in a later discussion on invention, conflicts yield only dismissal of the choice, rather than resolution of the conflict via an inventive problem solution.

2.4.2 EVALUATION CHARTS

The next level up in complexity to the above selection chart is an evaluation chart. This chart generally is used to assist in the ranking of various wishes, qualities, or other aspects of a proposed solution. Wishes or qualities are tabulated in a vertical column, each of these are assigned weights (importance) on an arbitrary scale, often ranging 1–10, for example. No zero values are assigned as this would dismiss this row as a valid choice. Each set of columns from this point on carry the value of the particular column's solution and the net weight of the product of the solution and the weighting given to that wish. The totals are then added for each proposed solution and the "winner" is normally the column with the highest total. A fabricated example for a Daddy Warbuck's transportation choices between New York City and Rome for vacation purposes is given in Figure 2.7.

	Choice			
Demand Number	Fountain Pen	Pencil	Chalk	Marker
1. Writes on paper	+	+	?	+
2. Won't stain hands	−	+	+	−
3. Damp paper tolerant	?	+	?	?
Summary	No go	Go	Recheck	No go

FIGURE 2.6 An example of a product selection matrix.

"Wish"	Weight	Commercial Airline		Ocean Liner	
		Value	Product	Value	Product
High speed	3	5	15	2	6
Convenience	5	4	20	5	25
Comfort	5	3	15	5	25
Low cost	2	2	4	3	6
Food and drink	5	1	5	5	25
Total			59		87

FIGURE 2.7   Transportation evaluation chart, maximum weight is equal to five.

In this example, the maximum possible score for a mode of transportation would be 100 (the total of the weights times the maximum weight of 5 each), thus the choice of an ocean liner meets 87% of the above persons wishes, versus the 59% figure for the commercial airliner. This method is subjective but is useful to help rank order a potentially large list of choices and wishes.

2.5 OBJECTIVE TREES

The final minor evaluation process involves the generation of an objectives tree. This is a formulation that involves the assignment of priorities to a series of objectives and subobjectives such that a determination of the value of each of several subobjectives is quantified in the designers' mind. At the first branching each subobjective is given a weight such that the sum of all weights is one. At every successive branching the weights continue to sum to unity, but the overall weight of each branch is the product of all branchings prior. An example objectives tree to illustrate this process is given in Figure 2.8.

The objectives tree is meant to be simple, but illustrative. In actual practice the tree may have many more branchings and levels of branching. The overall value of such diagrams is that they may give insight into overall priorities in a complicated situation. The drawback is that such trees are based upon the designer's personal bias as to the value of each branch, which will be borne out by the values in the final column.

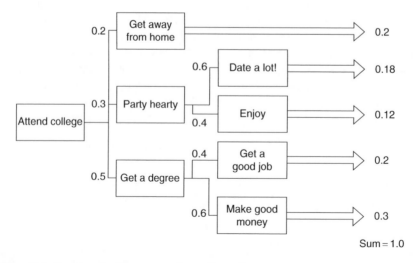

FIGURE 2.8   Objectives tree for the process of attending college.

2.6 INTRODUCTION TO QUALITY FUNCTION DEPLOYMENT DIAGRAMS

The process known as quality function deployment (QFD) is a structured approach to assist in translating customer requirements into realistic goals and sometimes even manufacturing specifications, dependent on the level and complexity of the structures used. This section will give an overview of only the first level of the QFD process, that of building most of the first, and most valuable diagram, which is intended to relate the qualities that a customer wants to the functions that the device or process must perform. The deployment part of the process involves (generally) the manufacturing or other specification processes; this will be left for a later discussion. The first process is in fact an extension of some of the material previously covered. A sample diagram for the process of evaluating a marriage partner will be used to introduce the topic (see Figure 2.9).

The QFD diagram below is similar to the ones that would normally be used in product comparison and product planning exercises. The function of the device is normally plotted on the vertical axis of a grid that is used for the comparison of "your product" and the competitors. In this case, we might posit that the function of marriage is to enable a couple to have legally recognized sex, reproduction, and companionship. The qualities desired in this partner as defined by your customers are elaborated on the left-hand axis. The relationship between qualities and functions is

Marriage partner selection — House of quality

Relationship: ● Strong   ○ Medium   △ Weak

Customer needs		Legal sex (1)	Companionship (2)	Reproduction (3)	Customer rating	Customer rating—US	Customer rating—Company 1	Customer rating—Company 2	Customer rating—Company 3	Planned level	Improvement ratio	Ranking factor	Importance weight	Relative weight
Likes to eat out	1		●		3.0	3	2	2	1	4	1.3	1.0	4.0	20.4
Likes to travel	2	○	○	○	2.0	4	3	2	1	2	0.5	1.0	1.0	5.1
Wants kids	3	●	○	●	5.0	4	3	1	1	4	1.0	1.0	5.0	25.5
Same religion	4		○	△	4.0	5	5	3	2	4	0.8	1.0	3.2	16.3
Intelligent	5	○	○	○	4.0	4	3	4	1	3	0.8	1.0	3.0	15.3
Good looking	6	●	●	●	4.0	5	3	2	1	3	0.6	1.0	2.4	12.2
Good job	7	○	△	△	3.0	3	1	1	1	1	0.3	1.0	1.0	5.1

	1	2	3
Importance weight total	416	486	422
Relative weight total	31	37	32
Target			

Project no: 1001   Date: July 23, 2007
Status: Complete   Distribution: Class notes
Issue: None   Originator: King

FIGURE 2.9 QFD diagram for the process of marriage partner selection.

designated with a symbol such as a filled in circle if they are highly correlated, an open circle if slightly so, a triangle if minimally so, etc. (e.g., likes to eat out correlates well with companionship). This ranking of functions (dependent on the software used) then allows a comparison of the relative importance of each of the functions (so that companionship is most highly desired in this example.) The right-hand side of the figure allows for entry of customer importance of each item (rating), the customer rating of our product (us) versus three competitors, a goal level (planned), a ranking factor for each of the planned levels, an improvement factor desired (planned over desired), a ranking factor for each of the improvements, and a calculation of relative weight (importance) of each of the improvements. One can now use this diagram to "select" a best partner, based, for example on calculations involving the summed product of customer rating times customer rating of "us" versus the three listed competitors. Alternately, the best "improvement" over "us" could be selected based upon the relative weights of improvements (final column) and the customer ratings relative to planned levels.

On occasion, the interactions between the functions are also plotted, using a triangular matrix attached to the top of the function columns. These interactions may also lead to design considerations that may need to be resolved, as there might be a conflict between two of the variables listed. This triangular matrix, when added to the above diagram, gives rise to the name "House of Quality" that is often attached to such a display. These negative interactions may be resolved using other techniques, such as TRIZ, to be discussed in Section 2.7.

2.7   INTRODUCTION TO TRIZ

It was mentioned earlier that the patent database is a site for initial idea solution searches using key word searches. An alternate method would be to look for distillations of patent materials in terms of methods for solving design problems. Such a method was developed initially in the late 1940s by Genrich Altshuller in the U.S.S.R., this method has been continually upgraded and added to in the intervening years. Altshuller held the position of patent clerk in the Russian Navy patent office, his work allowed him to study and distil design solution methodologies from inspection of thousands of patents. His work formed the basis for a series of papers on the theory of inventive problem solving; the Russian acronym for this is TRIZ.

Alshuller's goals were to codify what knowledge he could from the patent database and reduce the design process as much as possible to a step-by-step procedure. Some of his accomplishments are briefly listed below:

1. Altshuller recognized that problem solving ranged from the application of methodology that is commonly used in whatever specialty one is working in to true discovery entailing the development of new science or discovery of new principals. As the level of difficult increased, the number of solutions to be examined increased. Highly difficult problems generally involved the use of material outside of one's own specialty.
2. He recognized that most inventions went through different stages of development with a finite number of variations on different themes of transition. One such theme is that systems tend toward increasing ideality, another that systems tend toward Microsystems, another is that systems tend to less need human involvement.
3. Altshuller's first major observation was that most inventive problems involved solution of technical contradictions (negative interaction between desired functions or between desired qualities). This work gave rise to two useful devices, an inventive principles listing (aka engineering parameters) and a technical contradiction matrix. The principles list (Appendix 3) listed 40 different parameters that can be applied in the design of a system. This initial listing includes such terms as segmentation, asymmetry, extraction, nesting, and so forth. The technical contradiction matrix is a 39 by 39 matrix of contradictions

(Appendix 3), one axis lists all features that conceivably could be changed in a system (such as the weight of a moving object), the other axis repeats this list but is now labeled undesired result. The intersection of two features lists suggested solutions from the principles list. An example intersection point is weight of moving object (say an airplane wing, row 1) where increasing weight compromises strength (column 14) one solution technique is to use composites (solution 40), which is in fact in practice.

4. Altshuller went well beyond this level of work, studying and developing advanced inventive problem-solving techniques that are more algorithmic in nature. Object-action diagramming techniques and directed product evolution arose from this work. These subjects will be included later as specific topics in Chapter 6.

2.8 SUMMARY

This chapter introduced several simple design tools before an in-depth study of the overall process of design as applied to medical devices and processes. As such, it is generic in nature and may be applied to many design processes both in an engineering design environment and in personal decision-making.

EXERCISES

1. Perform a Web search with the search term "brainstorming." Evaluate several of the sites, try some of the software available and report on the usefulness of the program.
2. Do a Web search with the term concept map. Find and explore one or more example concept maps.
3. Draw a process diagram for the process of taking hamburger meat, grinding it, then flattening it and cutting out presized hamburger patties. The meat that is in between the patties is reinserted into the process just after the incoming meat is ground. What is wrong with this process? If necessary do a Web search to answer this question.
4. Visit the Web site www.jellybelly.com and find their process listing. Do a flowchart of this process, specifically identifying delays. Discuss means to speed up this process. Extra credit, request that samples be sent to your instructor.
5. Visit any Web site that has an example concept map that is of interest to you, print out the map, and comment on the value of it.
6. Pick two design terms or terms relating to a project you have worked on. Pick two different search engines and search on these two terms. What are the differences in yield? Would you recommend one search engine over the other? Why?
7. Do a Web search on the term "biomimetics," find a good example of this as applied to a design problem, print it out and discuss it.
8. Generate a simple process chart for the process of brushing teeth.
9. Generate a flowchart for the process of obtaining breakfast. Be sure to indicate delays and make suggestions to decrease same.
10. Generate a simple selection diagram to determine whom you will date for a formal dance.
11. Generate an evaluation chart to assist you in the determination between camping in the mountains or going to the beach for your vacation this year.
12. Generate a simple QFD chart for the selection of an automobile.
13. Generate a QFD diagram to help design a better device for closure of an atrial septal defect.
14. Problem that arose in the early use of long-barreled cannons was that they "wilted" during repeated use due to heating and uneven cooling, especially during rainstorms. Use brainstorming with one or two friends to help solve this problem. Reference the TRIZ contradiction matrix and attempt to find a solution. Document your choices.

BIBLIOGRAPHY

Ball, P., Life's lessons in design, *Nature*, 409, 2001:413–416.

Clarke, D., TRIZ: Through the eyes of an American TRIZ specialist, Ideation International (www.ideationtriz. com), 1997.

de Bono, E., *Six Thinking Hats*. New York: Little Brown and Company, 1985.

John T., *Step-by-Step QFD, Customer-Driven Product Design*, 2nd ed. Boca Raton, FL: St. Lucie Press, 1997.

Pahl, G. and Beitz, W., *Engineering Design, A Systematic Approach*. London: Springer-Verlag, 1988.

3 Design Team Management, Reporting, and Documentation

Never tell people how to do things. Tell them what to do and they will surprise you with their ingenuity.

George S. Patton, Jr.

Design management is a multistep process that is a necessary part of every product development process. Design management consists of

- Design team construction and management
- Documentation techniques and requirements
- Reporting techniques

All form an integral part of success in developing a product. All are interrelated and interdependent. All will be audited by the Food and Drug Administration and quality system auditors.

3.1 DESIGN TEAM CONSTRUCTION AND MANAGEMENT (INDUSTRY BASED)

The team is a basic unit of performance for most organizations. A team melds together the skills, experiences, and insights of several people. It is the natural complement to individual initiative and achievement because it engenders higher levels of commitment to common ends. Increasingly, management looks to teams throughout the organization to strengthen performance capabilities.

In any situation requiring the real-time combination of multiple skills, experiences, and judgments, a team inevitably gets better results than a collection of individuals operating within confined job roles and responsibilities. Teams are more flexible than larger organizational groupings because they can be more quickly assembled, deployed, refocused, and disbanded, usually in ways that enhance rather than disrupt more permanent structures and processes. Teams are more productive than groups that have no clear performance objective because their members are committed to deliver tangible performance results. Teams invariably contribute significant achievements in all areas of a business.

3.1.1 DEFINITION OF A TEAM

At the heart of a definition of team is the fundamental premise that teams and performance are inextricably connected. The truly committed team is the most productive performance unit management has at its disposal, provided there are specific results for which the team is collectively responsible, and provided the performance ethic of the company demands those results.

Within an organization, no single factor is more critical to the generation of effective teams than the clarity and consistency of the company's overall performance standards or performance ethic.

Companies with meaningful, strong performance standards encourage and support effective teams by helping them both tailor their own goals and understand how the achievement of those goals will contribute to the company's overall aspirations. A company's performance ethic provides essential direction and meaning to the team's efforts.

This crucial link between performance and teams is the most significant piece of wisdom learned from teams. It leads directly to the definition: a team is a small number of people with complementary skills that are committed to a common purpose, performance goals, and approach for which they hold themselves mutually accountable.

3.1.2 CHARACTERISTICS OF TEAMS

There are six basic characteristics of successful teams, including

1. Small number
2. Complementary skills
3. Common purpose
4. Common set of specific performance goals
5. Commonly agreed-upon working approach
6. Mutual accountability

The majority of teams who are successful have their membership range from 2 to 25. The most successful have numbered approximately 12. A larger number of people can theoretically become a team, but they usually break into subteams, rather than function as a single team. The main reason for this is that large numbers of people, by virtue of their size, have trouble interacting constructively as a group, much less agreeing on actionable specifics. Large groups also face logistical issues like finding enough physical space and time to meet together. They also confront more complex constraints like crowd or herd behaviors that prevent the intense sharing of viewpoints needed to build a team.

Teams must develop the right skills, that is, each of the complementary skills necessary to do the team's job. These team skill requirements fall into three categories:

1. Technical or functional expertise
2. Problem-solving and decision-making skills
3. Interpersonal skills

A team cannot get started without some minimum complement of skills, especially technical and functional ones. No team can achieve its purpose without developing all the skill levels required. The challenge for any team is in striking the right balance of the full set of complementary skills needed to fulfill the team's purpose over time.

A team's purpose and performance goals go together. The team's near-term performance goals must always relate directly to its overall purpose. Otherwise, team members become confused, pull apart, and revert to mediocre performance behaviors. Successful teams have followed the following premises:

- Common, meaningful purpose sets the tone and aspiration.
- Specific performance goals are an integral part of the purpose.
- Combination of purpose and specific goals is essential to performance.

Teams also need to develop a common approach, that is, how they will work together to accomplish their purpose. Teams should invest just as much time and effort crafting their working approach as shaping their purpose. A team's approach must include both an economic and administrative aspect as well as a social aspect. To meet the economic and administrative challenge, every member of the team must do equivalent amounts of real work that goes beyond commenting, reviewing, and deciding.

Team members must agree on who will do particular jobs, how schedules will be set and adhered to, what skills need to be developed, how continuing membership is earned, and how the group will make and modify decisions, including when and how to modify its approach in getting the job done. Agreeing on the specifics of work and how it fits together to integrate individual skills and advance team performance lies at the heart of shaping a common approach. Effective teams always have team members who, over time, assume different important social as well as leadership roles such as challenging, interpreting, supporting, integrating, remembering, and summarizing. These roles help promote the mutual trust and constructive conflict necessary to the team's success.

No group ever becomes a team until it can hold itself accountable as a team. Like common purpose and approach, this is a stiff test. Team accountability is about the sincere promises team members make to themselves and others, promises that underpin two critical aspects of teams:

1. Commitment
2. Trust

By promising to hold themselves accountable to the team's goals, each member earns the right to express their own views about all aspects of the team's effort and to have their views receive a fair and constructive hearing. By following through on such a promise, the trust upon which any team must be built is preserved and extended.

3.1.3 TEAM SUCCESS FACTORS

There are six team success factors inherent to any effective team:

1. Multifunctional involvement
2. Simultaneous full-time involvement
3. Colocation
4. Communication
5. Shared resources
6. Outside involvement

Multifunctional involvement means representation at least of the following stakeholders:

- Customers
- Dealers
- Suppliers
- Marketers
- Lawyers
- Manufacturing personnel
- Service personnel
- Engineers
- Designers
- Managers
- Nonmanagers

All personnel should be involved with the team from its inception.

Key team members—design, manufacturing, and marketing—must be represented full time from the start. The involvement of others should be full time for the duration of the most intense activity. Rewards should go to teams as a whole. Evaluation, even for members who are only full time for a short time, should be based principally on team performance.

Numerous studies indicate the astonishing exponential decrease in communication that ensues when thin walls or some distance exists between team members. For the most effective environment, team members must be in close proximity. (The best design teams on occasion are called Skunk Works, based upon a famous design team working for Lockheed Martin.)

Communication is everyone's panacea for everything, but nowhere more so than in teams. Examination of successful teams has shown that the most important element in ensuring a team's effectiveness and success is the constant communication across functional boundaries. Regular meetings with all functional areas represented and written status reports circulated to everyone are the norm for effective teams.

Duplication of every resource for every development project is not always a true possibility. However, team research has reported that the sharing of resources between new product/service teams and mainline activities, including manufacturing, marketing, and sales, is a leading cause of delayed product development and introduction efforts. One option is to devote areas of laboratories or manufacturing areas for the new product development efforts.

Suppliers, distributors, and ultimate customers must become partners in the development process from the start. Much, if not most, innovation will come from these constituents, if the team trusts them and vice versa.

3.1.4 TEAM LEADER

Successful team leaders instinctively know that their primary goal is team performance results instead of individual achievement, including their own. Unlike working groups, whose performance depends solely on optimizing individual contributions, real team performance requires impact beyond the sum of the individual parts. Hence, it requires a complementary mix of skills, a purpose that goes beyond individual tasks, goals that define joint work products, and an approach that blends individual skills into a unique collective skill, all of which produces strong mutual accountability.

Team leaders act to clarify purpose and goals, build commitment and self-confidence, strengthen the team's collective skills and approach, remove externally imposed obstacles, and create opportunities for others. Most important, like all members of the team, team leaders do real work themselves. They also believe that they do not have all the answers, so they do not insist on providing them. They believe they do not need to make all key decisions, so they do not do so. They believe they cannot succeed without the combined contributions of all the other members of the team to a common end, so they avoid any action that might constrain inputs or intimidate anyone on the team.

Team leaders must work hard to do the seven things necessary to good team leadership:

1. Keep the purpose, goals, and approach relevant and meaningful.
2. Build commitment and confidence.
3. Strengthen the mix and level of skills.
4. Monitor timing and schedules for planned activities.
5. Manage relationships with outsiders, including removing obstacles.
6. Create opportunities for others.
7. Do real work.

A team leader critically influences whether a potential team will mature into a real team or even a high-performance team. Unless a leader believes in a team's purpose and the people on the team, they cannot be effective.

3.1.5 DESIGN TEAM

The typical product design team is a collection of individuals from various departments within a company who come together for the specific purpose of designing and developing a new medical device. The design team is composed of two subteams:

1. Core product team
2. Working design team

3.1.5.1 Core Product Team

Core product teams are responsible for performing the research required to reduce risks and unknowns to a manageable level, to develop the product specification, and to prepare the project plan. They are responsible for all administrative decisions for the project, regulatory and standards activity, as well as planning for manufacturing and marketing the device.

The core product team is composed of individuals representing the following functions:

- Marketing
- Engineering
- Electrical
- Mechanical
- Biomedical
- Chemical
- Software
- Reliability engineering
- Human factors
- Safety engineering
- Manufacturing
- Service
- Regulatory
- Quality assurance
- Finance

The leader of the core team is usually from engineering or marketing. The leader is responsible for conducting periodic team meetings, ensuring minutes of such meetings are recorded and filed, establishing and tracking time schedules, tracking expenses and comparing then to budgeted amounts, presenting status reports to the senior staff, and ensuring sufficient resources in all areas are supplied. The leader will also provide performance evaluation of each member of the team to line managers.

The approximate amount of time required of each participant as well as incremental expenses, such as model development, simulation software, travel for customer verification activities, laboratory supplies, market research, and project status reviews, should also be estimated.

3.1.5.2 Working Design Team

The members of the working team, primarily engineers, take the product specification and develop the more detailed design specification. Working teams exist in all areas of engineering, including electrical, mechanical, and software. Working team members are responsible for developing designs from the design specification, ensuring all requirements are verified through testing, providing test reports. Certain members may also be responsible for verifying requirements and validating the system as a whole. Individual working teams may be divided into subteams to address individual design assignments.

3.2 STUDENT DESIGN TEAM CONSTRUCTION AND MANAGEMENT

Section 3.1 represents an ideal that must be modified to be workable with student design teams. This section elaborates on several of these differences.

3.2.1 DEFINITION OF A STUDENT TEAM

The definition of a student team must again satisfy the description regarding accountability. The typical student team will need a mentor who is also accountable and willing to supervise a given project, donating an average of an hour or more per week to direct supervision of the design project. Additionally, an overall course supervisor should be available to intervene if necessary in the conduct of the design work, overall evaluation of a class, and coordination of judging exercises.

3.2.2 CHARACTERISTICS OF STUDENT TEAMS

By definition, the six characteristics of successful teams will again apply. Teams should generally be of a small number, team limits of five per team are fairly common. The pressures due to the scheduling of meeting times for larger groups often prove to be insurmountable. For many biomedical engineering problems, an interdisciplinary team (ME + EE + BME, e.g., reflecting the needs of the project) is more likely to succeed to the development (at least) of a prototype than a single major group. For long-term projects, a mixture of upperclassmen and underclassmen (if possible) can provide continuity of effort.

3.2.3 TEAM SUCCESS FACTORS

The six team factors for success apply directly to student design teams. Item six, "outside involvement", especially on the part of both the course instructor and the project mentor, is critical due to the general inexperience of the group with the (typical) material at hand. Communication must be positive; collocation should be strived for at least for weekly planning/work delegation meetings. Overall effort levels should be approximately the same for each member, ranging from 2 to 3 h per week per credit hour for the course.

3.2.4 TEAM LEADER

The initial choice of a team leader should be based upon having the requisite academic and social skill set to supervise a particular design project and team. Some design instructors will select the initial leader; others will allow the group to elect a member. With either method, the design group should not feel constrained to keep with a particular leader; the best leader may depend on the phase of the given project.

3.2.5 SPECIAL CONSIDERATIONS FOR STUDENT DESIGN TEAMS

Student design teams will often go through team developmental phases without realizing that the interrelationships between members and within the project are evolving. These phases may be summarized as follows:

- In phase 1, the group dynamic is one of forming, the group is setting its purpose, its structure, and its membership, and individual duties.
- There is almost universally a phase 2 labeled "storming," in which conflicts arise regarding group purpose, individual expectations, leadership is called into question, the whole idea of doing a design project seems overwhelming, etc. This is a critical phase, which will be addressed again below.

- Phase 3 is generally termed "norming," wherein management of relationships and tasks once again become normalized, and real work begins.
- Phase 4 is often called performing; tasks are completed, the project is evaluated, and there is a sense of completion of the project.
- On occasion, with the completion of a project, there is a sense of loss, occasionally termed "adjourning."

It is useful to question the above listing, as not all groups go through all phases. Most groups will go through a period of crisis, where an internal or external assist may be needed.

Some of the options available to groups include the following:

1. Disband the group—if it is early enough new projects may be found and graduation requirements therefore may be met, otherwise there is a major penalty.
2. Request intervention on the part of your advisor or instructor.
3. Try to talk it out in a nonconfrontal manner.
4. Use peer evaluation tools to provide feedback, both positive and negative, to each of your team members.
5. Check with your instructor to see if there is a mechanism in place to give different grades to different students, based upon peer evaluation as well as instructor and supervisor grading. If there is, be sure to understand the ground rules and be sure all team members agree to abide by these rules.
6. Do nothing.
7. Reorganize.

Items 3 and 4 are useful if the project is early on (early one third of a project). Constructively addressing problem areas in a relaxed manner can be rewarding. Peer evaluation forms may be found that allow one to rank performance ($5 =$ outstanding, $1 =$ unsatisfactory) versus skills and behaviors (planning, teamwork, time management, motivation, etc.) Many will additionally prompt with questions such as what are their strengths?, what areas need improvement?, etc. If the group is large enough, responses can be anonymous. These same forms may later be used with item 5, if possible. Disbanding a project is seldom warranted, unless the project is very early on. Doing nothing is not recommended. If conflicts may be simply solved with reorganization, try this method.

3.3 REPORTING TECHNIQUES: WRITTEN AND ORAL, POSTERS, WEB SITES

Reporting methods vary considerably dependent on the nature of the project (industrial vs. academic), the size of the team and of the project, and the expectations of the person(s) to whom the report is being made. Typical reporting techniques involve oral presentations using transparencies or PowerPoint Slides, Poster presentations (especially in academic settings), and formal reports of progress or results (Web or hard copy). For the student and advisor, a combination of these techniques with the addition of a Web site can be very useful.

3.3.1 PROGRESS REPORTS: WRITTEN

Progress reports, at the lowest level, are fairly simple documentation, generally on paper or in a Web section, listing the following items: current status, work completed, current work, and future work. An example progress report might read like the following:

Progress report: EKG transmitter project—week 7 of 11
Current status: We recently completed our library and Web search for applicable EKG transmitter designs. It appears that the transmitter system designed by Goldman et al. (see references) is both out of patent and a prime candidate for reverse engineering.

Work completed: We were able to x-ray the transmitter we borrowed from Dr. Sachs; we were able to do a component count and figure out what brand of battery system they used.

Current work: B. will meet with EE Professor Heller this week and go over a chip level redesign of the amplifier system. C. will review chip level transmitter systems, and encapsulation materials.

Future work: We appear to be on schedule for the planned end date of next month. Our earlier request for funding has been adequate to obtain the necessary supplies and equipment.

An example progress report (NOT!) follows as an antithesis:

Progress report: Transmitter project, week 12 of 11 (sorry this is late.)

Current status: We have been unable to meet with Dr. Wilson as he is out of town. We came by your office but did not find you in to explain our problem. It will not be our fault if we do not complete this project.

Work completed: We had exams last week and next week we have spring break so we have not had time to do anything on this project. It can wait.

Current work: Packing my bags for spring break.

Future work: We will place our order for parts when we get back. We are sure X university will quickly get the purchase order in the mail and that the advertised 6-week delivery time is a gimmick. Got to go catch some rays!

Obviously, this is fairly low-level reporting, and should serve as a minimum reference. More elaborate reporting schemes would involve additional line items from the original listing of the problem specification (the who, what, where, why list) and elaborations on the details of current budget levels, current interactions with all interested parties (design, manufacturing, sales, etc.) and a good discussion of status with respect to the original detailed timelines and specifications.

In general, these progress reports should include enough material that your advisor and course instructor can properly evaluate your progress and give you feedback (and funding) as necessary.

In an industrial setting, often in the interest of brevity and time, memos may be constrained to one page or less for status meetings.

3.3.2 ORAL REPORTING

Oral reporting of progress includes the above terms at whatever level of complexity is required to convey the information to the audience. PowerPoint presentations will generally convey information better than those using transparencies if and only if they are properly done. Some general rules to follow are

- Use your slide area well, but place no more than six to eight lines of information on a page. Text needs to be used sparingly; your job is to fill in the blanks, not read material to your audience.
- Use color and specific colors judiciously. If possible, use color to make a point.
- Use motion (PowerPoint) sparingly. Overuse of materials flying in from different areas will quickly lose the audience.
- Learn your style of lecturing; determine if you are a one slide per minute or a one slide per 3 min speaker. Any talk faster than two to three slides per minute tends to be irritating and likely will lose the audience.
- Use graphics if they assist in understanding the talk.
- If your talk is more than 15 min, consider some way to interest (awaken) the group, via a personal account or a clean joke.
- Be sure to tell the group what you are going to tell them; then, tell them and summarize what you told. (Someone may have slept through two of the three!)

- Be sure that you cover your bases, double-check that you have done the who/what/why . . . material.
- Practice your talk. Do not overpractice your talk. Give a dry run if possible in front of a coworker or fellow student whom you can trust to give you valid feedback.
- Consider whether or not to give your audience handouts of your slides to retain your message(s).
- If necessary, to avoid nervousness, visualize your audience in their underwear. They are now the ones to be embarrassed.

Sufficient detail should be given in the assigned time frame to allow your course advisor or mentor to properly evaluate your current and pending work. You should expect to get feedback from your advisors and other audience members, and you will be expected to actively participate in other student feedback sessions.

3.3.3 POSTER PRESENTATIONS

In academic circles, if you really want to meet people that are interested in your work, and want a one-to-one discussion with them, poster presentations are a good method. For reporting of student work in an academic environment, there seems to be about an even split in the reporting schemes (oral vs. written) used in courses. Some general rules for poster presentations follow:

- Know the size of the poster you are going to place your work on. Typically, a board will be provided which measures 6 or 8 ft wide by 4 ft tall, elevated so that the bottom of the board is approximately 2 ft off the ground. However, if you are reporting in a foreign country, sizes may be 1 m wide by 2 m high.
- Plan ahead and lay out your poster presentation on a marked off floor area before packing it. Or, as is now the more prevalent, design your entire poster using software such as PowerPoint, and try different color schemes and layout without having to print and test.
- Check to see what method of attachment is allowed. Some situations call for pushpins, others for double-sided tape, etc. Bring your own if unsure.
- Title of your poster should appear at the top in capital letters, at least 4 to 6 in. high, readable at a distance of 6 ft. Use upper and lowercase; block lettering using all caps is not a good form. If you can do so, put the title on a continuous sheet of paper, rather than on pasted together single sheets. Author names and affiliations are best placed below this banner, using a slightly smaller font, and perhaps italics. If you have access to a poster printer, use it. The visual appeal of a well-designed single sheet poster is generally far superior to that composed of series of 8½ in. by 11 in. sheets of paper.
- Subsections should also be legible at a distance and should be abstractions of the primary points of the poster, rather than a text or textbook presentation. Be sure all key points are covered (abstract, introduction, . . . , conclusion, references) as necessary.
- Bold text is generally easier to read, but check this, as not all fonts are easy to read bolded.
- Use color in your text to make a point, if appropriate.
- Your poster should read from the top left, in vertical columns, to the bottom right. If you need to change this, be sure to use arrows to interconnect your panels to help guide your readers.
- Color highlighting of your text blocks, which may also be on colored paper, will make your poster more attractive, if the colors are complementary.
- Use pictures, diagrams, cartoons, figures, etc., rather than text wherever possible. If you are allowed to do so, prepare handout material to supplement your poster. Be sure to include contact information.
- Your poster should be self-explanatory. You may be talking to one person; another can be reading and deciding if they wish to wait to get additional information.

- If appropriate and allowed, bring in additional materials, such as a computer to give a visual demonstration of your work. Use sound appropriately, if at all. Do not induce headaches in yourself and the adjacent poster presenters with inappropriately loud or obnoxious sounds.
- Prepare brief comments for questioners.
- Retrieve your materials at the time stated; otherwise you may lose the effort put into their development.
- Test drive your poster with fellow students, if possible.
- As appropriate, carry and hand out your business card. Your next job may come of your interactions here.

3.3.4   WEB SITES

For student design projects, the development of a Web site for the project is a good communication tool, both for the students and the advisors. If done appropriately, the site may also prove of value in a job hunt. Items that should be posted on a Web site generally include weekly progress reports, monthly oral PowerPoint slide shows, end of term poster, term paper, etc. Optional items, dependent on the course requirements, might be an initial project proposal, Gantt charts, 510(k) drafts, patent drafts, safety analyses, design notes, design history file, a corrected proposal, design notes and calculations, and the like. Your instructor should give you guidelines as to content and timing of each of the above items. Design considerations, paralleling those above for posters, should also be considered in the layout of the site, such that it is easy to read and to navigate.

If a project has a possibility of patentability or may involve intellectual property of any kind (or may be considered potentially offensive to some group), the project Web site might be password protected. This should be set up so that your team, instructor, and advisor have access. As most Web site software can also can time stamp each of the page entries, this information may prove of value if a patent is sought on the material contained on your site.

3.3.5   EXPECTATIONS FOR COMMUNICATIONS

The above communication tools involve reporting of your efforts to others. It is important that you understand, before you engage a topic fully, the criteria under which your work will be judged. Each of the above sections will be discussed in turn.

Weekly written reports, unless made very formal and lengthy, should generally serve as an indicator that the team is on track. If there are immediate needs (financial or otherwise), these should be indicated here or via e-mails to the instructor/supervisor. A weekly report might count toward 1% of the term grade, and generally will be a function of being posted on time and complete (binary).

Oral reports may also be graded, with instructor feedback on such items as presentation skills and organization, technical content, and determination that the solution being pursued (dependent on the phase of the term) is correct given the demands and constraints of the project at hand. Graded (or not) the oral sessions should provide a means for feedback to the group regarding conduct of their project. As oral reports will in general demand more student effort, they typically will weigh in at about three times the value of a written (short) progress report.

In industry, and to a limited extent in academia, the term "at expectations" is often used. For the academic, it means (generally) that the person has published papers, done service work, done research, and has taught at some expected level. In industry, in the context of this text, the term at expectations will mean that the person performed at a level of competence expected of the experience and pay level in a given situation. In both cases, the term applies to such matters as promotions and pay raises. Poster presentations and final paper evaluations may use this same terminology in an attempt to describe very different outputs from different teams on different tasks. Such a rubric will be discussed next.

When grading design projects, there are several main topical items that must be considered. These will typically include such major topics such as engineering goals, creative ability, thoroughness, overall competence in design, clarity of expression, and ethical/societal/political considerations. These topics may then be subdivided (and will be below) and judged individually. Evaluations on the subdivisions of each of these may include the terms such as at expectations, above expectations, below expectations, not applicable, and failure. Failure is a grade reserved for an item that is mandatory (given the project at hand) but very poorly done (or not at all). The term "not applicable" would apply in a case where an item does not need to be considered, at least in depth (e.g., a market analysis on a device custom made for an individual).

Each of the above topical items and their subdivisions will now be discussed in detail.

The engineering goals section will typically include three mandatory sections. First is a problem statement, where the evaluator asks the questions: Did the project have a clear problem definition, identifying constraints and alternatives? Is the work properly based upon customer requirements (often termed demands and wishes)? Second, did the team prototype, or at least test or predict the performance of their solution? Has the group considered manufacturability of the proposed solution? If the product is a database, has it satisfied the requirements of the situation? Is there at least a proof of concept? Third, is the problem solution, at whatever level, properly documented? Is there a prototype (preferred)? Is there at least a feasibility analysis? Is the design valid? Does it meet standards? A fourth section, if needed, would include safety health, and risk analysis for the project outcome. An estimate of environmental impact should be done here, if necessary. Some level of mathematical analysis of risk should be done. A fifth, and final section, if needed, should include economic and market considerations. Questions such as—can it be made? Can people afford it? What is the number of potential users? What are the outcomes of a market survey?—should be answered.

The second section of the poster or report grading form considers creative ability, with a mandatory section evaluating the team's approach to the problem at hand. The questions, did the team use a logical analysis of the problem at hand, were alternative solutions considered and evaluated? A second, but not mandatory consideration would involve determining the originality of the solution, with potential contributions to engineering knowledge, and perhaps to the patent literature.

A third section addresses thoroughness of the report. It requires documentation of effort in the form of a project notebook, design file, or other file, such as a Web site. A second requirement is that a thorough literature and patent search be done, as necessary. Third, as necessary, a review of applicable standards is optional, as needed.

A fourth section addresses the team's overall competence in design. Required first is proof that proper engineering skills tool, and techniques were applied to the problem at hand. A second requirement is a demonstrated ability to design a process, system, or component. An optional judgment may involve the scope of the problem tackled, and whether the original problem was difficult or not.

The fifth set of judging terms involve clarity of presentation. A first criteria involves the engineering layout of the poster or paper—is it well laid out, easy to read, a good mix of text and pictures? Second, is it a good stand-alone poster or paper, or is additional information needed? Third, does it convince the reader that it is a good solution? Last, and optionally, is there evidence of teamwork in this presentation?

The final, but very important judging criteria (mandatory) asks about the ethical and political considerations (if any) involved in the solution, in an attempt to judge the aspects of universal design that might not be addressed elsewhere.

The above list is a codification of experience in judging design posters and presentations over a period of years. Weighting given to each section is 20%, 20%, 20%, 25%, 10%, and 5%. "At expectations" is defined as 85%; an oval grade of 85% is therefore a B in a design course. This scoring system puts emphasis on overall outcome, rather than perceived quality, such as for a refereed journal. A review form for a journal might include such terms as content (20%), degree of novelty or originality (10%), structure of paper (10%), quality of text (10%), and reviewers' general opinion and comments (50%), which stresses the reviewer's personal biases.

3.4 INTRODUCTION TO DATABASES

Throughout the design process, data will be generated that may need to be managed in one of the two ways: storage in an Excel spreadsheet for later documentation or analysis purposes or storage in a database for similar purposes. Design projects may also involve the design of such a spreadsheet and an overlay of software for data analysis, or design of a database for data storage and subsequent data querying and reporting. At the graduate level, this latter analysis may include such techniques as knowledge discovery, an assembly of techniques used to derive rules from data collected in an environment. This section introduces the needed concepts involved here before your use of them, which may be mandated by the problem at hand.

3.4.1 EXCEL® SPREADSHEETS

Excel spreadsheets are useful in situations where data fields are essentially flat; data can be managed adequately in a simple two-dimensional array or arrays (aka multiple worksheets). Data that is nonrepetitive can be easily managed using a spreadsheet; data that is repetitive, such as individual patients' demographics for each clinic visit is better handled with a database. Excel spreadsheets are useful for data sets that do not exceed 32,000 data points in length, after this typically data must be chunked in multiple spreadsheets or put into databases. Excel spreadsheets are optimized for simple statistical and other analysis of data and for easily generated plots of data sets. With the use of the Visual Basic editor, some very useful data entry/calculation programs may be generated.

Such programs may include elementary electrocardiogram analyses, simple lab test statistics and documentation, real-time display of data, clinic utilization statistics, what-if analyses, etc. More mundane applications include the storage of design specifications and change orders, verification and validation documentation, and straightforward safety process documentation. The need for these databases will be covered in later chapters.

3.4.2 DATABASES

Databases are very much in use in the field of design; modern society probably could not junction without this invention. Databases are simply a convenient and (should be an) efficient method of storing data, with a high-level language that allows convenient manipulation of the data. Properly designed databases are efficient in storage of data and in fact can reduce costs due to rapid retrieval of data. Redundant entry of information, such as the address of a supplier, is entered only once in a table, rather than in multiple occurrences when the supplier is referenced. Commercial databases include DB2, SQL (Structured Query Language) server, FoxPro, Access, Oracle, Sybase, Informix, and Paradox. With competition, this field will likely narrow soon. Each has advantages, dependent on your background and the size of the problem you are trying to solve. Access, for example, might work well in an initial design for a small clinic database, but growth to a larger clinic or the use of multiple simultaneous data entry points would push a designer to SQL or better server systems.

Most databases have the following in common. Data that would otherwise be repeated is keyed in once into a structure termed a table. Data that is entered in a table column (field) generally has a given structure (date, alphanumeric, number, etc.); this can be checked for integrity as well as check for reality during entry. The structure of the database allows for relationships between tables, for example, one table may link to several others (one patient links to multiple cases) or may link to only one other table (one patient, one home address). Tables link through keys; such a key might be a patient's social security number, or a patient encounter number generated by a clinic. Data entry techniques can involve the generation and utilization of forms. Data extraction techniques involve the use of a query and the reporting using another form generated for this use.

Data from databases may be exported for use in spreadsheets, and vice versa; thus mastery of databases is not necessary for some work in data analysis.

3.4.3 EXAMPLE FOR DATABASE DEVELOPMENT

In the early 1990s, the pain clinic in a major hospital was simply using paper forms to capture all information. The forms were used for an initial patient interaction/interview/evaluation (five pages), subsequent psychological patient evaluations if needed (one page each), and subsequent physical/medical evaluations as needed (multiple pages). The initial interaction form held the expected patient demographics information, hospital identification number, referring physician information, pain history, and medical examination information, assessment, diagnosis, and plan of treatment. The psychological assessment plan included a current psychological evaluation, testing results, treatment plan, and other activities. The medical follow-up paperwork included an evaluation of the pain history since the initial visit, treatment for the pain (such as injections, medications, counseling, physical or occupational therapy, biofeedback, relaxation technique training, biofeedback, or transcutaneous electrical nerve stimulation), effects of the treatment, drugs prescribed, pain evaluation, diagnosis, and follow-up plans.

As is common in a circumstance such as this, three main driving forces pushed this clinic toward the use of a database. First, their record-keeping system was entirely dependent on preserving and accessing the paper records generated, lost or incomplete paperwork meant lost revenue. Second, the patient population in the clinic was gradually increasing, putting more of a demand on the one secretary/filing clerk available. Third, and most important, was the pressure from insurers to adequately document and report consistently all interventions performed and the reasons for the interventions, at the risk of nonpayment of billed services.

The above scenario led to the initial development of an Access database. Three tables were created, one for the initial visit record system, one for all psychological evaluation/treatment interventions, and one for all medical treatment interventions. The link between all three tables (key) was set to be the patient identifier; the second identifier to keep visits unique was set to be the date of service. An initial system was developed with a paper form tool for data entry (teleforms) such that information was entered using block letters and numbers in a standard fill in the blank method. A later version was based upon the same information layout, but using a direct computer entry method. The disadvantage of the paper-based forms was the need for a paper intermediate step to data entry, and the resultant errors due to lack of data entry and misread data forms due to sloppy form copying techniques. The direct terminal entry of data allows the programmer to validate each data entry field as entered and to warn of incomplete data entry on a given patient.

Two main advantages can be gained by development of this database. All insurance information can be adequately entered, documented, and billed for using the report functions in Access. Patient summary letters can be generated for the referring physician also using the report writing functions of Access.

EXERCISES

1. Take the material written just above on the rules for a poster session; generate a PowerPoint presentation that conveys the same thoughts in a more vital fashion.
2. For the material on PowerPoint presentations, demonstrate several of the points made using a PowerPoint presentation.
3. Perform a Web search for optical character recognition. Comment on the range of uses for this method of data entry. Comment on some of the disadvantages of this method.
4. Draft a design for a computer method that would contain the relevant information needed to catalog equipment used in a medium-sized biomedical engineering department. At what point would you consider the use of Access over Excel?

5. You are in charge of developing a database for a drop-in clinic for a medium-sized city. What would be some of the key parameters you would need to enter on every patient? Discuss briefly.
6. Perform a Web search using the term "teleforms"; comment on the uses outlined.
7. Construct a design team exercise during or after class to tackle a design exercise. Reporting will be done orally by one of the team members. Members must take one of the following roles: marketing, manufacturing/distribution, legal/safety, engineering, or team leader; members are responsible for assuming their roles on the design team. Design topics could include any one of the following:

- Design a device to detect SIDS in an infant.
- Design an automated EEG electrode placement system.
- Design a device to track Alzheimer's patients locations.
- Design a system to track asthmatics location and sample the environment for noxious stimuli.
- Design a head restraint system for race car drivers.
- Design a pain clinic database.
- Design a system to quantify male or female arousal in an MRI machine.
- Any other design suggested by your instructor.

BIBLIOGRAPHY

Carr, D. et al., *The Team Learning Assistant Workbook*. Boston, FL: McGraw Hill, 2005.
Chevlin, D.H. and Joseph, J. III, Medical device software requirements: Definition and specification, *Medical Instrumentation*, 30(2), March/April 1996.
Davis, A.M. and Pei, H., Giving voice to requirements engineering, *IEEE Software*, 11(2), March 1994.
Dubnicki, C., Building high-performance management teams, *Healthcare Forum Journal*, May–June 1991.
Fairly, R.E., *Software Engineering Concepts*. New York: McGraw-Hill, 1985.
Fries, R.C., *Reliable Design of Medical Devices*. New York: Marcel Dekker, 1997.
Gause, D.C. and Gerald, M.W., *Exploring Requirements: Quality Before Design*. New York: Dorset House, 1989.
Goodman, P.A. and Associates, *Designing Effective Work Groups*. San Francisco, CA: Jossey-Bass, 1986.
Hackman, J.R., Ed., *Groups That Work (And Those That Don't)*. San Francisco, CA: Jossey-Bass, 1990.
Hirschhorn, L., *Managing in the New Team Environment*. Reading, MA: Addison-Wesley, 1991.
Hoerr, J., The payoff from teamwork, *Business Week*. July 10, 1989.
House, C.H., The return map: Tracking product teams, *Harvard Business Review*. January–February, 1991.
Katz, R., High performance research teams, *Wharton Magazine*. Spring, 1982.
Katzenbach, J.R. and Douglas K.S., *The Wisdom of Teams*. New York: Harper Collins, 1999.
Kidder, T., *The Soul of a New Machine*. Boston, FL: Atlantic-Little, Brown, 1981.
Kotkin, J., The smart team at Compaq computer, *Inc.*, February, 1986.
Larson, C.E. and Frank M.L., *Teamwork: What Must Go Right/What Can Go Wrong*. Newbury Park, CA: Sage Publications, 1989.
Pantages, A., The new order at Johnson wax, *Datamation*. March 15, 1990.
Peters, T, *Thriving on Chaos*. New York: Alfred A. Knopf, 1988.
Potts, C., Kenji T., and Annie, I.A., Inquiry-based requirements analysis, *IEEE Software*, 11(2), March 1994.
Smith, P.G. and Donald G.R., *Developing Products in Half the Time*. New York: Van Nostrand Reinhold, 1991.
Weisbord, M.R., *Productive Workplaces*. San Francisco, CA: Jossey-Bass, 1989.

4 Product Definition

The greatest mistake in the treatment of diseases is that there are physicians for the body and physicians for the soul, although the two cannot be separated.

Plato

Medical devices are an important part of health care. Yet they are an extraordinarily heterogeneous category of products. The term "medical device" includes such technologically simple articles as ice bags and tongue depressors on one end of the continuum and very sophisticated articles such as pacemakers and surgical lasers on the other end. Perhaps it is this diversity of products coupled with the sheer number of different devices that makes the development of an effective and efficient regulatory scheme a unique challenge for domestic and international regulatory bodies.

The patient is the ultimate consumer of medical devices, from the simplest cotton swab to the most sophisticated monitoring devices. However, with the exception of some over-the-counter products, the medical device manufacturer rarely has a direct relationship with the patient in the marketplace. Unlike many other consumer products, a host of intermediaries influence the demand for medical devices. These intermediaries include policy makers, providers, and payers of health care services.

4.1 WHAT IS A MEDICAL DEVICE?

There are as many different definitions for a medical device as there are regulatory and standards organizations. Though the definitions may differ in verbiage, they have a common thread of content. Two of the more popular definitions are reviewed below.

4.1.1 FOOD AND DRUG ADMINISTRATION DEFINITION

Section 201(h) of the Federal Food, Drug, and Cosmetic Act defines a medical device as

an instrument, apparatus, implement, machine, contrivance, implant, in vitro reagent, or other similar or related article, including any component, part, or accessory, which is

- recognized in the official National Formulary, or the United States Pharmacopeia, or any supplement to them
- intended for use in the diagnosis of disease or other conditions, or in the cure, mitigation, treatment, or prevention of disease, in man or other animals, or
- intended to affect the structure or any function of the body of man or other animals, and

which does not achieve any of its principal intended purposes through chemical action within or in the body of man or other animals and which is not dependent upon being metabolized for the achievement of any of its principal intended purposes.

The Medical Device Amendments of 1976 expanded the definition to include

- Devices intended for use in the diagnosis of conditions other than disease, such as pregnancy
- In vitro diagnostic products, including those previously regulated as drugs

A significant risk device is a device that presents the potential for serious risk to the health, safety, or welfare of a subject and is (1) intended as an implant, (2) used in supporting or sustaining human life, and (3) of substantial importance in diagnosing, curing, mitigating, or treating disease or otherwise preventing impairment of human health. A nonsignificant risk device is a device that does not pose a significant risk.

4.1.2 MEDICAL DEVICE DIRECTIVES DEFINITION

The various Medical Device Directives (MDD) define a medical device as

any instrument, appliance, apparatus, material or other article, whether used alone or in combination, including the software necessary for its proper application, intended by the manufacturer to be used for human beings for the purpose of

- Diagnosis, prevention, monitoring, treatment or alleviation of disease
- Diagnosis, monitoring, alleviation of or compensation for an injury or handicap
- Investigation, replacement or modification of the anatomy or of a physiological process
- Control of conception

and which does not achieve its principal intended action in or on the human body by pharmacological, immunological or metabolic means, but which may be assisted in its function by such means.

One important feature of the definition is that it emphasizes the "intended use" of the device and its "principal intended action." This use of the term "intended" gives manufacturers of certain products some opportunity to include or exclude their product from the scope of the particular Directive.

Another important feature of the definition is the inclusion of the term software. The software definition will probably be given further interpretation, but is currently interpreted to mean that (1) software intended to control the function of a device is a medical device, (2) software for patient records or other administrative purposes is not a device, (3) software which is built into a device, e.g., software in an electrocardiographic monitor used to drive a display, is clearly an integral part of the medical device, and (4) software update sold by the manufacturer, or a variation sold by a software house is a medical device in its own right.

4.2 PRODUCT DEFINITION PROCESS

Numerous methods of obtaining new product information exist. They include various ways of collecting data, such as internal sources, industry analysis, and technology analysis. Then the information is screened and a business analysis is conducted (Figure 4.1). Regardless of the method of obtaining the information, there are certain key questions:

- Where are we in the market now?
- Where do we want to go?
- How big is the potential market?
- What does the customer really want?
- How feasible is technical development?
- How do we get where we want to go?
- What are the chances of success?

4.2.1 SURVEYING THE CUSTOMER

The customer survey is an important tool in changing an idea into a product. The criticality of the survey is exhibited by an estimate that, on an average, it takes 58 initial ideas to get one

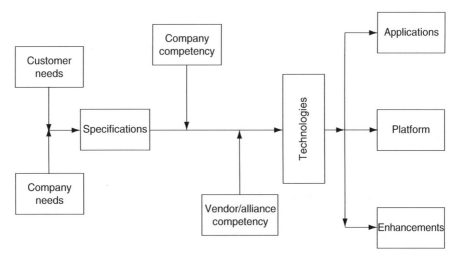

FIGURE 4.1   The product definition process.

commercially successful new product to the market. It is therefore necessary to talk with various leaders in potential markets to build a credible database of product ideas.

The goal of the customer survey is to match the needs of the customer with the product concept. Quality has been defined as meeting the customer needs. So a quality product is one that does what the customer wants it to do. The objective of consumer analysis is to identify segments or groups within a population with similar needs so that marketing efforts can be directly targeted to them. Several important questions must be asked to find that market which will unlock untold marketing riches:

- What is the "need" category?
- Who is buying and who is using the product?
- What is the "buying" process?
- Is what I'm selling a high- or low-involvement product?
- How can the market be segmented?

4.2.2   Defining the Company's Needs

While segmentation analysis focuses on consumers as individuals, market analysis takes a broader view of potential consumers to include market sizes and trends. Market analysis also includes a review of the competitive and regulatory environment. Three questions are important in evaluating a market:

- What is the "relevant" market?
- Where is the product in its product life cycle?
- What are the key competitive factors in the industry?

4.2.3   What Are the Company's Competencies?

Once a market segment has been chosen, a plan to beat the competition must be chosen. To accomplish this, a company must look at itself with the same level of objectivity it looks at its competitors. Important questions to assist in this analysis include

- What are our core competencies?
- What are our weaknesses?
- How can we capitalize on our strengths?
- How can we exploit the weaknesses of our competitors?

- Who are we in the marketplace?
- How does my product map against the competition?

4.2.4   What Are the Outside Competencies?

Once a company has objectively looked at itself, it must then look at others in the marketplace:

- What are the strengths of the competition?
- What are their weaknesses?
- What are the resources of the competition?
- What are the market shares of the industry players?

4.2.5   Completing the Product Definition

There are many other questions that need to be answered to complete the product definition. In addition to those mentioned earlier, an organization needs to determine

- How does the potential product fit with our other products?
- Do our current technologies match the potential product?
- How will we differentiate the new product?
- How does the product life cycle affect our plans?

It is also important to consider the marketing mix of products, distribution networks, pricing structure, and the overall economics of the product plan. These are all important pieces of the overall product plan as developed in a business proposal. However, the needs and wants of the customer remain the most important information to be collected. One method of obtaining the required customer requirements is quality function deployment (QFD).

4.3   OVERVIEW OF QFD

QFD is a process in which the "voice of the customer" is first heard and then deployed through an orderly, four-phase process in which a product is planned, designed, made and then made consistently. It is a well-defined process that begins with customer requirements and keeps them evident throughout the four phases. The process is analytical enough to provide a means of prioritizing design trade-offs, to track product features against competitive products, and to select the best manufacturing process to optimize product features. Moreover, once in production, the process affords a means of working backward to determine what a prospective change in the manufacturing process or in the product's components may do to the overall product attributes.

The fundamental insight of QFD from an engineering perspective is that customer wants and technical solutions do not exist in a one-to-one correspondence. Though this sounds simplistic, the implications are profound. It means that product features are not what customers want; instead, they want the "benefits" provided by those features. To make this distinction clear, QFD explicitly distinguishes between customer attributes that the product may have and technical characteristics that may provide some of the attributes the customer is looking for. Taking a pacemaker, as an example, the customer attribute might be that the patient wants to extend their life, while the technical characteristic is that the pacemaker reduces arrhythmias.

4.4   QFD PROCESS

The QFD process begins with the wants of the customer, because meeting these is essential to the success of the product. Product features should not be defined by what the developers think their customers want. For a clear product definition that will lead to market acceptance, manufacturers must spend both time and money learning about their customer's environments, their constraints,

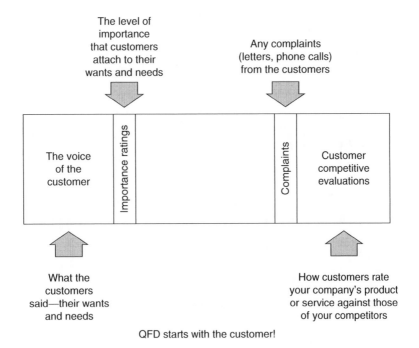

FIGURE 4.2   The customer information portion of the matrix.

and the obstacles they face in using the product. By fully understanding these influencers, a manufacturer can develop products that are not obvious to its customers or competitors at the outset, but will have high customer appeal.

QFD should be viewed from a very global perspective as a methodology that will link a company with its customers and assist the organization in its planning processes. Often, an organization's introduction to QFD takes the form of building matrices. A common result is that building the matrix becomes the main objective of the process. The purpose of QFD is to get in touch with the customer and use this knowledge to develop products that satisfy the customer, not to build matrices.

QFD uses a matrix format to capture a number of issues pertinent and vital to the planning process. The matrix represents these issues in an outline form that permits the organization to examine the information in a multidimensional manner. This encourages effective decisions based on a team's examination and integration of the pertinent data.

The QFD matrix has two principal parts. The horizontal portion of the matrix contains information relative to the customer (Figure 4.2). The vertical portion of the matrix contains technical information that responds to the customer inputs (Figure 4.3).

4.4.1   Voice of the Customer

The voice of the customer is the basic input required to begin a QFD project. The customer's importance rating is a measure of the relative importance that customers assign to each of the voices. The customer's competitive evaluation of the company's products or services permits a company to observe how its customers rate its products or services on a numerical scale. Any complaints that the customers have personally registered with the company serve as an indication of dissatisfaction.

4.4.2   Technical Portion of the Matrix

The first step in developing the technical portion of the matrix is to determine how the company will respond to each voice. The technical or design requirements that the company will use to describe

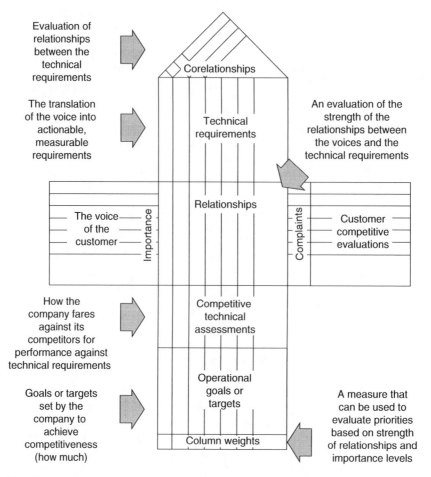

FIGURE 4.3 The technical portion of the matrix.

and measure each customer's voice are placed across the top of the matrix. For example, if the voice of the customer stated "want the control to be easy to operate," the technical requirement might be "operating effort." The technical requirements represent how the company will respond to its customers' wants and needs.

The center of the matrix, where the customer and technical portion intersect, provides an opportunity to record the presence and strength of relationships between these inputs and action items. Symbols may be used to indicate the strength of these relationships. The information in the matrix can be examined and weighed by the appropriate team. Goals or targets can be established for each technical requirement. Trade-offs can be examined and recorded in the triangular matrix at the top of Figure 4.3. This is accomplished by comparing each technical requirement against the other technical requirements. Each relationship is examined to determine the net result that changing one requirement has on the others.

4.4.3 OVERVIEW OF THE QFD PROCESS

The QFD process is a nine-step process consisting of

1. Determining the voice of the customer
2. Customer surveys for importance ratings and competitive evaluation
3. Developing the customer portion of the matrix

4. Developing the technical portion of the matrix
5. Analyzing the matrix and choose priority items
6. Comparing proposed design concepts and synthesize the best
7. Developing a part planning matrix for priority design requirements
8. Developing a process planning matrix for priority process requirements
9. Developing a manufacturing planning chart

In planning a new project or revisions to an old one, organizations need to be in touch with the people who buy and use their products and services. This is vital for hard issues, such as a product whose sales are dependent on the customers' evaluation of how well their needs and wants are satisfied. It is equally crucial for softer issues, such as site selection and business planning.

Once the customers' wants and needs are known, the organization can obtain other pertinent customer information. Through surveys, it can establish how its customers feel about the relative importance of the various wants and needs. It can also sample a number of customers who use its products and competitors' products. This provides the customers evaluation of both the organization's performance and that of its chief competitors.

Records can be examined to determine the presence of any customer complaint issues. This can be the result of letters of complaint, phone complaints, reports to the FDA, or other inquiries and comments.

Once this information is available, it can be organized and placed in the horizontal customer information portion of the QFD matrix. The voices of the customers represent their wants and needs—their requirements. These are the inputs to the matrix, along with importance ratings, competitive evaluations, and complaints.

The appropriate team can then begin developing the technical information portion of the matrix. The customers' voices must be translated into items that are measurable and actionable within the organization. Companies use a variety of names to describe these measurable items, such as design requirements, technical requirements, product characteristics, and product criteria.

The relationship between the inputs and the actionable items can then be examined. Each technical requirement is analyzed to determine if action on the item will affect the customer's requirements. A typical question would be "Would the organization work on this technical requirement to respond favorably to the customers' requirements?"

For those items in which a relationship is determined to exist, the team must then decide on the strength of the relationship. Symbols are normally used to denote a strong, moderate, or weak relationship. Some of the symbols commonly used are double circle, single circle, and triangle, respectively. The symbols provide a quick visual impression of the overall relationship strengths of the technical requirements and the customers' wants and needs.

The team must instigate testing to develop technical data showing the performance of the parent company and its competitors for each of the technical requirements. Once this information is available, the team can begin a study to determine the target value that should be established for each technical requirement. The objective is to ensure that the next-generation product will be truly competitive and satisfy its customers' wants and needs. A comparison of the customers' competitive ranges and the competitive technical assessments helps the organization determine these targets.

Additional information can be added to the matrix depending on the team's judgment of value. Significant internal and regulatory requirements may be added. Measure of organizational difficulty can be added. Column weights can be calculated. These can serve as an index for highlighting those technical requirements that have the largest relative effect on the product.

Once this matrix is complete, the analysis stage begins. The chief focus should be on the customer portion of the matrix. It should be examined to determine which customer requirements need the most attention. This is an integrated decision involving the customers' competitive evaluation, their importance ratings, and their complaint histories. The number of priority items selected will be a balance between their importance and the resources available within the company.

Items selected for action can be treated as a special project or can be handled by use of the QFD matrix at the next level of detail. Any items so selected can become the input to the new matrix. While the first matrix is a planning matrix for the complete product, this new matrix is at the lower level. It concerns the subsystem or assembly that affects the requirement.

The challenge in the second-level matrix (Figure 4.4) is to determine the concept that best satisfies the deployed requirement. This requires evaluation of some design concept alternatives. Several techniques are available for this type of comparative review. The criteria or requirements for the product or service are listed on the left of the matrix. Concept alternatives are listed on the top. The results of the evaluation of each concept versus criteria can be entered in the center portion.

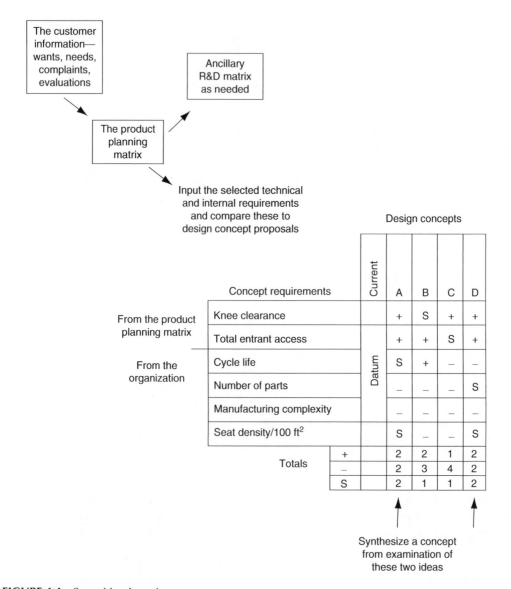

Concept requirements	Current	A	B	C	D
Knee clearance		+	S	+	+
Total entrant access		+	+	S	+
Cycle life	Datum	S	+	−	−
Number of parts		−	−	−	S
Manufacturing complexity		−	−	−	−
Seat density/100 ft²		S	−	−	S
Totals +		2	2	1	2
−		2	3	4	2
S		2	1	1	2

FIGURE 4.4   Second-level matrix.

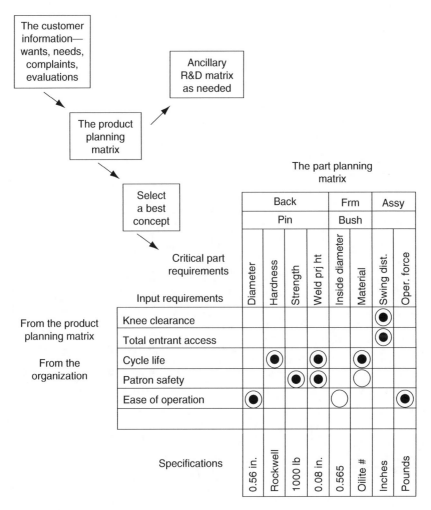

FIGURE 4.5 Part planning matrix.

Once the best concept alternative is selected, a QFD part planning matrix can be generated for the component level (Figure 4.5). The development of this matrix follows the same sequence as that of the prior matrix. Generally, less competitive information is available at this level and the matrix is simpler. The technical requirements from the prior matrix are the inputs. Each component in the selected design concept is examined to determine its critical part requirements. These are listed in the upper portion. Relationships are examined and symbols are entered in the center portion. The specifications are then entered for these selected critical part requirements in the lower portion of the matrix.

The part planning matrix should then be examined. Experience with similar parts and assemblies should be a major factor in this review. The analysis should involve the issue of which of the critical part requirements listed are the most difficult to control or ensure continually. This review will likely lead to the selection of certain critical part requirements that the team believes deserve specific follow-up attention.

If a team believes the selected critical part characteristics are best handled through the QFD process, a matrix should be developed (Figure 4.6) for process planning. The critical part concerns from the part planning matrix should be used as inputs in the left area of the matrix. The critical process requirements are listed across the top. Relationships are developed and examined in the

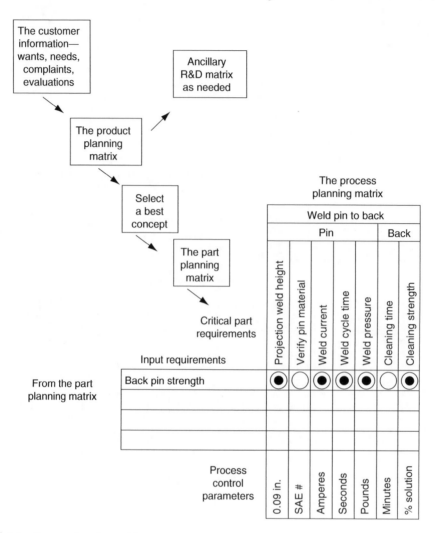

FIGURE 4.6 Process planning matrix.

central area. The specifications for operating levels for each process requirement are recorded in the lower area of the matrix. For example, if a critical part requirement was spot-weld strength, one critical process parameter would be weld current. The amount of current would be a critical process parameter to ensure proper spot-weld strength. The specification for this critical process requirement would be the amperes of current required to ensure the weld strength.

Upon completion of the planning at the part and process levels, the key concerns should be deployed to the manufacturing level. Most organizations have detailed planning at this level and have developed spreadsheets and forms for recording their planning decisions. The determinations from the prior matrices should become inputs to these documents. Often, the primary document at this level is a basic planning chart (Figure 4.7). Items of concern are entered in the area farthest left. The risk associated with these items is assessed and recorded in the next column. In typical risk assessments, the level of the concern and the probability of its occurrence are listed, as are the severity of any developing problems and the probability of detection. These items, along with other concerns, can be used to develop an index to highlight items of significant concern. Other areas in the chart can be used to indicate issues such as the general types of controls, frequency of checking, measuring devices, responsibility, and timing.

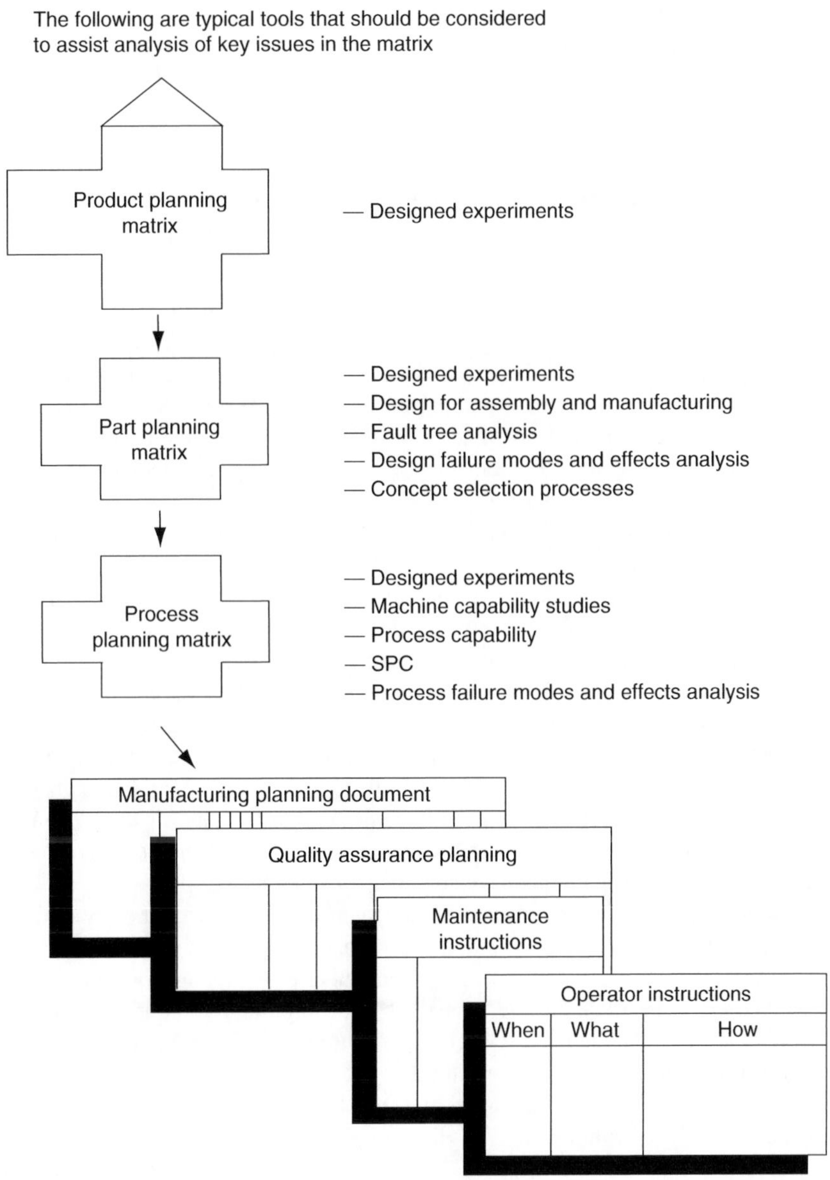

The following are typical tools that should be considered
to assist analysis of key issues in the matrix

Product planning matrix
— Designed experiments

Part planning matrix
— Designed experiments
— Design for assembly and manufacturing
— Fault tree analysis
— Design failure modes and effects analysis
— Concept selection processes

Process planning matrix
— Designed experiments
— Machine capability studies
— Process capability
— SPC
— Process failure modes and effects analysis

Manufacturing planning document

Quality assurance planning

Maintenance instructions

Operator instructions

When	What	How

FIGURE 4.7   Manufacturing planning chart.

4.5   SUMMARY OF QFD

The input to the QFD planning matrix is the voice of the customer. The matrix cannot be started until the customers' requirements are known. This applies to internal planning projects as well as products and services that will be sold to marketplace customers. Use of the QFD process leads an organization to develop a vital customer focus.

The initial matrix is usually the planning matrix. The customers' requirements are inputs. Subsequent matrices may be used to deploy or flow down selected requirements from the product planning matrix for part planning and process planning. Some forms of a manufacturing chart or matrix can be used to enter critical product and process requirements from prior matrices.

The principal objective of the QFD process is to help a company organize and analyze all the pertinent information associated with a project and to use the process to help it select the items demanding priority attention. All companies do many things right. The QFD process will help them focus on the areas that need special attention.

4.6 REQUIREMENTS, DESIGN, VERIFICATION, AND VALIDATION

As medical products encompass more features and technology, they will grow in complexity and sophistication. The hardware and software for these products will be driven by necessity to become highly synergistic and intricate which will in turn dictate tightly coupled designs. The dilemma is whether to tolerate longer development schedules to achieve the features and technology, or to pursue shorter development schedules. There really is no choice given the competitive situation of the marketplace. Fortunately, there are several possible solutions to this difficulty. One solution that viably achieves shorter development schedules is a reduction of the quantity of requirements that represent the desired feature set to be implemented. By documenting requirements in a simpler way, the development effort can be reduced by lowering the overall product development complexity. This would reduce the overall hardware and software requirements which in turn reduces the overall verification and validation time.

The terms verification and validation are sometimes confused and used interchangeably, when in reality they are very different. Verification is the process of evaluating the products of a given phase of development to ensure correctness and consistency with respect to the products and standards provided as input to that phase. Verification ensures that all requirements for the device have been tested and proven to be correct. Verification is performed during the development process and is accomplished on subsystems as well as the system. Validation is the process of evaluating a product to ensure compliance with specified and implied requirements. Validation ensures that the product, as designed, is the device the customer requested and meets all the customer's needs. Validation is performed at the end of the design cycle and is accomplished on the actual device as manufactured according to all manufacturing specifications and standards.

The issue is how to reduce the number of documented requirements without sacrificing feature descriptions. This can be achieved by limiting the number of product requirements, being more judicious about how the specified requirements are defined, or by recognizing that some requirements are really design specifications. A large part of requirements definition should be geared toward providing a means to delay making decisions about product feature requirements that are not understood until further investigation is carried out.

As stated above, verification and validation must test the product to assure that the requirements have been met and that the specified design has been implemented. At worst, every requirement will necessitate at least one test to demonstrate that it has been satisfied. At best, several requirements might be grouped such that at least one test will be required to demonstrate that they all have been satisfied. The goal for the design engineer is to specify the requirements in such a manner as to achieve as few requirements as are absolutely necessary and still allow the desired feature set to be implemented. Several methods for achieving this goal are refinement of requirements, assimilation of requirements, and requirements versus design.

4.6.1 REFINEMENT OF REQUIREMENTS

As an example, suppose a mythical device has the requirement "the output of the analog to digital converter (ADC) must be accurate to within plus or minus 5%." Although conceptually this appears to be a straight forward requirement, to the software engineer performing the testing to demonstrate satisfaction of this requirement, it is not as simple as it looks. As stated, this requirement will necessitate at least three independent tests and most likely five tests. One test will have to establish that the ADC is outputting the specified nominal value. The second and third tests will be needed to confirm that the output is within the $\pm 5\%$ range. Being a good software engineer, the 5% limit is not as arbitrary as it may seem due to the round-off error of the percent calculation with the ADC output units. Consequently, the fourth and fifth test will be made to ascertain the sensitivity of the round-off calculation.

A better way to specify this requirement is to state "the output of the ADC must be between X and Y," where X and Y values correspond to the original requirement of $\pm 5\%$. This is a better requirement statement because it simplifies the testing that occurs. In this case, only two tests are required to demonstrate satisfaction of this requirement. Test one is for the X value and test two is for the Y value. The requirement statements are equivalent but the latter is more effective because it has reduced the test set size, resulting in less testing time and consequently a potential for the product to reach the market earlier.

4.6.2 Assimilation of Requirements

Consider the situation where several requirements can be condensed into a single equivalent requirement. In this instance, the total test set can be reduced through careful analysis and an insightful design. Suppose that the user interface of a product is required to display several fields of information that indicate various parameters, states, and values. It is also required that the user be able to interactively edit the fields, and that key system critical fields must flash or blink so that the user knows that a system critical field is being edited. Further assume that the software requirements document specifies that "all displayed fields can be edited. The rate field shall flash while being edited. The exposure time shall flash while being edited. The volume delivered field shall flash while being edited."

These statements are viable and suitable for the requirements specification but they may not be optimum from an implementation and test point of view. There are three possible implementation strategies for these requirements. First, a "monolithic" editor routine can be designed and implemented that handles all aspects of the field editing, including the flash function. Second, a generic field editor can be designed which is passed a parameter that indicates whether or not the field should flash during field editing. Third, an editor executive could be designed such that it selects either a nonflashing or flashing field editor routine depending on whether the field was critical or not. Conceptually, based on these requirements statements, the validation team would ensure that (1) only the correct fields can be displayed, (2) the displayed fields can be edited, (3) critical fields blink when edited, and (4) each explicitly named field blinks.

The first "monolithic" design option potentially presents the severest test case load and should be avoided. Since it is monolithic in structure and performs all editing functions, all validation tests must be performed within a single routine to determine whether the requirements are met. The validation testing would consist of the four test scenarios presented above.

The second design option represents an improvement over the first design. Because the flash/no flash flag is passed as a parameter into the routine, the testing internally to the routine is reduced because part of the testing burden has been shifted to the interface between the calling and called routines. This is easier to test because the flash/no flash discrimination is made at a higher level. It is an inherent part of the calling sequence of the routine and therefore can be visually verified without formal tests. The validation testing would consist of test situations one, two, and four as presented above.

The third design option represents the optimum from a test standpoint because the majority of the validation testing can be accomplished with visual inspections. This is possible because the flash/no flash discrimination is also implemented at a higher level and the result of the differentiation is a flashing field or a nonflashing field. The validation testing would consist of test situations two and four as presented above.

Based on the design options, the requirements could be rewritten to simplify testing even further. Assume that the third design option in fact requires less testing time and is easier to test. The requirement statements can then be written to facilitate this situation even more. The following requirements statements are equivalent to those above and in fact tend to drive the design in the direction of the third design option. "All displayed fields can be edited. All critical items being edited shall flash to inform the user that editing is in progress." In this instance, the third design can be augmented by creating a list or look-up table of the fields required to be edited and a flag can be associated with each that indicates whether the field should flash or not. This approach allows a completely visual inspection to replace the testing because the field is either in the edit list or it is not, and if it is, then it either flashes or it does not. Testing within the routine is still required, but it is now associated with debug testing during development and not with formal validation testing after implementation.

4.6.3 Requirements versus Design

There is agreement that there is a lot of overlap between requirements and design, yet the division between these two is not a hard line. Design can itself be considered a requirement. Many individuals, however, do not appreciate that the distinction between them can be used to simplify testing and consequently shorten overall software development times. Requirements and their specification concentrate on the functions that are needed by the system or product and the users. Requirements need to be discussed in terms of what has to be done, and not how it is to be done.

The requirement "hardcopy strip chart analysis shall be available" is a functional requirement. The requirement "hardcopy strip chart analysis shall be from a pull down menu" has design requirements mixed with the functional requirements. Consequently, there may be times when requirements specifications will contain information that can be construed as design. When developing a requirements specification, resist placing the "how to" design requirements in the system requirements specification and concentrate on the underlying "what" requirements.

As more "how" requirements creep into the requirements specification, more testing must occur on principally two levels. First, there is more detail to test for and second, but strategically more important, there is more validation than verification that needs to be done. Since verification is qualitative in nature and ascertains that the process and design were met, low-key activities have been transferred from the visual and inspection methods into validation testing which is more rigorous and requires formal proof of requirements fulfillment. The distinction of design versus requirements is difficult, but a careful discrimination of what goes where is of profound benefit. As a rule of thumb, if it looks like a description of "what" needs to be implemented, then it belongs in the requirements specification. If it looks like a "how to" description, if a feature can be implemented in two or more ways and one way is preferred over another, or if it is indeterminate as to whether it is requirements or design, then it belongs in the design specification.

There is another distinct advantage to moving as many "how" requirements to design as possible. The use of computer-aided software engineering (CASE) tools has greatly automated the generation of code from design. If a feature or function can be delayed until the design phase, it can then be implemented in an automated fashion. This simplifies the verification of the design because the automation tool has been previously verified and validated so that the demonstration that the design was implemented is simple.

4.7 PRODUCT SPECIFICATION

The product specification is the first step in the process of transforming product ideas into approved product development efforts. It details the results of the customer survey and subsequent interface between the marketing, design engineering, reliability assurance, and regulatory affairs personnel. It specifies what the product will do, how it will do it, and how reliable it will be. To be effective, it must be as precise as possible.

The product specification should be a controlled document, that is, subject to revision level control, so that any changes that arise are subjected to review and approval before implementation. It prevents the all too typical habit of making verbal changes to the specification, without all concerned personnel informed. This often leads to total confusion in later stages of development, as the current specification is only a figment of someone's imagination or a pile of handwritten papers in someone's desk.

The specification should also have joint ownership. It should only be written after all concerned departments have discussed the concept and its alternatives and have agreed on the feasibility of the design. Agreement should come from marketing, design engineering, manufacturing, customer service, reliability assurance, and regulatory affairs.

The specification is a detailed review of the proposed product and includes

- Type of product
- Market it addresses
- Function of the product
- Product parameters necessary to function effectively
- Accuracy requirements
- Tolerances necessary for function
- Anticipated environment for the device
- Cautions for anticipated misuse
- Safety issues
- Human factors issues
- Anticipated life of the product
- Reliability goal
- Requirements from applicable domestic or international standards

Each requirement should be identified with some form of notation, such as brackets and a number. For traceability purposes, each numbered subsection of the specification should start numbering its requirements with the number 1. For example:

5.3.1 Analog to Digital Converter
The output of the analog to digital converter must be between X and Y [1].

In parsing the requirements, this particular one would be referred to as 5.3.1-1. Subsequent requirements in this paragraph would be numbered in consecutive order. Requirements in the next paragraph would restart the numbering with number 1. See Figure 4.8 for an example.

Requirement Number	Requirement	Paragraph Number	Requirement Number	Author	Requirement Responsibility	Test Type
1221	The machine shall contain no burrs or sharp edges	3.1	1	Smith	System	Visual
1222	The maximum height of the machine shall be 175 cm	3.1	2	Smith	System	Valid
1223	The maximum height of the shipping package shall be 185 cm	3.1	3	Smith	System	Valid
1224	The power supply shall have a maximum inrush current of 7.3 V	3.2	1	Jones	Subsystem B	Verification
1225	The power supply shall provide currents of +5, +15, and −15 V	3.2	2	Jones	Subsystem B	Verification
1226	The check valve shall withstand a pressure of 150 psi	3.3	1	Thomas	Subsystem C	Verification

FIGURE 4.8   An example section of a specification database.

Software programs are available to assist in the parsing process. The software establishes a database of requirements for which a set of attributes are developed that help trace each requirement. Some attributes which might be established include

- Paragraph number
- Requirement number
- Author of the requirement
- System or subsystem responsible for the requirement
- Type of verification or validation test

EXERCISES

1. Do a Web search on QFD, report the number and geographical distribution of the information found, comment on these results.
2. QFD can be used for technical as well as social system development. Find and report on an example of an improved clinic or other system based on QFD principles.
3. Related term is called six-sigma. Do a Web search to define this term, then comment on its relationship with QFD.
4. Find and report on any QFD application to a technical problem.
5. Develop the first level QFD diagram for your next car purchase.
6. Develop the second level QFD for the car purchase.
7. Develop the third level QFD for the car purchase.
8. You are an employee of Sleep-EZ Inc. You are charged with the development of an inexpensive anesthesia machine for use in third world countries. Develop a three-level QFD matrix for this task.
9. Develop a set of requirements for the above anesthesia machine.

BIBLIOGRAPHY

Aston, R., *Principles of Biomedical Instrumentation and Measurement*. Columbus, OH: Merrill, 1990.

Banta, H.D., The regulation of medical devices, *Preventive Medicine*, 19(6), November 1990.

Croswell, D.W., The evolution of biomedical equipment technology, *Journal of Clinical Engineering*, 20(3), May–June, 1995.

Day, R.G., *Quality Function Deployment—Linking a Company with Its Customers*. Milwaukee, WI: ASQC Quality Press, 1993.

Food and Drug Administration, *Medical Device Good Manufacturing Practices Manual*. Washington, DC: U.S. Department of Health and Human Services, 1991.

Fries, R.C., *Reliability Assurance for Medical Devices, Equipment and Software*. Buffalo Grove, IL: Interpharm Press, 1991.

Fries, R.C., *Handbook of Medical Device Design*. New York: Marcel Dekker, 2001.

Fries, R.C., *Reliable Design of Medical Devices*, 2nd ed. New York: Marcel Dekker, 2006.

Fries, R.C. et al., Software regulation, in *The Medical Device Industry—Science, Technology, and Regulation in a Competitive Environment*. New York: Marcel Dekker, 1990.

Gause, D.C. and Gerald, M.W., *Exploring Requirements: Quality before Design*. New York: Dorset House Publishing, 1989.

Guinta, L.R. and Nancy, C.P, *The QFD Book: The Team Approach to Solving Problems and Satisfying Customers*. American Management Association, 1993.

Hamrell, M.R., The history of the United States Food, Drug and Cosmetic Act and the development of medicine, *Biochemical Medicine*, 2(33), April 1985.

Jorgens, J. III, Computer hardware and software as medical devices, *Medical Device and Diagnostic Industry*, Medical Device and Diagnostic Industry Magazine, 5(5), May 1983, 62–67.

King, P.H. and Richard CF, *Design of Biomedical Devices and Systems*. New York: Marcel Dekker, 2002.

Kriewall, T.J. and Gregory, P.W., An application of quality function deployment to medical device development, in *Case Studies in Medical Instrument Design*. New York: The Institute of Electrical and Electronics Engineers, 1991.

Potts, C., Kenji, T., and Annie, I.A., Inquiry-based requirements analysis, *IEEE Software*, 11(2), March 1994.

Pressman, R.S., *Software Engineering, A Practitioner's Approach*, 2nd ed. New York: McGraw-Hill, 1987.

Sheretz, R.J. and Stephen, A.S., Medical devices: Significant risk vs nonsignificant risk, *The Journal of the American Medical Association*, 272(12), September 26, 1994.

Siddiqi, J. and Shekaran, M.C., Requirements engineering: The emerging wisdom, *Software*, 13(2), March 1996.

Taylor, J.W., *Planning Profitable New Product Strategies*. Radnor, PA: Chilton Book Company, 1984.

5 Product Documentation

> I love being a writer, what I can't stand is the paperwork.
>
> **Peter de Vries**
>
> Documentation is mandatory, resistance is futile.
>
> **Paul H. King**

This chapter covers product documentation requirements in great detail, primarily from a medical device industry viewpoint. Documentation in the medical device and pharmaceutical industry is mandated in the United States under Title 21 of the *Code of Federal Regulations* (*CFR*). The *CFR* is a codification of the general and permanent rules published in the *Federal Register* by the executive departments and agencies of the federal government. Title 21 of the *CFR* is reserved for rules of the Food and Drug Administration. Each title (or volume) of the *CFR* is revised once each calendar year. A revised title 21 is issued on approximately April 1 of each year.

The additions and revisions to the *CFR* governing food and drugs used in humans and animals, biologics, cosmetics, medical devices, radiological health, and controlled substances are published in the following volumes:

- Volume 1: Parts 1–99 (FDA, General)
- Volume 2: Parts 100–169 (FDA, Food for Human Consumption)
- Volume 3: Parts 170–199 (FDA, Food for Human Consumption)
- Volume 4: Parts 200–299 (FDA, Drugs: General)
- Volume 5: Parts 300–499 (FDA, Drugs for Human Use)
- Volume 6: Parts 500–599 (FDA, Animal Drugs, Feeds and Related Products)
- Volume 7: Parts 600–799 (FDA, Biologics; Cosmetics)
- Volume 8: Parts 800–1299 (FDA, Medical Devices)
- Volume 9: Parts 1300–End (DEA and Office of National Drug Control Policy)

21 CFR part 820, for example, defines medical device quality system (QS) regulation; a section of this part of the act (part M) defines general and specific record-keeping requirements for medical devices.

All documents and records required by the quality system regulation and the Medical Device Directives (MDD; Europe) must be maintained at the manufacturing establishment or other location that is reasonably accessible to responsible officials of the manufacturer and to auditors. They must be legible and stored so as to minimize deterioration and to prevent loss. Those stored in computer systems must be backed up and have a disaster plan in effect.

Documents and records deemed confidential by the manufacturer may be marked to aid the auditor in determining whether information may be disclosed. All records must be retained for a period equivalent to the design and expected life of the device, but not less than 2 years from the date of release of the product by the manufacturer.

There are several types of documents that must be kept by every medical device manufacture. These types include

- Business proposal
- Product specification

- Design specification
- Software quality assurance plan (SQAP) (where applicable)
- Software requirements specification (SRS) (where applicable)
- Software design description (SDD) (where applicable)

There are four primary types of records which must be kept by every medical device manufacturer. These types are

1. Design history file (DHF)
2. Device master record (DMR)
3. Device history record (DHR)
4. Technical documentation file (TDF)

Each type of record is discussed in the following sections.

5.1 PRODUCT DOCUMENTATION DOCUMENTS

5.1.1 BUSINESS PROPOSAL

The purpose of the business proposal is to identify and document market needs, market potential, the proposed product and product alternatives, risks and unknowns, and potential financial benefits. The business proposal also contains a proposal for further research into risks and unknowns, estimated project costs, schedule, and a request to form a core team to carry out needed research, to define the product and to prepare the project plan.

The business proposal usually contains

- Project overview, objectives, major milestones, schedule
- Market need and market potential
- Product proposal
- Strategic fit
- Risk analysis and research plan
- Economic analysis
- Recommendation to form a core project team
- Supporting documentation

5.1.1.1 Project Overview, Objectives, Major Milestones, and Schedule

This portion of the business proposal contains a statement of overall project objectives and major milestones to be achieved. The objectives clearly define the project scope and provide specific direction to the project team.

The major milestones and schedule follow the statement of objectives. The schedule anticipates key decision points and completion of the primary deliverables throughout all phases of development and implementation. The schedule contains target completion dates; however, it must be stressed that these dates are tentative and carry an element of risk. Events contingent upon achievement of the estimated dates should be clearly stated. Examples of milestones include

- Design feasibility
- Patent search completed
- Product specification verified by customers
- Design concept verified through completion of subsystem functional model completed
- Process validation completed

- Regulatory approval obtained
- Successful launch into territory A (for example)
- Project assessment complete, project transferred to manufacturing and sustaining engineering

This information will generally be in the form of a Gantt chart (Figure 5.1). The left-hand side of the Gantt chart lists (in this case) specific tasks to be accomplished, the horizontal axis denotes, via bars, the expected time line for the particular task. Tasks that are dependent on each other, such as "find project," must precede "begin project," and are linked with an arrow. Single events, such as "final exam," occur only at a specific time, and do not necessarily have to do with "find project," except that the two events occur once at the same time. Gantt charts are useful for project scheduling, if the number of elements is not large. They are also a good initial planning tool for use when outlining an overall task, such as a redesign of a system, etc.

5.1.1.2   Market Need and Market Potential

This section defines the customer and clinical need for the product or service and, identifies the potential territories to be served. Specific issues that are to be addressed include, but should not be limited, to the following:

- What is the market need for this product, that is, what is the problem to be solved?
- What clinical value will be delivered?
- What incremental clinical value will be added over existing company or competitive offerings?
- What trends are occurring, which predict this need?
- In which markets are these trends occurring?
- What markets are being considered, what is the size of the market, and what are the competitive shares?
- What are the market size and the estimated growth rate for each territory to be served?
- What are the typical selling prices and margins for similar products?
- When must the product be launched to capture the market opportunity?
- If competitors plan to launch similar products, what is our assessment of their launch date?
- Have competitors announced a launch date?
- What other similar products compose the market?
- Will the same product fit in all markets served? If not, what are the anticipated gross differences and why? What modifications will be required?
- Is the target market broad-based and multifaceted or a focused niche?
- What are the regulatory requirements, standards, and local practices which may impact the product design for every market to be served?

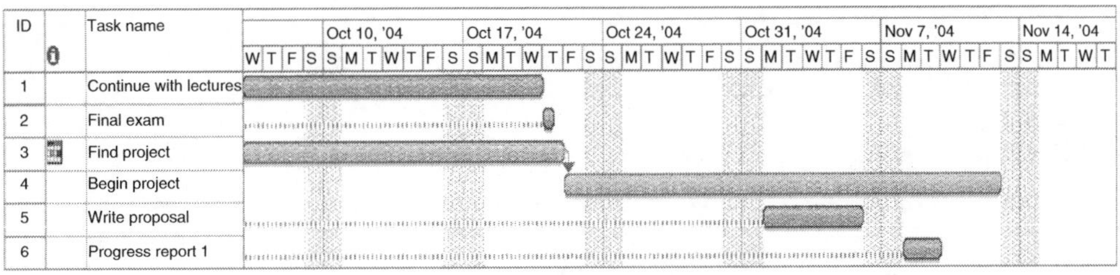

FIGURE 5.1   Gantt chart for a section of a design course.

5.1.1.3   Product Proposal

This section proposes the product idea that fulfills the market need sufficiently well to differentiate its features and explain how user or clinical value will be derived. The product specification is neither written nor does design commence during this phase. It may be necessary to perform some initial feasibility studies, construct nonworking models, perform simulations, and conduct research to have a reasonable assurance that the product can be designed, manufactured, and serviced. Additionally, models, simulations, and product descriptions will be useful to verify the idea with customers. If central to your development effort, the elements of a quality function diagram must be developed and evaluated. It is also recommended that several alternative product ideas be evaluated against the base case idea. Such evaluation will compare risks, development time lines, costs, and success probabilities.

5.1.1.4   Strategic Fit

This section discusses how the proposed product conforms with (or departs from) stated strategy with respect to product, market, clinical setting, technology, design, manufacturing, and service.

5.1.1.5   Risk Analysis and Research Plan

This section contains an assessment of risks and unknowns, an estimate of the resources needed to reduce the risks to a level whereby the product can be designed, manufactured, and serviced with a reasonable high level of confidence. The personnel resource requirement should be accompanied by the plan and timetable for addressing, researching, and reducing the risks.

The following categories of risks and unknowns should be addressed. Not all of these categories apply for every project. Select those which could have a significant impact on achieving project objectives.

- Technical
  - Feasibility (proven, unknown, or unfamiliar?)
  - New technology
  - Design
  - Manufacturing process
  - Accessibility to technologies
  - Congruence with core competencies
  - Manufacturing process capability
  - Cost constraints
  - Component and system reliability
  - Interface compatibility
- Market
  - Perception of need in market place
  - Window of opportunity; competitive race
  - Pricing
  - Competitive positioning and reaction
  - Cannibalization of existing products
  - Customer acceptance
- Financial
  - Margins
  - Cost to develop
  - Investment required
- Regulatory
  - Filings and approvals (FDA and other regulations)
  - Compliance with international standards

- Clinical studies; clinical trials
- Clinical utility and factors, unknowns
- Intellectual property
  - Patents
  - Licensing agreements
  - Software copyrights
- Requisite skill sets available or needed to design and develop
  - Electrical
  - Biomedical
  - Mechanical
  - Software
  - Industrial design
  - Human factors
  - Reliability
- Manpower availability
  - Workload of potential members of the team
  - Priorities of this and other projects
- Vendor selection
  - Quality system
  - Documentation controls
  - Process capability
  - Component reliability
  - Business stability
- Schedule
  - Critical path
  - Early or fixed completion date
  - Resource availability
- Budget

The critical path mentioned above may be derived from the Gantt chart; it is the path of activities that are dependent on each other such that the project cannot be completed in any shorter a time than is fixed by their dependencies.

5.1.1.6   Economic Analysis

This section includes a rough estimate of the costs and personnel required to specify, design, develop, and launch each product variant into the market place.

5.1.1.7   Core Project Team

This section discusses the formation of a core project team to perform the research required to reduce risks and unknowns to a manageable level, to develop and verify the user specification and to prepare the project plan.

The requisite skills of the proposed team members should also be outlined. To the extent possible, the following functions should be involved in research, preparation of the user specification, and the preparation of the project plan.

- Marketing
- Engineering
- Human factors
- Reliability assurance
- Manufacturing

- Service
- Regulatory
- Quality assurance
- Finance

Approximate amount of time required of each participant as well as incremental expenses should also be estimated. Some examples of incremental expenses include model development, simulation software, travel for customer verification activities, laboratory supplies, market research, and project status reviews.

5.1.2 PRODUCT SPECIFICATION

Product specification is the first step in the process of transforming product ideas into approved product development efforts. It details the results of the customer survey and subsequent interface between the marketing, design engineering, reliability assurance, and regulatory affairs personnel. It specifies what the product will do, how it will do it, and how reliable it will be. To be effective, it must be as precise as possible.

The product specification should be a controlled document, that is, subject to revision level control, so that any changes that arise are subjected to review and approval before implementation. It prevents the all too typical habit of making verbal changes to the specification, without all concerned personnel informed. This often leads to total confusion in later stages of development, as the current specification is only a figment of someone's imagination or a pile of handwritten papers in someone's desk.

The specification should also have joint ownership. It should only be written after all concerned departments have discussed the concept and its alternatives and have agreed on the feasibility of the design. Agreement should come from marketing, design engineering, manufacturing, customer service, reliability assurance, quality assurance, and regulatory affairs.

The specification is a detailed review of the proposed product and includes

- Type of product
- Market it addresses
- Technology to be used
- Function of the product
- Product parameters necessary to function effectively
- Accuracy requirements
- Tolerances necessary for function
- Anticipated environment for the device
- Cautions for anticipated misuse
- Safety issues
- Human factors issues
- Anticipated life of the product
- Reliability goal
- Requirements from applicable domestic or international standards

Each requirement should be identified with some form of notation, such as brackets and a number. For traceability purposes, each numbered subsection of the specification should start numbering its requirements with the number 1. For example,

5.3.1   Analog to Digital Converter
The output of the analog to digital converter must be between X and Y [1].

In parsing the requirements, this particular one would be referred to as 5.3.1-1. Subsequent requirements in this paragraph would be numbered in consecutive order. Requirements in the next paragraph would restart the numbering with number 1.

Software programs are available to assist in the parsing process. The software establishes a database of requirements for which a set of attributes are developed that help trace each requirement. Some attributes which might be established include

- Paragraph number
- Requirement number
- Author of the requirement
- System or subsystem responsible for the requirement
- Type of verification or validation test

These packages are generally Excel based.

5.1.3 DESIGN SPECIFICATION

The design specification is a document, which is derived from the product specification. Specifically, the requirements found in the product specification are partitioned and distilled down into specific design requirements for each subassembly. The design specification should address the following areas for each subsystem:

- Reliability budget
- Service strategy
- Manufacturing strategy
- Hazard consideration
- Environmental constraints
- Safety
- Cost budgets
- Standards requirements
- Size and packaging
- Power budget
- Heat generation budget
- Industrial design/human factors
- Controls/adjustments
- Material compatibility

In addition, all electrical and mechanical inputs and outputs and their corresponding limits under all operating modes must be defined.

Each performance specification should be listed with nominal and worst-case requirements under all environmental conditions. Typical performance parameters to be considered include

- Gain
- Span
- Linearity
- Drift
- Offset
- Noise
- Power dissipation
- Frequency response

- Leakage
- Burst pressure
- Vibration
- Long-term stability
- Operation forces/torques

As in the product specification, the requirements in the design specification should be identified by a notation such as a bracket and numbers. The parsing tool works well for focusing on these requirements.

5.1.4 Software Quality Assurance Plan

The term software quality assurance (SQA) is defined as a planned and systematic pattern of activities performed to assure the procedures, tools, and techniques used during software development and modification are adequate to provide the desired level of confidence in the final product. The purpose of an SQA program is to assure the software is of such quality that it does not reduce the reliability of the device. Assurance that a product works reliably has been classically provided by a test of the product at the end of its development period. However, because of the nature of software, no test appears sufficiently comprehensive to adequately test all aspects of the program. SQA has thus taken the form of directing and documenting the development process itself, including checks and balances.

Specifying the software is the first step in the development process. It is a detailed summary of what the software is to do and how it will do it. The specification may consist of several documents, including the SQAP, the SRS and the software design specification. These documents serve not only to define the software package, but are the main source for requirements to be used for software verification and validation.

A typical SQAP includes the following 16 sections.

5.1.4.1 Purpose

This section delineates the specific purpose and scope of the particular SQAP. It lists the names of the software items covered by the SQAP and the intended use of the software. It states the portion of the software life cycle covered by the SQAP for each software item specified.

5.1.4.2 Reference Documents

This section provides a complete list of documents referenced elsewhere in the text of the SQAP.

5.1.4.3 Management

This section describes the organizational structure that influences and controls the quality of the software. It also describes the portion of the software life cycle covered by the SQAP, the tasks to be performed with special emphasis on SQA activities, and the relationships between these tasks and the planned major checkpoints. The sequence of the tasks shall be indicated as well as the specific organizational elements responsible for each task.

5.1.4.4 Documentation

This section identifies the documentation governing the development, verification and validation, use, and maintenance of the software. It also states how the documents are to be checked for adequacy.

5.1.4.5 Standards, Practices, Conventions, and Metrics

This section identifies the standards, practices, conventions, and metrics to be applied as well as how compliance with these items is to be monitored and assured.

5.1.4.6   Review and Audits

This section defines the technical and managerial reviews and audits to be conducted, states how the reviews and audits are to be accomplished, and states what further actions are required and how they are to be implemented and verified.

5.1.4.7   Test

This section identifies all the tests not included in the software verification and validation plan and states how the tests are to be implemented.

5.1.4.8   Problem Reporting and Corrective Action

This section describes the practices and procedures to be followed for reporting, tracking, and resolving problems identified in software items and the software development and maintenance processes. It also states the specific organizational responsibilities for your company.

5.1.4.9   Tools, Techniques, and Methodologies

This section identifies the special software tools, techniques, and methodologies that support SQA, states their purpose, and describes their use.

5.1.4.10   Code Control

This section defines the methods and facilities used to maintain, store, secure, and document controlled versions of the identified software during all phases of the software life cycle.

5.1.4.11   Media Control

This section states the methods and facilities used to identify the media for each computer product and the documentation required to store the media and protect computer program physical media from unauthorized access or inadvertent damage or degradation during all phases of the software life cycle.

5.1.4.12   Supplier Control

This section states the provisions for assuring that software provided by suppliers meets established requirements. It also states the methods that will be used to assure that the software supplier receives adequate and complete requirements.

5.1.4.13   Records Collection, Maintenance, and Retention

This section identifies the SQA documentation to be retained, states the methods and facilities to be used to assemble, safeguard, and maintain this documentation, and designates the retention period.

5.1.4.14   Training

This section identifies the training activities necessary to meet the needs of the SQAP (Software Quality Assurance Plan).

5.1.4.15   Risk Management

This section specifies the methods and procedures employed to identify, assess, monitor, and control areas of risk arising during the portion of the software life cycle covered by the SQAP.

5.1.4.16   Additional Sections as Required

Some material may appear in other documents. Reference to these documents should be made in the body of the SQAP. The contents of each section of the plan shall be specified either directly or by reference to another document.

5.1.5   SOFTWARE REQUIREMENTS SPECIFICATION

SRS is a specification for a particular software product, program, or set of programs that perform certain functions. The SRS must correctly define all of the software requirements, but no more. It should not describe any design, verification, or project management details, except for required design constraints. A good SRS is unambiguous, complete, verifiable, consistent, modifiable, traceable, and usable during the operation and maintenance phase.

Each software requirement in an SRS is a statement of some essential capability of the software to be developed. Requirements can be expressed in several of ways:

- Through input/output specifications
- By use of a set of representative examples
- By specification of models

A typical SRS includes the following 11 sections.

5.1.5.1   Purpose

This section should delineate the purpose of the particular SRS and specify the intended audience.

5.1.5.2   Scope

This section should identify the software product to be produced by name, explain what the software product will, and if necessary, will not do, and describe the application of the software being specified.

5.1.5.3   Definitions, Acronyms, and Abbreviations

This section provides the definitions of all terms, acronyms, and abbreviations required to properly interpret the SRS.

5.1.5.4   References

This section should provide a complete list of all documents referenced elsewhere in the SRS or in a separate specified document. Each document should be identified by title, report number if applicable, date, and publishing organization. It is also helpful to specify the sources from which the references can be obtained.

5.1.5.5   Overview

This section should describe what the rest of the SRS contains and explain how the SRS is organized.

5.1.5.6   Product Perspective

This section puts the product into perspective with other related products. If the product is independent and totally self-contained, it should be stated here. If the SRS defines a product that is a component of a larger system then this section should describe the functions of each subcomponent of the system, identify internal interfaces, and identify the principal external interfaces of the software product.

5.1.5.7 Product Functions

This section provides a summary of the functions that the software will perform. The functions should be organized in a way that makes the list of functions understandable to the customer or to anyone else reading the document for the first time. Block diagrams showing the different functions and their relationships can be helpful. This section should not be used to state specific requirements.

5.1.5.8 User Characteristics

This section describes those general characteristics of the eventual users of the product that will affect the specific requirements. Certain characteristics of these people, such as educational level, experience, and technical expertise impose important constraints on the system's operating environment. This section should not be used to state specific requirements or to impose specific design constraints on the solution.

5.1.5.9 General Constraints

This section provides a general description of any other items that will limit the developer's options for designing the system. These can include regulatory policies, hardware limitations, interfaces to other applications, parallel operation, control functions, higher order language requirements, and criticality of the application, or safety and security considerations.

5.1.5.10 Assumptions and Dependencies

This section lists each of the factors that affect the requirements stated in the SRS. These factors are not design constraints on the software, but include any changes to them that can affect the requirements.

5.1.5.11 Specific Requirements

This section contains all the details the software developer needs to create a design. The details should be defined as individual specific requirements. Background should be provided by cross referencing each specific requirement to any related discussion in other sections. Each requirement should be organized in a logical and readable fashion. Each requirement should be stated such that its achievement can be objectively verified by a prescribed method.

The specific requirements may be classified to aid in their logical organization. One method of classification would include

- Functional requirements
- Performance requirements
- Design constraints
- Attributes
- External interface requirements

This section is typically the largest section within the SRS.

5.1.6 SOFTWARE DESIGN DESCRIPTION

SDD is a representation of a software system that is used as a medium for communicating software design information. The SDD is a document that specifies the necessary information content and recommended organization for an SDD. The SDD shows how the software system will be structured to satisfy the requirements identified in the SRS. It is a translation of requirements into a description of the software structure, software components, interfaces, and data necessary for the implementation phase. In essence, the SDD becomes a detailed blueprint for the implementation activity. In a complete SDD, each requirement must be traceable to one or more design entities.

The SDD should contain the following six items of information:

1. Introduction
2. References
3. Decomposition description
4. Dependency description
5. Interface description
6. Detailed design

5.1.6.1   Introduction

The introduction should describe the software being documented, defining the program, and its uses in general terms.

5.1.6.2   References

The reference section should allow one to refer to any standards by name (such as IEEE Std 1016–1998) that are necessary to understand the beginning point for analyses that follow.

5.1.6.3   Decomposition Description

The decomposition description records the division of the software system into design entities. It describes the way the system has been structured and the purpose and function of each entity. For each entity, it provides a reference to the detailed description via the identification attribute.

The decomposition description can be used by designers and maintainers to identify the major design entities of the system for purposes such as determining which entity is responsible for performing specific functions and tracing requirements to design entities. Design entities can be grouped into major classes to assist in locating a particular type of information and to assist in reviewing the decomposition for completeness. In addition, the information in the decomposition description can be used for planning, monitoring and control of a software project. Both hierarchical diagrams and natural language may be used.

5.1.6.4   Dependency Description

The dependency description specifies the relationships among entities. It identifies the dependent entities, describes their coupling, and identifies the required resources. This design view defines the strategies for interactions among design entities and provides the information needed to easily perceive how, why, where, and at what level system actions occur. It specifies the type of relationships that exist among the entities.

The dependency description provides an overall picture of how the system works to assess the impact of requirements and design changes. It can help maintenance personnel to isolate entities causing system failures or resource bottlenecks. It can aid in producing the system integration plan by identifying the entities that are needed by other entities and that must be developed first. This description can also be used by integration testing to aid in the production of integration test cases.

5.1.6.5   Interface Description

The entity interface description provides everything designers, programmers, and testers need to know to correctly use the functions provided by an entity. This description includes the details of external and internal interfaces not provided in the SRS.

The interface description serves as a binding contract among designers, programmers, customers, and testers. It provides them with an agreement needed before proceeding with the detailed design of entities. In addition, the interface description may be used by technical writers to produce customer documentation or may be used directly by customers.

5.1.6.6 Detailed Design Description

The detailed design description contains the internal details of each design entity. These details include the attribute descriptions for identification, processing, and data. The description contains the details needed by programmers before implementation. The detailed design description can also be used to aid in producing unit test plans.

5.2 RECORDS

5.2.1 DESIGN HISTORY FILE

DHF is a compilation of records, which describes the design history of a finished device. It covers the design activities used to develop the device, accessories, major components, labeling, packaging, and production processes.

The DHF contains or references the records necessary to demonstrate that the design was developed in accordance with the approved design plans and the requirements of the quality system regulation.

The design controls in CFR 21 820.30(j) require that each manufacturer establish and maintain a DHF for each type of device. Each type of device means a device or family of devices that are manufactured according to one DMR. That is, if the variations in the family of devices are simple enough that they can be handled by minor variations on the drawings then only one DMR exists. It is a common practice to identify device variations on drawings by dash numbers. For this case, only one DHF could exist because only one set of related design documentation exists. Documents are never created just to go into the DHF.

The QS regulation also requires that the DHF shall contain or reference the records necessary to demonstrate that the design was developed in accordance with the approved design plan and the requirements of this part. As noted, this requirement cannot be met unless the manufacturer develops and maintains plans that meet the design control requirements. The plans and subsequent updates should be part of the DHF. In addition, the QS regulation specifically requires that

- Results of a design review, including identification of the design, the date, and the individual(s) performing the review, shall be documented in the DHF.
- Design verification shall confirm that the design output meets the design input requirements. The results of the design verification, including identification of the design, method(s), the date, and the individual(s) performing the verification, shall be documented in the DHF.

Typical documents that may be in, or referenced in, a DHF include

- Design plans
- Design review meeting information
- Sketches
- Drawings
- Procedures
- Photos
- Engineering notebooks
- Component qualification information
- Biocompatibility (verification) protocols and data
- Design review notes
- Verification protocols and data for evaluating prototypes
- Validation protocols and data for initial finished devices
- Contractor/consultants information

- Parts of design output/DMR documents that show plans were followed
- Parts of design output/DMR documents that show specifications were met

The DHF contains documents such as the design plans and input requirements, preliminary input specs, validation data and preliminary versions of key DMR documents. These are needed to show that plans were created, followed, and specifications were met. The DHF is not required to contain all design documents or to contain the DMR; however, it will contain historical versions of key DMR documents that show how the design evolved.

The DHF also has value for the manufacturer. When problems occur during redesign and for new designs, the DHF has the institutional memory of previous design activities. The DHF also contains valuable verification and validation protocols that are not in DMR. This information may be very valuable in helping to solve a problem; pointing to the correct direction to solve a problem; or, most important, preventing the manufacturer from repeating an already tried and found to be useless design.

5.2.2 DEVICE MASTER RECORD

DMR is a compilation of those records containing the specifications and procedures for a finished device. It is set up to contain or reference the procedures and specifications that are current on the manufacturing floor. The DMR for each type of device should include or refer to the location of the following information:

- Device specifications including appropriate drawings, composition, formulation, component specifications, and software specifications
- Production process specifications including the appropriate equipment specifications, production methods, production procedures, and production environment specifications
- Quality assurance procedures and specifications including acceptance criteria and the quality assurance equipment used
- Packaging and labeling specifications, including methods and processed used
- Installation, maintenance, and servicing procedures and methods

It is more important to construct a document structure that is workable and traceable than to worry about whether something is contained in one file or another.

5.2.3 DEVICE HISTORY RECORD

DHR is the actual production records for a particular device. It should be able to show the processes, tests, rework, etc. that the device went through from the beginning of its manufacture through distribution. The DHR should include or refer to the location of the following information:

- Dates of manufacture
- Quantity manufactured
- Quantity released for distribution
- Acceptance records which demonstrate the device is manufactured in accordance with the DMR
- Primary identification label and labeling used for each production unit
- Any device identification and control numbers used

5.2.4 TECHNICAL DOCUMENTATION FILE

TDF contains all the relevant design data by means of which the product can be demonstrated to satisfy the essential safety requirements, which are formulated in the MDD. In liability proceedings or a control procedure; it must be possible to turn over the relevant portion of this file. For this reason, the file must be compiled in a proper manner and must be kept for a period of 10 years after the production of the last product.

The TDF must allow assessment of the conformity of the product with the requirements of the MDD. It must include

- General description of the product, including any planned variants.
- Design drawings, methods of manufacture envisaged and diagrams of components, sub-assemblies, circuits, etc.
- Descriptions and explanations necessary to understand the above mentioned drawings and diagrams and the operations of the product.
- Results of the risk analysis and a list of applicable standards applied in full or in part, and descriptions of the solutions adopted to meet the essential requirements of the directives if the standards have not been applied in full.
- For products placed on the market in a sterile condition, a description of the method is used.
- Results of the design calculations and of the inspections carried out. If the device is to be connected to other device(s) to operate as intended, proof must be provided that it conforms to the essential requirements when connected to any such device(s) having the characteristics specified by the manufacturer.
- Test reports and, where appropriate, clinical data.
- Labels and instructions for use.

The manufacturer must keep copies of European Community type-examination certificates and/or the supplements thereto in the TDF. These copies must be kept for a period ending at least 5 years after the last device has been manufactured.

5.3 COMPARISON OF THE MEDICAL DEVICE RECORDS

A manufacturer will accumulate a large amount of documentation during the typical product development process. The primary question then becomes which documentation is kept and where is it kept? Table 5.1 is an attempt to summarize the typical types of documentation and where they are kept. This is not an exclusive list, but serves only as guidance.

TABLE 5.1
Comparison of Record Storage

Record	Inclusion			
	DHF	DMR	DHR	Technical File
Agency submittals		X		X
Assembly inspection records		X	X	
Bills of material		X		
Calibration instructions/records		X		
Certificate of vendor compliance			X	
Certificates of compliance	X			X
Check sheets			X	
Clinical trial information	X			X
Combined product analysis				X
Component specifications		X		
Declarations of conformity				X

(continued)

TABLE 5.1 (continued)
Comparison of Record Storage

Record	DHF	DMR	DHR	Technical File
		Inclusion		
Design review records	X			
Design specification	X	X		
Design test protocols	X			
Design test results	X			X
Design validation plans	X			
Design validation protocols	X			
Design validation results	X	X		
Design verification plans	X			
Design verification protocols	X			
Design verification results	X			X
Engineering drawings	X	X		
Essential requirements checklists				X
Evaluations of potential vendors	X			
Evaluations of contractors	X			
Evaluations of consultants	X			
Field action reports			X	
Field service reports			X	
Final inspection instructions		X		
Incoming material quality records		X		
Inspection instructions		X		
Inspection plans		X		
Installation instructions		X		
Labeling requirements	X	X		X
Lab notebooks	X	X		
Letters of transmittal		X		
Listings of applicable standards				X
Machining inspection records			X	
Maintenance procedures		X		
Maintenance service reports			X	
MDD design specifications				X
Medical device reports (MDRs)			X	
Medical device vigilance reports			X	
Nonconforming material reports			X	
Packaging instructions		X		
Packaging specifications		X		
Postrelease design control change records		X		
Prerelease design control change records	X			
Primary inspection records		X		
Process change control records		X		
Process validation records		X		
Product complaints			X	
Product descriptions		X		X
Product environmental specs		X		
Product manuals		X		
Product routings		X		
Product specifications	X	X		
Product test specifications	X	X		
Production release documentation		X		
Project plans	X			

TABLE 5.1 (continued)
Comparison of Record Storage

Record	Inclusion			
	DHF	DMR	DHR	Technical File
Project team minutes	X			
Promotional materials		X		
Purchase orders			X	
Quality inspection audit reports		X		
Quality problem reporting sheets			X	
Quality memorandums		X		
Rationale for deviation from standards/regulations				X
Receipt vouchers			X	
Regulatory submittals		X		
Rework plans	X			
Risk analysis	X			X
Sales order reports			X	
Service specifications		X		
Shipping orders			X	
Software source code	X	X		
Tooling specs/revision log		X		
Work orders			X	

EXERCISES

1. Write a one page business proposal for your design project. Rough out a product specification page and design specification page if applicable.
2. You are going into competition with Johnson & Johnson; you plan to capture 30% of the market for band-aids. Do the needed Web search to determine your market potential in terms of the U.S. market.
3. Web sites medicaldesignonline.com has daily columns discussing new medical developments. Go to this Web site (or a related one) and peruse the industry news section. For one of the recent developments listed, discuss and document the market need. Identify what was obtained from this site versus what you obtain from other site searches.
4. Improper record keeping and other poor practices have bankrupted several medically related firms. Do a Web or library search to find such a case. Briefly discuss the case.
5. You are assigned to investigate the consequences of prostatectomy. Identify the current market for this operation and the consequences of the operation. Identify a need for improvement relating to your observations.
6. Do a Web search using the term "medical device." Detail how many hits are really consulting firms that assist in the structuring of a business proposal or product specification. Print out documentation on two or three of these companies and discuss what the product really is in terms of this chapter. The use of a good search engine (such as Go Network) is recommended, most of the single search engines are not powerful enough.
7. There are a few Web sites that specialize in determining the market for devices or treatments that target a complex of consequences of lung disease or the like. Most charge a high fee for identifying opportunities for entrepreneurship in the field. Find such a site, document it, and discuss the perceived value of the information.

BIBLIOGRAPHY

ANSI/IEEE Standard 730, *IEEE Standard for Software Quality Assurance Plans*. New York: The Institute of Electrical and Electronics Engineers, 1989.

ANSI/IEEE Standard 830, *IEEE Guide to Software Requirements Specifications*. New York: The Institute of Electrical and Electronics Engineers, 1984.

ANSI/IEEE Standard 1016, *IEEE Recommended Practice for Software Design Descriptions*. New York: The Institute of Electrical and Electronics Engineers, 1987.

Brodie, C.H. and Gary, B., *Voices Into Choices—Acting on the Voice of the Customer*. Madison, WI: Joiner Associates, 1997.

Chevlin, D.H. and Joseph, J. III, Medical device software requirements: Definition and specification, *Medical Instrumentation*, 30(2), March/April 1996.

Davis, A.M. and Pei, H., Giving voice to requirements engineering, *IEEE Software*, 11(2), March 1994.

Day, R.G., *Quality Function Deployment—Linking a Company with Its Customers*. Milwaukee, WI: American Society for Quality Control Quality Press, 1993.

Fairly, R.E., *Software Engineering Concepts*. New York: McGraw-Hill Book Company, 1985.

Fries, R.C., *Reliable Design of Medical Devices*. New York: Marcel Dekker, 1997.

Gause, D.C. and Gerald, M.W., *Exploring Requirements: Quality Before Design*. New York: Dorset House Publishing, 1989.

Guinta, L.R. and Nancy, C.P., *The QFD Bool: The Team Approach to Solving Problems and Satisfying Customers Through Quality Function Deployment*. New York: AMACOM Books, 1993.

Higson, G.R., *The Medical Devices Directives—A Manufacturer's Handbook*. Brussels: Belgium, Medical Technology Consultants (Europe), 1993.

Keller, M. and Ken, S., *Software Specification and Design*: *A Disciplined Approach for Real-Time Systems*. New York: John Wiley & Sons, 1992.

Kriewall, T.J. and Gregory, P.W., An application of quality function deployment to medical device development, in *Case Studies in Medical Instrument Design*. New York: The Institute of Electrical and Electronics Engineers, 1991.

Potts, C., Kenji, T., and Annie, I.A., Inquiry-based requirements analysis, *IEEE Software*, 11(2), March 1994.

Pressman, R.S., *Software Engineering*: *A Practitioner's Approach*. New York: McGraw-Hill, 1987.

Schoenmakers, C.C.W., *CE Marking for Medical Devices*. New York: The Institute of Electrical and Electronic Engineers, 1997.

Siddiqi, J. and Chandra Shekaran, M., Requirements engineering: The emerging wisdom, *Software*. 13(2), March 1996.

Silbiger, S., *The Ten Day MBA*. New York: William Morrow and Company, 1993.

Taylor, J.W. *Planning Profitable New Product Strategies*. Radnor, PA: Chilton Book, 1984.

Trautman, K.A., *The FDA and Worldwide Quality System Requirements Guidebook for Medical Devices*. Milwaukee, WI: American Society for Quality Press, 1997.

21 CFR §820. Washington, DC: Food and Drug Administration, 1996.

6 Product Development

In nothing do men more nearly approach the gods than in giving health to men.

Cicero

A product development process ensures that the design, development, and transfer of a new or modified medical device will result in a product that is safe, effective, and meets user needs and intended use requirements. As shown in Figure 6.1, design controls begin with the approval of product requirements. Product requirements include the needs of the users, patients, and intended use of the device. A design and development plan is developed to describe the design and development activities. The product requirements are converted into technical design inputs (system requirements specification [SRS]) that serve as a basis for the design of a medical device. Iterations of the design process result in design outputs that are verified against the design inputs to ensure that the design outputs adequately address the technical design inputs. The finished device is validated to ensure that all product requirements have been addressed. Final product and process specifications are transferred to production. In the course of the design process, documentation pertaining to the design of the finished device is maintained in a design history file (DHF). Changes to the device design are managed and controlled both before and postdesign transfer until retirement. Risk management is performed simultaneously with device design and development. Formal design reviews are conducted at appropriate points to evaluate the adequacy of the design to fulfill all requirements.

6.1 PRODUCT REQUIREMENTS

The product concept must be documented. This can range from a brief description for products similar to existing ones to a formal document, such as marketing requirements document, for new and complex products.

Product requirements include the needs of the users and patients and address the intended use of the device. They also include the following requirements, if applicable:

- User/patient/clinical performance characteristics
- Privacy and security
- Safety
- Regulatory
- Quality
- Reliability
- Compatibility with accessories/auxiliary devices or products
- Compatibility with the intended environment
- Human factors
- Physical characteristics
- Sterility
- Manufacturability
- Serviceability
- Labeling, packaging, and storage
- Requirements for intended markets (domestic or international)

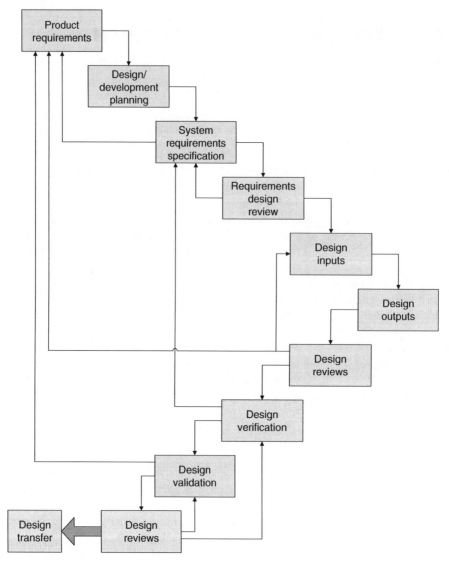

FIGURE 6.1 Product development process.

The types of information described above may come from a variety of sources such as market research studies, customer complaints, field failure analysis, service records, regulatory needs, user interviews, and customer satisfaction analysis. Input sources used shall be documented. Requirements that are essential to quality, safety, and proper function must be identified. Product requirements are reviewed, approved, and documented in the DHF.

6.2 DESIGN AND DEVELOPMENT PLANNING

Each product program must establish and maintain a plan(s) that describes or references the design and development activities and defines responsibility for implementation. It identifies and describes the interfaces with different groups or activities that provide, or result in, input to the design and development process. The design and development plan is reviewed, updated, and approved as the design and development of a product evolves.

The design and development plan describes how the different design control requirements are to be met. It includes all major activities, design deliverables, responsibilities, resources, and associated timelines for the development of a product. The program team creates the design and development plan and reviews, updates, and approves the plan as design and development evolves. The design and development plan resides in the DHF and any changes made to the plan also reside in the DHF.

6.2.1   Design and Development Plan

The following elements are addressed, if applicable, in the design and development plan. The applicability of these elements is determined by the program team and justification is provided for elements deemed not applicable.

6.2.1.1   Program Goals

High-level goals and objectives of the product are described, i.e., what is to be developed and other considerations that communicate the size, scope, and complexity of the product development project.

6.2.1.2   Design and Development Elements

Design and development elements refer to different categories of activities performed in the design of a medical device from design inputs through design transfer to manufacturing and service. The design and development plan describes the different elements including their scope and planned approach to fulfill the requirements of each element. Timeline for the activities associated with the different elements is incorporated in the design and development schedule. Required design and development elements include the following:

- Design input: Identify the design inputs that will be used during design and development. Identify the activities for translating user needs and product requirements into technical design inputs.
- Design activities: Identify the design activities anticipated to develop the product including those performed by suppliers and contractors. Include anticipated design iterations and contingencies. Design activities shall include, if applicable:
  - Development of new technologies
  - Reuse of existing technologies
  - Definition of system, subsystem, and module architectures
  - Design characterization and definition of design parameters
  - Component selection and supplier quality
  - Development and testing of subsystem prototypes and modules
  - System integration and testing
  - Design for reliability and risk analyses
  - Software design and development (including configuration management)
  - Activities to develop other design outputs
  - Technical assessments
  - Regulatory strategy and submissions
- Design outputs: Identify the design output elements that will be developed and the activities for developing them.
- Formal design reviews: Identify the timing, intended content, and the reviewers for the formal design review(s) that will be conducted during the product program. Each product program should have at least one formal design review. Formal design review(s) should be conducted to review, at a minimum, the following:
  - Completed design inputs
  - Completed design outputs
  - Completed design validation

- Design verification: Identify and provide an overview of the verification activities, for developing objective evidence that design input requirements have been met, including activities for the development of verification plans, test methods, testing, reporting, and reviewing results.
- Design validation: Identify and provide an overview of the validation activities for developing objective evidence that the device design meets product requirements, including activities for the development of validation plans, test methods, testing, reporting, and reviewing results.
- Design transfer: Identify the activities for translating the device design to production and service specifications and for transferring it to the manufacturing and service operations. Identify the requirements to be considered in selecting a manufacturing site or identify the manufacturing site, if known.
- Design change control: Identify the mechanism(s) and responsibilities for reviewing and approving design changes.
- Design history file: Identify the location of the product program DHF contents to allow ease of access. Reference other DHFs (and their locations) that may be leveraged for the product being developed. Identify key milestones at which all the documents in the DHF shall be brought up to date and be revision controlled, as appropriate.
- Risk management: Summarize the methods and activities that will be used to address potential product and process hazards to customers through risk management.

6.2.1.3   Organizational and Key Interfaces

Identify the key individuals/functions responsible for performing the design and development tasks, including cross-functional program team members and external resources, such as suppliers, contractors, or partners. At a minimum, define the roles for R&D, marketing, manufacturing, quality, reliability, regulatory, and service.

6.2.1.4   Deliverables and Responsibilities

Identify the design control deliverables for the product program and indicate the personnel responsible for completing them. The deliverables to be addressed are dependent on the size, scope, and complexity of the product program and must be defined by the program team leader and program team.

6.2.1.5   Design and Development Schedule

Based on the size, scope, and complexity of the product program; design and development elements; and list of deliverables prepare a design and development schedule. The schedule is specified at the level of detail necessary for carrying out major activities, completing program deliverables, and addressing design control requirements. Identify these activities, deliverables, the responsible individual/function, resources required, and the associated due dates. Indicate which activities are concurrent, sequential, and dependent on other activities. Identify the major milestones and formal design reviews.

6.2.1.6   Approve Design and Development Plan

The plan is completed and approved by the program team before the commencement of detailed design.

6.2.1.7   Incorporate Updates to Design and Development Plan

Changes to the design and development plan are reviewed and approved at key milestones as determined by the program team. The design and development plan identifies the number and timing of plan reviews by the program team. The plan is revision controlled.

6.3   SYSTEM REQUIREMENTS SPECIFICATION

Product requirements are translated into the SRS that specifies what the design must do to an engineering level of detail. Inputs from results of risk management are included.

The SRS includes the following types of requirements:

- Functional requirements: These requirements specify what the device does, focusing on the operational capabilities of the device and processing of inputs and the resultant outputs.
- Physical and performance requirements: These requirements specify how much or how well the design must perform, addressing issues such as speed, strength, size, weight, response times, accuracy, precision, limits of operation, device safety, and reliability.
- Interface requirements: These requirements specify characteristics that are critical to compatibility with external systems (including user and patient interface).
- System architecture: These requirements specify relationships among logical functions, physical systems/subsystems, and interfaces.
- Software requirements (if applicable): These requirements specify product functionality to be implemented through software and the functional, performance, interface, and safety requirements for the software subsystem(s).

Where appropriate, the SRS should include additional design details in areas such as specification limits and tolerance, risk management, toxicity and biocompatibility, electromagnetic compatibility (EMC), human factors, software, chemical characteristics, reliability, regulatory requirements, manufacturing processes, service design requirements, and testing. If the design logically decomposes into subsystems, the SRS may be used to generate subsystem level requirements. Traceability of the SRS to product requirements and design outputs is maintained. Requirements that are essential to the quality, safety, and proper function are identified.

Incomplete, ambiguous, or conflicting requirements are identified and resolved using the following mechanism:

- Program team reviews design inputs to identify and resolve incomplete, ambiguous, or conflicting requirements.
- Any remaining incomplete, ambiguous, or conflicting requirements are addressed in a formal design review.

The SRS is reviewed, approved, and documented in the DHF.

6.4   DESIGN INPUT

Each product program must establish design inputs to ensure that design requirements relating to a device are appropriate and address the intended use of the device, including the needs of the user and patient. There should be a mechanism for addressing incomplete, ambiguous, or conflicting requirements. The design input requirements are documented, reviewed, and approved by a designated individual(s). The approval, including the date and signature of the individual(s) approving the requirements is documented.

Each product program establishes product requirements. Product requirements include the needs of the users and patients and intended use of the device. Product requirements are translated into technical design inputs that are specified at an engineering level of detail.

Product requirements and the SRS obtained from the translation of product requirements constitute design input for the product program. Traceability is maintained to ensure that product requirements are linked to the corresponding SRS and design outputs.

6.5   DESIGN OUTPUT

Design outputs are the results of the design effort. Initial design activities result in intermediate design outputs. As design and development progresses, intermediate design outputs evolve into final design outputs that form the basis of the device master record (DMR).

The following general requirements apply to design outputs:

- Design outputs are maintained and documented such that they can be evaluated for conformance to design inputs. Traceability of design outputs to design inputs shall be maintained.
- Acceptance criteria for design outputs are established to enable verification and validation. Acceptance criteria related to device performance, such as accuracy and reliability are defined with tolerance limits.
- Design outputs that are essential to the quality, safety, and proper functioning of the device are identified. These outputs are identified by design and risk analysis.

6.5.1   INTERMEDIATE DESIGN OUTPUT

Intermediate design outputs are deliverables, which define and characterize the design. The following intermediate design outputs are created and recorded in the DHF as applicable:

- Preliminary design specifications
- Models and prototypes
- Software source code
- Risk analysis results
- Traceability documents
- Biocompatibility and bioburden test results
- Other intermediate design outputs as appropriate

6.5.2   FINAL DESIGN OUTPUT

Final design outputs form the basis of the DMR, which are recorded in the DHF, and shall include the following elements:

- Device specifications
- Device drawings
  - Component
  - Assembly
  - Finished device
- Composition, formulation, and component specifications
  - Subassembly specifications (if applicable)
  - Component and material specifications
  - Product configuration documents
  - Parts list
  - Bill of materials
- Software specifications (if applicable)
- Software machine code, such as a diskette or master EPROM
- Production process specifications
- Critical production process specifications
- Equipment specifications
- Production methods and procedures
  - Test protocols
  - Work instructions

- Production environmental specifications
- Quality assurance procedures and specifications
  - Acceptance criteria
  - Purchasing and acceptance requirements
  - Quality assurance equipment to be used
- Packaging and labeling specifications including methods and processes used
- Installation, maintenance, and servicing procedures and methods
  - Installation instructions
  - Service and maintenance instructions

6.6   FORMAL DESIGN REVIEW

Formal documented reviews of design results should be planned and conducted at appropriate stages of device design and development. Participants at these reviews include representatives of all functions concerned with the design stage being reviewed and an individual(s) who does not have direct responsibility for the design stage being reviewed, as well as any necessary specialists. The results of these reviews are documented in the DHF and include identification of the design, date, and individual(s) performing the review.

Formal design reviews are performed at major decision points or milestones in the design process as specified by the design and development plan. They are intended to be a systematic assessment of design results and to provide feedback to designers on existing or emerging problems. Each formal design review must ensure that design outputs meet design inputs.

6.6.1   ACTION TRACKING AND ISSUE RESOLUTION

Action items identified in formal design reviews are tracked to completion. Objective evidence of completion is documented. Resolution of issues may involve a design change, requirements change, or analysis justifying no action. The program team is responsible for ensuring that all issues and differences identified during the formal design review are resolved. Unresolved issues are escalated to management for resolution, guidance, or additional resources.

6.7   DESIGN VERIFICATION

Design verification is performed to confirm that the design output meets design input requirements. The results of design verification are documented in the DHF and include the identification of the design, test methods, date, and individual(s) performing the verification.

6.7.1   DESIGN VERIFICATION PLAN

Plans for subsystem and system level verification activities need to be developed. Typically, subsystem level verification activities, if applicable, are performed before system level verification activities. The plan identifies the timing and types of verification activities to be performed, the personnel performing the activities, and equipment to be used. Design verification includes

- Verification of requirements (system and subsystem level where appropriate)
- Verification of labeling, packaging, on-screen displays, printouts, and any other similar specifications

There must be confirmation that acceptance criteria have been established before the performance of verification. As appropriate, necessary statistical techniques to confirm the acceptance criteria must be identified.

Traceability is maintained between design outputs, their corresponding design inputs, and verification activities to confirm that design outputs meet the SRS. Verification plans must be reviewed and approved.

6.7.2 DESIGN VERIFICATION TEST METHODS

Test and inspection methods (protocols/scripts/procedures) for design are developed, documented, and approved before use. Verification methods include the following, if applicable:

- Integration testing
- Functional testing
- Accuracy testing
- System and subsystem performance testing
- Software testing such as unit/module, integration, system level, regression testing
- Package integrity tests
- Biocompatibility testing of materials
- Bioburden testing of products to be sterilized

Verification may be done by analysis where testing is not appropriate or practical, such as

- Tolerance analysis
- Worst case analysis of an assembly to verify that components are derated properly and not subject to overstress during handling and use
- Thermal analysis of an assembly to assure that internal or surface temperatures do not exceed specified limits
- Fault tree analysis of a process or design
- Failure modes and effects analysis of a process or design
- Finite element analysis
- Software source code evaluations such as code inspections and walkthroughs
- Comparison of a design to a previous product having an established history of successful use
- Clinical evaluation analysis

Test methods are based on generally acceptable practices for the technologies employed in similar products, such as compendia methods (e.g., ASTM, IEC, IEEE, and NIST). Test methods include defined conditions for testing. The test equipment used for verification must be calibrated and controlled according to quality system requirements. Repeatability and reproducibility of test procedures are determined. Technical comments about any deviations or other events that occur during testing shall be documented.

6.7.3 DESIGN VERIFICATION REPORT

A Design Verification Report summarizes the results of verification activities. Detailed verification results, such as original data, are contained or referenced in the report. The Design Verification Report and referenced documents are included in the DHF. Documentation of the results includes identification of the design, method(s), date, and the individual(s) performing the verification. Review and approve verification results to ensure that acceptance criteria have been met and all discrepancies identified by verification are resolved.

6.8 DESIGN VALIDATION

Design validation is performed to ensure that the device design conforms to user needs and intended uses. Design validation is performed under defined operating conditions on initial production units, lots, batches, or their equivalents and shall include testing of production units under actual or simulated use conditions. Design validation includes software validation and risk analysis, where appropriate. The results of validation are documented in the DHF and include the identification of the design, test methods, date, and individual(s) performing the validation.

6.8.1 Design Validation Plan

The design validation plan identifies the timing and types of validation activities to be performed, performance characteristics to be assessed, personnel performing the tests, and equipment to be used in validating the device.

Design validation includes the following, if applicable:

- Software validation
- External evaluations
- Process validation
- Risk analysis
- Validation of labeling and packaging

There must be confirmation that acceptance criteria have been established before the performance of validation. As appropriate, identify necessary statistical techniques to confirm the acceptance criteria.

Validation needs to be performed on initial production units, lots, batches, or their equivalents and done under actual or simulated use conditions. Where equivalent materials are used for design validation, such materials must be manufactured using the same methods and specifications to be used for commercial production. Justification needs to be provided to establish why the results are valid and must include a description of any differences between the manufacturing process used for the equivalent device and the process intended to be used for routine production. Validation must be complete before commercial distribution of the product.

Traceability must be maintained between design outputs, their corresponding design inputs, and validation activities to confirm that design outputs meet product requirements. Validation plans are reviewed for appropriateness, completeness, and to ensure that user needs and intended use(s) are being addressed.

6.8.2 Design Validation Test Methods

Test and inspection methods (protocols/scripts/procedures) for design validation must be developed, documented, and approved before use. Validation methods include the following, if applicable:

- Simulated use testing
- Testing confirming product data sheets, users' manual, product labels, and user interface screens
- Safety testing

Validation may be done by analysis where testing is not appropriate or practical, such as

- Historical comparisons to older devices
- Scientific literature review
- Failure modes and effects analysis of a design or process
- Workload analysis
- Alternative calculations
- Auditing design output
- Comparison of a design to a previous product having an established history of successful use

Validation is performed according to a written protocol that includes defined conditions for testing and simulations of expected environmental conditions such as temperature, humidity, shock, and vibration, and environmental stresses encountered during shipping and installation.

The test methods identified in the plan are based on generally acceptable practices for the technologies employed in similar products. The test equipment used for validation is calibrated and controlled according to quality system requirements. Repeatability and reproducibility of test procedures is determined.

6.8.3 Design Validation Report

A Design Validation Report summarizing the results of validation activities is developed. Detailed validation results, such as original data, are contained or referenced in the report. The Design Validation Report and referenced documents are included in the DHF.

Documentation of the results includes identification of the design, method(s), date, and the individual(s) performing the validation. Validation results are reviewed and approved to ensure that acceptance criteria have been met and all discrepancies identified by verification are resolved.

6.9 DESIGN TRANSFER

Design transfer ensures that the device design is correctly translated into production specifications and that the finished device is successfully transferred from design to production and service. Production specifications ensure that devices are repeatedly and reliably produced within product and process capabilities.

EXERCISES

1. From the QFD developed for the anesthesia machine in Chapter 4, develop a list of requirements for the device.
2. Develop a list of design inputs for the anesthesia machine above, based on the requirements.
3. Develop a list of risks involved in the use of the anesthesia machine.
4. How would the activities for the software portion of the anesthesia machine differ from the hardware portion of the device?
5. Develop a list of design outputs for the anesthesia machine and the activities necessary to accomplish them.
6. What verification activities would be necessary to prove your requirements?
7. Identify the activities for translating the anesthesia machine design to production and service.
8. Identify the requirements to be considered in selecting a manufacturing site for the anesthesia machine.

BIBLIOGRAPHY

Crow, K.A., *Characterizing and Improving the Product Development Process*. DRM Associates, Palos Verdes, C.A., 2001. http://www.npd-solutions.com/pdprocess.html, last accessed December 5, 2008.
Fries, R.C., Ed., *Handbook of Medical Device Design*. New York: Marcel Dekker, 2001.
Fries, R.C. and Andre, E.B., Medical device software development, in *Clinical Engineering Handbook*. Burlington, MA: Elsevier Academic Press, 2004.
Mallory, S.R., *Software Development and Quality Assurance for the Healthcare Manufacturing Industries*. Boca Raton, FL: CRC Press, 2002.
Oberg, P.A., Medical device research and design, in *Clinical Engineering Handbook*. Burlington, MA: Elsevier Academic Press, 2004.
Whitmore, E., *Development of FDA-Regulated Medical Products: Prescription Drugs, Biologics, and Medical Devices*. Milwaukee, WI: ASQ Quality Press, 2003.
Wiklund, M.E. and Stephen, B.W., *Designing Usability into Medical Products*. Boca Raton, FL: CRC Press, 2005.

7 Hardware Development Methods and Tools

The future of the aircraft industry is still the responsibility of the engineer. Money alone never did and never will create anything.

Aviation Week

Design input provides the foundation for product development. The objective of the design input process is to establish and document the design input requirements for the device. The design input document is as comprehensive and precise as possible. It contains the information necessary to direct the remainder of the design process. It includes design constraints, but does not impose design solutions.

Once the documentation describing the design and the organized approach to the design is complete, the actual design work begins. As the design activity proceeds, there are several failure-free or failure-tolerant principles that must be considered to make the design more reliable. Each is important and has its own place in the design process.

7.1 SIX SIGMA

Six sigma is a revolutionary business process geared toward dramatically reducing organizational inefficiencies that translate into bottom-line profitably. It started in the 1980s at Motorola and spread to organizations such as Allied Signal, Seagate, and General Electric. The process consists of five steps known as DMAIC:

1. Define
2. Measure
3. Analyze
4. Improve
5. Control

By systematically applying these steps, with the appropriate tools, practitioners of this approach have been able to save substantial dollars.

The basis of six sigma is measuring a process in terms of defects. The statistical concept of six sigma means your processes are working nearly perfectly, delivering only 3.4 defects per million opportunities (DPMO). Most organizations in the United States are operating at a 3–4 sigma quality level. This means they could be losing up to 25% of their total revenue due to processes that deliver too many defects, defects that take up time and effort to repair as well as generating unhappy customers.

The central idea of six sigma management is that if you can measure the defects in a process, you can systematically figure out ways to eliminate them, thus approaching a quality level of zero defects. The goal is to get the maximum return on your six sigma investment by spreading it throughout your company, continuing to train employees in the six sigma methodology and tools to lead process improvement teams, and sustaining the exponential gains you achieve by continuing to improve. One area the methodology of six sigma can be extended to is product design.

7.1.1 Design for Six Sigma

Design for six sigma (DFSS) is an approach to designing or redesigning product and services to meet or exceed customer requirements and expectations. Like its parent six sigma initiative, DFSS uses a disciplined methodology and set of tools to bring high quality to product development. It begins by conducting a gap analysis of your entire product development system. This analysis finds the gaps in your processes that are negatively affecting new product performance. It also addresses a highly significant factor, the voice of the customer (VOC). Every new product decision must be driven by the VOC. Otherwise, what basis is there for introducing it? By learning how to identify that voice and respond to it, the designer is in a far better position to deliver a new product or service that the customer actually wants.

7.1.2 Methodologies

Once the gap analysis is completed and the VOC is defined, DFSS applies its own version of the six sigma DMAIC methodology. The steps in the DFSS methodology, known as DMADV, include

- Define
- Measure
- Analyze
- Design
- Verify

The define step determines the project goals and the requirements of both internal and external customers. The measure step assesses customer needs and specifications. The analyze step examines process options to meet customer requirements. The design step develops the process to meet the customer requirements. The verify step checks the design to ensure that it meets customer requirements.

There are other methodologies for DFSS that have been used, including

- DMADOV
- IDEAS
- IDOV
- DMEDI
- DCCDI

DMADOV is a slight modification of the DMADV methodology mentioned above. The addition to DMADV is the optimize step, where the design is optimized.

IDEAS is a methodology with the following steps:

- Identify
- Design
- Evaluate
- Affirm
- Scale up

IDOV is a well-known design methodology, especially in the manufacturing world. The identify step identifies the customer and the critical to quality specifications. The design step translates the customer specifications into functional requirements (FRs) and into solution alternatives. A selection process brings the list of solutions down to the "best" solution. The optimize step used advanced statistical tools and modeling to predict and optimize design and performance. The Validate step ensures the design that was developed will meet the customer specifications.

DMEDI is a methodology with the following steps:

- Define
- Measure
- Explore
- Develop
- Implement

DCCDI is a methodology that is fairly new. The define step defines the project goals. The customer step ensures the analysis of the potential customer and their requirements is complete. The concept step is where ideas are developed, reviewed, and selected. The design step is performed to meet the customer and business specifications. And the implementation step is completed to develop and commercialize the product or service.

7.1.3 STRUCTURE

The DFSS approach can utilize any of the many possible methodologies. The fact is that all of these methodologies use the same advanced design tools, such as quality function deployment (QFD), failure modes and effects analysis (FMEC), benchmarking, design of experiments, simulation, robust design, etc. Each methodology primarily differs in the name of each phase and the number of phases.

DFSS packages, methods, and tools in a framework promotes cultural change under a recognized brand name that helps overcome an initial resistance to change. It is the most useful if it generates permanent behavior changes that outlast its own life as a brand. Given the DFSS toolset is not substantially new, the rationale for DFSS should not focus on tools. Over time, DFSS should emerge as a scientific approach to product development that leverages the six sigma culture. It will become a means to re-instill rigorous deductive and inductive reasoning in product development processes. It requires

- Identifying customer desires
- Developing validated transfer functions that describe product performance through objective measures
- Correlating these objective measures to customer desires
- Effectively assessing the capability to meet those desires well before product launch
- Applying transfer function knowledge to optimize designs to satisfy customer desires and avoid failure modes

Six sigma culture aids implementation of these steps by providing

- Cross company common language for problem resolution and prevention
- Mind-set that demands the use of valid data in decision-making
- Expectation across the organization that results should be measurable
- Disciplined project management system to help achieve timely results

None of the elements of this approach are revolutionary, but together they provide a template for success.

7.1.4 DESIGN FOR SIX SIGMA TOOLS

The use of six sigma tools and techniques should be introduced in a well thought out manner at various phases of the project. Tools that should be considered during a product development process include

- Robust design
- QFD
- Design FMEA (DFMEA)
- Axiomatic design

7.1.4.1 Robust Design

Robust design method, also called the Taguchi Method, pioneered by Dr. Genichi Taguchi, greatly improves engineering productivity. By consciously considering the noise factors (environmental variation during the product's usage, manufacturing variation, and component deterioration) and the cost of failure in the field the robust design method helps ensure customer satisfaction. Robust design focuses on improving the fundamental function of the product or process, thus facilitating flexible designs and concurrent engineering. Indeed, it is the most powerful method available to reduce product cost, improve quality, and simultaneously reduce development interval.

7.1.4.1.1 Why Use the Robust Design Methodology?

During the last five years many leading companies have invested heavily in the six sigma approach aimed at reducing waste during manufacturing and operations. These efforts have had great impact on the cost structure and hence on the bottom line of those companies. Many of them have reached the maximum potential of the traditional six sigma approach. What would be the engine for the next wave of productivity improvement?

Brenda Reichelderfer of ITT Industries reported on their benchmarking survey of many leading companies, "design directly influences more than 70% of the product life cycle cost; companies with high product development effectiveness have earnings three times the average earnings; and companies with high product development effectiveness have revenue growth two times the average revenue growth." She also observed, "40% of product development costs are wasted!" These and similar observations by other leading companies are compelling them to adopt improved product development processes under the banner DFSS. The DFSS approach is focused on (1) increasing engineering productivity so that new products can be developed rapidly and at low cost, and (2) value-based management.

Robust design method is central to improving engineering productivity. Pioneered by Dr. Genichi Taguchi after the end of the World War II, the method has evolved over the last five decades. Many companies around the world have saved hundreds of millions of dollars by using the method in diverse industries: automobiles, xerography, telecommunications, electronics, software, etc.

7.1.4.1.2 Typical Problems Addressed by Robust Design

A team of engineers was working on the design of a radio receiver for ground to aircraft communication requiring high reliability, i.e., low bit error rate, for data transmission. On the one hand, building series of prototypes to sequentially eliminate problems would be forbiddingly expensive. On the other hand, computer simulation effort for evaluating a single design was also time-consuming and expensive. Then, how can one speed up development and yet assure reliability?

In an another project, a manufacturer had introduced a high speed copy machine to the field only to find that the paper feeder jammed almost 10 times more frequently than what was planned. The traditional method for evaluating the reliability of a single new design idea used to take several weeks. How can the company conduct the needed research in a short time and come up with a design that would not embarrass the company again in the field?

The robust design method has helped reduce the development time and cost by a factor of two or better in many such problems.

In general, engineering decisions involved in product/system development can be classified into two categories:

- Error-free implementation of the past collective knowledge and experience
- Generation of new design information, often for improving product quality/reliability, performance, and cost

While CAD/CAE tools are effective for implementing past knowledge, robust design method greatly improves productivity in generation of new knowledge by acting as an amplifier of engineering skills. With robust design, a company can rapidly achieve the full technological potential of their design ideas and achieve higher profits.

7.1.4.1.3 Robustness Strategy

Variation reduction is universally recognized as a key to reliability and productivity improvement. There are many approaches to reducing the variability, each one having its place in the product development cycle. By addressing variation reduction at a particular stage in a product's life cycle, one can prevent failures in the downstream stages. The six sigma approach has made tremendous gains in cost reduction by finding problems that occur in manufacturing or white-collar operations and fixing the immediate causes. The robustness strategy is to prevent problems through optimizing product designs and manufacturing process designs.

The manufacturer of a differential op-amplifier used in coin telephones faced the problem of excessive offset voltage due to manufacturing variability. High offset voltage caused poor voice quality, especially for phones further away from the central office. So, how to minimize field problems and associated cost? There are many approaches:

1. Compensate the customers for their losses.
2. Screen out circuits having large offset voltage at the end of the production line.
3. Institute tighter tolerances through process control on the manufacturing line.
4. Change the nominal values of critical circuit parameters such that the circuit's function becomes insensitive to the cause, namely, manufacturing variation.

The approach is the robustness strategy. As one moves from approach 1 to 4, one progressively moves upstream in the product delivery cycle and also becomes more efficient in cost control. Hence it is preferable to address the problem as upstream as possible. The robustness strategy provides the crucial methodology for systematically arriving at solutions that make designs less sensitive to various causes of variation. It can be used for optimizing product design as well as for manufacturing process design.

The robustness strategy uses five primary tools:

1. Parameter diagram (P-diagram) is used to classify the variables associated with the product into noise, control, signal (input), and response (output) factors.
2. Ideal function is used to mathematically specify the ideal form of the signal–response relationship as embodied by the design concept for making the higher-level system work perfectly.
3. Quadratic loss function (also known as quality loss function) is used to quantify the loss incurred by the user due to deviation from target performance.
4. Signal to noise (S/N) ratio is used for predicting the field quality through laboratory experiments.
5. Orthogonal arrays are used for gathering dependable information about control factors (design parameters [DPs]) with a small number of experiments.

7.1.4.1.3.1 P-Diagram

P-diagram is a must for every development project. It is a way of succinctly defining the development scope. It is discussed in detail in Section 7.1.4.3.2.

7.1.4.1.3.2 Quality Measurement

In quality improvement and design optimization the metric plays a crucial role. Unfortunately, a single metric does not serve all stages of product delivery. It is common to use the fraction of products outside the specified limits as the measure of quality. Though it is a good measure of the loss due to scrap, it miserably fails as a predictor of customer satisfaction. The quality loss function serves that purpose very well.

Let us define the following variables:

m is the target value for a critical product characteristic
$+/- \Delta_0$ is the allowed deviation from the target
A_0 is the loss due to a defective product

Then the quality loss, L, suffered by an average customer due to a product with y as value of the characteristic is given by the following equation:

$$L = k(y - m)^2$$

where

$$k = \left(A_0/\Delta_0{}^2\right)$$

If the output of the factory has distribution of the critical characteristic with mean μ and variance σ^2, then the average quality loss per unit of the product is given by

$$Q = k\left\{(\mu - m)^2 + \sigma^2\right\}$$

7.1.4.1.3.3 S/N Ratios

The product/process/system design phase involves deciding the best values/levels for the control factors. The S/N ratio is an ideal metric for that purpose. The equation for average quality loss, Q, says that the customer's average quality loss depends on the deviation of the mean from the target and also on the variance. An important class of design optimization problem requires minimization of the variance while keeping the mean on target.

Between the mean and standard deviation, it is typically easy to adjust the mean on target, but reducing the variance is difficult. Therefore, the designer should minimize the variance first and then adjust the mean on target. Among the available control factors most of them should be used to reduce variance. Only one or two control factors are adequate for adjusting the mean on target.

The design optimization problem can be solved in two steps:

1. Maximize the S/N ratio, η, defined as

$$\eta = 10 \ \log_{10}(\eta^{2\sim}/\sigma^2)$$

   This is the step of variance reduction.
2. Adjust the mean on target using a control factor that has no effect on h. Such a factor is called a scaling factor. This is the step of adjusting the mean on target.

One typically looks for a scaling factor to adjust the mean on target during design and another for adjusting the mean to compensate for process variation during manufacturing.

7.1.4.1.3.4   Static versus Dynamic S/N Ratios
In some engineering problems, the signal factor is absent or it takes a fixed value. These problems are called static problems and the corresponding S/N ratios are called static S/N ratios. The S/N ratio described in Section 7.1.4.1.3.3 is a static S/N ratio.

In other problems, the signal and response must follow a function called the ideal function. In the cooling system example described earlier, the response (room temperature) and signal (set point) must follow a linear relationship. Such problems are called dynamic problems and the corresponding S/N ratios are called dynamic S/N ratios. The dynamic S/N ratio will be illustrated in a later section using a turbine design example. Dynamic S/N ratios are very useful for technology development, which is the process of generating flexible solutions that can be used in many products.

7.1.4.1.3.5   Steps in Robust Parameter Design
Robust parameter design has the following four main steps:

1. Problem formulation: This step consists of identifying the main function, developing the P-diagram, defining the ideal function and S/N ratio, and planning the experiments. The experiments involve changing the control, noise, and signal factors systematically using orthogonal arrays.
2. Data collection/simulation: The experiments may be conducted in hardware or through simulation. It is not necessary to have a full-scale model of the product for the purpose of experimentation. It is sufficient and more desirable to have an essential model of the product that adequately captures the design concept. Thus, the experiments can be done more economically.
3. Factor effects analysis: The effects of the control factors are calculated in this step and the results are analyzed to select optimum setting of the control factors.
4. Prediction/confirmation: To validate the optimum conditions we predict the performance of the product design under baseline and optimum settings of the control factors. Then we perform confirmation experiments under these conditions and compare the results with the predictions. If the results of confirmation experiments agree with the predictions, then we implement the results. Otherwise, the above steps must be iterated.

7.1.4.2   Quality Function Deployment

QFD was discussed in Chapter 4.

7.1.4.3   Robust Design Failure Mode and Effects Analysis

Failure mode and effects analysis (FMEA) is a methodology that has been used in the medical industry for many years. It is usually developed early in the product development cycle, in conjunction with a risk analysis. Risk by definition is the probable rate of occurrence of a hazard causing harm. Risk can be associated with device failure and also can be present in a normally operating device. The FMEA is an enhancement to the risk analysis by analyzing the potential failure down to the component level. Robust DFMEA, the subject of this paper, is an enhancement to the normal FMEA by anticipating safety and reliability failure modes through use of P-diagram.

Given the fact that product design responsibility starts at concept phase and ends when the product is obsolete, special emphases should be implemented to achieve design reliability and robustness. Robust DFMEA fits very well into this methodology. It is an invaluable tool to shorten product development times.

Robust DFMEA fits very well into the concept of concurrent engineering. It necessitates a close and continuous working relationship between design, manufacturing, service, suppliers, and customers. Robust DFMEA is best generated for the system level and used to derive through analysis the system's key subsystems and key components. Robust DFMEA preparation should incorporate inputs from a cross-functional team with expertize in design, human factors, manufacturing, testing, service, quality, reliability, clinical, regulatory, supplier or other fields, as appropriate.

The robust DFMEA is an integral part of the robust design methodology (RDM) currently being used in Europe and is an essential tool of the DFSS process. Robust DFMEA should be generated to analyze device design through a comprehensive and structured approach using the concept of a P-diagram. Robust DFMEA takes the failure mode analysis into a structural five dimensional failure–cause brainstorming approach, including:

- Total design and manufacturing variation: design variability refers to the ability of the design to allow a misuse (i.e., design symmetry, can be installed upside down). Manufacturing variability refers to the special design characteristics that are sensitive to variation in manufacturing/assembly processes.
- Changes over time refers to changes over time in dimensions or strength such as wearout or degradation.
- Customer usage refers to customer misuse and abuse of the product.
- External environment refers to external environmental conditions.
- System Interaction refers to the interaction of the various subsystems and components.

7.1.4.3.1 Benefits of a Robust DFMEA
There are many benefits when using the robust DFMEA, including:

1. Improved design reliability through a detailed analysis of system, subsystems, and components.
2. Traceability back to customer needs (VOC) for validation.
3. Ability to recognize and evaluate potential design failure modes and their effects.
4. Ability to recognize and evaluate potential special design characteristics.
5. Assure the implementation of proper mitigation, before the event action, to improve product reliability and robustness.
6. Improve or modify design verification/validation planning.
7. Analysis of interactions among various subsystems/components as well as interfaces to external systems.
8. Analysis of all interfaces and interactions with the customer and environment.
9. Definition of all stresses needed for testing.

7.1.4.3.2 P-Diagram
The P-diagram (Figure 7.1) is a block diagram used to facilitate the understanding of robust design as a concept. The P-diagram shows factors that affect a product. It models the product as a box affected by three types of parameters or factors that affect the response of the product, i.e., how the product performs its intended function:

- Signal factors
- Noise factors
- Control factors

The following two types of factors are controllable, while noise factors are uncontrollable in natural conditions of use.

1. Signal factors: Set by the user at a level that corresponds to the desired response
2. Control factors: Set by the designer

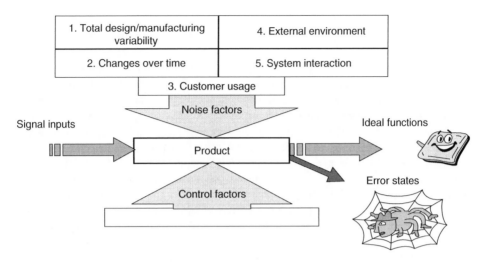

FIGURE 7.1   P-diagram.

There are five elements to every P-diagram:

1. Inputs: Any or all of the energy, material, and information that the product requires to deliver the desirable or undesirable output.
2. Ideal functions: Also called desirable output is referred to as the physical and performance requirements of an item that are used as a basis for design. Those requirements are stated in engineering terms that are unambiguous, complete, verifiable, and not in conflict with one another.
3. Error states: Also called undesirable output or failure modes are referred to as the ways in which the product may fail to meet the ideal function. Error states occur in one or all of the four states listed below:
   (a) No function
   (b) Over/under/degraded function
   (c) Intermittent function
   (d) Unintended function
4. Noise factors: Also called potential cause/mechanism of failure. Noise factors are the source of variation that can cause the error states/failure modes to occur. Noise factors are categorized into five, any or all of the below-mentioned five categories may cause the error states/failure modes to occur:
   Noise 1: Total design/manufacturing variability
      Design variability refers to the ability of the design to allow a misuse (i.e., design symmetry, can be installed upside down).
      Manufacturing variability refers to the key design characteristics that are sensitive to variation in manufacturing.
   Noise 2: Changes over time
      It is the changes over time in dimensions or strength such as wearout or degradation.
   Noise 3: Customer usage
      It is customer misuse and abuse of the product.
   Noise 4: External environment
      It is external environmental conditions.
   Noise 5: System interaction
      It is the interaction of the various subsystems and components.
5. Control factors: These are the DPs used to optimize performance in the presence of noise factors.

7.1.4.3.3 Performing a Robust DFMEA

The DFMEA form is illustrated in Figure 7.2. The form contains the following sections:

1. Number (no.): Enter ideal function number, start with ideal function number 1.
2. Item/function: Enter the name of the product being analyzed. Use the nomenclature and show the design level as indicated on the engineering drawing/specification. Robust DFMEA is best generated in the following order system, key subsystems then key components. Product under analysis in the P-diagram corresponds to item in item/function column. Enter, as concisely as possible, the function of the product being analyzed to meet the design intent. If the system has more than one function, list all the functions separately. Ideal functions in the P-diagram corresponds to function in item/function column.
3. Potential failure mode: Lists each potential failure mode for the particular product function. A recommended starting point is a review of product quality history, complaint reports, and group brainstorming. Remember that a hierarchical relationship exists between the components, subsystems, and system levels.
4. Potential effect(s) of failure: These effects are defined as the effects of the failure mode on the function, as perceived by the customer. Describe the effects of the failure in terms of what the customer might notice or experience, remembering that the customer may be an internal customer as well as the ultimate end user. State clearly if the function could impact safety or noncompliance to regulations. Remember that a hierarchical relationship exists between the components, subsystems, and system levels.
5. Severity (S): Severity is an assessment of the seriousness of the effect of the potential failure mode to customer if it occurs. Severity is rated and recorded for the worst-case scenario potential effect. To ensure continuity, the robust DFMEA team should use a consistent severity ranking system.
6. Potential cause(s)/mechanism(s) of failure: Causes are the source of variation that causes the failure modes/error states to occur. Noise factors in the P-diagram correspond to potential cause(s)/mechanism of failure column.
7. Occurrence (O): Occurrence is the likelihood that a specific cause/noise factor will occur and cause the potential failure during the design life. The likelihood of occurrence ranking number has a meaning rather than a value, below are some guidelines for defining an occurrence value:

   - What is the field experience with similar components, subsystems, or system?
   - Is it carryover or similar to a previous level component, subsystem, or system?
   - How significant are the changes from a previous level component, subsystem, or system?
   - Is it radically different from a previous level design?
   - Is it completely new?
   - Has its application/use changed?
   - What are the environmental changes?
   - Has any analysis (e.g., simulation) been used to estimate the expected comparable occurrence rate?
   - Have prevention mitigation been put in place?

   To ensure continuity, the robust DFMEA team should use a consistent occurrence ranking system.
8. Classification: This column may be used to classify any special product characteristics (safety and key design characteristics) for components, subsystems, and system that require mitigations. This column may also be used to highlight high priority failure modes for assessment.

POTENTIAL FAILURE MODE AND EFFECTS ANALYSIS IN DESIGN
(Robust DESIGN FMEA)
Baxter Confidential

Project Number: Product Number: Page   of

Project Name: Product Name: FMEA Number:
                                                                Rev.

No.	Item/Function	Potential Failure Mode	Potential Effect(s) of Failure	S E V	Potential Cause(s) Mechanism(s) of Failure	O C C	C L A S S	Current Mitigations	Verification	D E T	Recommended Actions	Acting results S E V	O C C	D E T	C L A S S
1	Ideal functions	Error states			Noise factors										
2															
3															

FIGURE 7.2 DFMEA diagram.

Occurrence	Sev. / Occ.	Severity				
		None	Low	Moderate	High	Very high
		1	2	3	4	5
Remote: Failure is unlikely, improbable	1	AO	AO	AO	AO	AO
Low: Relatively few failures	2	AO	AO	AO	AO	SC
Moderate: Occasional failures	3	AO	AO	KC	KC	SC
High: Repeated/frequent failures	4	AO	KC	KC	KC	SC
Very High: Failure is almost inevitable, frequent, persistent failures	5	AO	KC	KC	KC	SC

FIGURE 7.3 Classification codes for DFMEA.

The classification codes are illustrated in Figure 7.3.

9. Current mitigations: Current mitigations are the activities that will assure the design adequacy for the failure mode and cause under consideration. Those activities will prevent the cause/mechanism or failure mode/effect from occurring, or reduce the rate of occurrence, such as

- Proven modeling/simulation (e.g., finite element analysis)
- Tolerance stack up study (e.g., geometric dimensional tolerance)
- Material compatibility study (e.g., thermal expansion, corrosion)
- Subjective design and manufacturing reviews
- Redundancy
- Labeling
- Design of experiments studies
- Parameter design studies
- Tolerance design studies

10. Verification: Verify the design adequacy against cause/mechanism of failure or verify the design adequacy against failure mode, either by analytical or physical methods, such as

Classification		
Code	To Indicate	Criteria
SC	A potential safety characteristics	Severity = 5; occurrence = 2–5
KC	A potential key design characteristics	Severity = 4; occurrence = 3–5
		Severity = 3; occurrence = 3–5
		Severity = 2; occurrence = 4–5
AO	Action is optional	Not SC nor KC

- Tests on preproduction samples or prototype samples
- Analytical tests
- Design verification plan tests

Manufacturing tests or inspections conducted as part of the manufacturing and assembly process are "not" acceptable verification in design phase.

11. Detection (D): Detection is the ability (detection likelihood) of the current mitigations/verification to detect a potential cause/mechanism or failure mode and lead to corrective actions. Timeliness of current mitigations and verification application such as early in design concept stage or just before release for production plays a major role in ranking the detection level. To ensure continuity, the robust DFMEA team should use a consistent detection ranking system.

12. Recommended actions: Recommended actions intent is to reduce any one or all of the severity, occurrence, and detection rankings. Only a design revision/technology change can bring a reduction in the severity ranking. Occurrence reduction can be achieved by removing or controlling the cause/mechanism of the failure mode, where detection reduction can be achieved by increasing the design validation/verification actions. Additional mitigations and recommended actions shall be implemented, as illustrated in Figure 7.4. If no actions are recommended for a specific cause, indicate this by entering "none" or "none at this time" in this column.

13. Action results: Estimate and record the resulting severity, occurrence, and detection rankings and also assess classification. If no actions result, indicate this by entering a "NR" in the severity, occurrence, and detection columns.

As a result of performing the step-by-step robust DFMEA one should be able to define the special design characteristics (safety and reliability product characteristics) that contribute directly

Code	To indicate	Criteria	Additional mitigation and/or recommended actions
SC	A potential safety characteristics	Severity = 5 and Occurrence = 2 to 5	Risk must be mitigated and all current mitigations must be traced back to requirement. All recommended actions must be tracked via issues tracking system until Occurrence brought to less or equal to 1 and Detection brought to less or equal to 2.
KC	A potential key design characteristics	Severity = 4 and Occurrence = 3 to 5	All current mitigations must be traced back to requirement. All recommended actions must be tracked via issues tracking system until Occurrence brought to less or equal to 2 and Detection brought to less or equal to 3.
		Severity = 3 and Occurrence = 3 to 5	All current mitigations must be traced back to requirement. All recommended actions must be tracked via issues tracking system until Occurrence brought to less or equal to 2 and Detection brought to less or equal to 3.
		Severity = 2 and Occurrence = 4 to 5	All current mitigations must be traced back to requirement. All recommended actions must be tracked via issues tracking system until Occurrence brought to less or equal to 3 and Detection brought to less or equal to 3.
AO	Action optional	Not SC nor KC	Project team decides actions required.

FIGURE 7.4 Recommended actions for DFMEA.

to a failure mode/error state of the medical device under analysis. Special design characteristics (safety and reliability product characteristics) defined for the product under analysis are dependent on the robust DFMEA scope and boundary, when performed on a system, subsystem, or component. For example, in a system level robust DFMEA system level characteristics are defined, in a subsystem level robust DFMEA subsystem level characteristics are defined, and in a component level robust DFMEA component level characteristics are defined.

In many cases a system contains purchased subsystems and components, robust DFMEA is capable of defining all appropriate safety and reliability product characteristics that need to be reached to an agreement with the purchased subsystems and components suppliers.

Adding to all that, all safety and reliability product characteristics that are sensitive to manufacturing process defined in the robust DFMEA (designed in house or purchased subsystems and components) need to derive the process FMEAs and control plans to achieve product reliability and robustness.

7.1.4.4 Axiomatic Design

Axiomatic design, a theory and methodology developed at Massachusetts Institute of Technology (MIT; Cambridge, Massachusetts) 20 years ago, helps designers focus on the problems in bad designs. Says the theory's creator, Professor Nam Suh, "The goal of axiomatic design is to make human designers more creative, reduce the random search process, minimize the iterative trial-and-error process, and determine the best design among those proposed." Axiomatic design applies to designing all sorts of things: software, business processes, manufacturing systems, work flows, general systems with constraints, etc. What's more, it can be used for diagnosing and improving existing designs.

7.1.4.4.1 What Is Axiomatic Design?

While "MIT" and "axiomatic" might suggest some lofty academic theory, axiomatic design is well grounded in reality. It is a systematic, scientific approach to design. It guides de-signers through the process of first breaking up customer needs into FRs, then breaking up these requirements into DPs, and then finally figuring out a process to produce those DPs. In MIT-speak, axiomatic design is a decomposition process going from customer needs to FRs, to DPs, and then to process variables (PVs), thereby crossing the four domains of the design world: customer, functional, physical, and process. The fun begins in decomposing the design. A designer first "explodes" higher-level FRs into lower-level FRs, proceeding through a hierarchy of levels until a design can be implemented. At the same time, the designer "zigzags" between pairs of design domains, such as between the functional and physical domains. Ultimately, zigzagging between "what" and "how" domains reduces the design to a set of FR, DP, and PV hierarchies.

Along the way, there are these two axioms: the independence axiom and the information axiom. (From these two axioms come a bunch of theorems that tell designers "some very simple things," says Suh. "If designers remember these, then they can make enormous progress in the quality of their product design.") The first axiom says that the FRs within a good design are independent of each other. This is the goal of the whole exercise: Identifying DPs so that "each FR can be satisfied without affecting the other FRs," says Suh.

The second axiom says that when two or more alternative designs satisfy the first axiom, the best design is the one with the least information. That is, when a design is good, information content is zero (i.e., "information" as in the measure of one's freedom of choice, the measure of uncertainty, which is the basis of information theory). "Designs that satisfy the independence axiom are called uncoupled or decoupled," explains Robert Powers, president of Axiomatic Design Software, Inc. (Boston, Massachusetts), developers of Acclaro, a software application that prompts designers through the axiomatic design process. "The difference is that in an uncoupled design, the DPs are

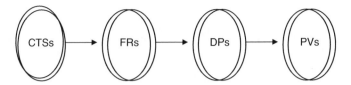

FIGURE 7.5   Axiomatic design process mapping.

totally independent; while in a decoupled design, at least one DP affects two or more FRs. As a result, the order of adjusting the DPs in a decoupled design is important."

The approach for design is to spend time upfront understanding customer expectations and delights (customer attributes) together with corporate and regulatory requirements. Then the following mappings are necessary:

- Perform QFD by mapping critical to satisfaction (CTS) to FRs.
- Perform mapping of axiomatic design between the FRs and DPs.
- Perform mapping of axiomatic design between the DPs and the PVs.

The design process involves three mappings between four domains as shown in Figure 7.5. The first mapping involves the mapping from CTS metrics to the FRs and then to DPs. The last mapping occurs between DPs and the PVs.

7.1.4.4.2   Mapping of Axiomatic Design

The axiomatic design method provides the process as a means to define physical and process structures. The design is first identified in terms of its FRs and then progressively detailed in terms of its lower-level FRs and DPs. Hierarchy is build by the decomposing design into a number of FRs and DPs. The principles that are use as guidance are

- Principle of independence—maintain the independence of the FRs
- Principle of information—minimize the information content in a design: reduce complexity

The principle of independence states that the optimal design maintains the independence of the FRs. An acceptable design will have the FRs and DPs related in such a way that a specific DP can be adjusted to satisfy a corresponding FR without affecting other FRs.

There are three possible mappings—uncoupled (optimal), decoupled (semi-optimal), and coupled. These mappings can be explained using the following matrices:

$$\{FR's\} = [A]\{DP's\}$$

The elements of the design matrix, A, indicate the effects of changes of DPs on the FRs, as an example, consider the design equation shown below:

$$\begin{Bmatrix} FR_1 \\ FR_2 \\ FR_3 \end{Bmatrix} = \begin{bmatrix} a_{11} & 0 & a_{13} \\ a_{21} & a_{22} & 0 \\ a_{31} & 0 & a_{33} \end{bmatrix} \begin{bmatrix} DP_1 \\ DP_2 \\ DP_3 \end{bmatrix}$$

Uncouple design is represented as follows showing the independence of FRs

$$\begin{Bmatrix} FR_1 \\ FR_2 \\ FR_3 \end{Bmatrix} = \begin{bmatrix} a_{11} & 0 & 0 \\ 0 & a_{22} & 0 \\ 0 & 0 & a_{33} \end{bmatrix} \begin{bmatrix} DP_1 \\ DP_2 \\ DP_3 \end{bmatrix}$$

FRs are represented as

$$FR_1 = a_{11} \times DP_1$$
$$FR_2 = a_{22} \times DP_2$$
$$FR_3 = a_{33} \times DP_3$$

Decouple design is represented as follows showing the semi-independence of FRs

$$\begin{Bmatrix} FR_1 \\ FR_2 \\ FR_3 \end{Bmatrix} = \begin{bmatrix} a_{11} & 0 & 0 \\ a_{21} & a_{22} & 0 \\ a_{31} & a_{32} & a_{33} \end{bmatrix} \begin{bmatrix} DP_1 \\ DP_2 \\ DP_3 \end{bmatrix}$$

FRs are represented as

$$FR_1 = a_{11} \times DP_1 + 0 \times DP_2 + 0 \times DP_3$$
$$= a_{11} \times DP_1$$
$$FR_2 = a_{21} \times DP_1 + a_{22} \times DP_2 + 0 \times DP_3$$
$$= a_{21} \times DP_1 + a_{22} \times DP_2$$
$$FR_3 = a_{31} \times DP_1 + a_{32} \times DP_2 + a_{33} \times DP_3$$

Coupled design is represented as follows showing the interdepencies of FRs

$$\begin{Bmatrix} FR_1 \\ FR_2 \\ FR_3 \end{Bmatrix} = \begin{bmatrix} a_{11} & 0 & a_{13} \\ a_{21} & a_{22} & 0 \\ a_{31} & 0 & a_{33} \end{bmatrix} \begin{bmatrix} DP_1 \\ DP_2 \\ DP_3 \end{bmatrix}$$

The FRs are highly interdependent that lead to a mediocre design.

$$FR_1 = a_{11} \times DP_1 + 0 \times DP_2 + a_{13} \times DP_3$$
$$= a_{11} \times DP_1 + a_{13} \times DP_3$$
$$FR_2 = a_{21} \times DP_1 + a_{22} \times DP_2 + 0 \times DP_3$$
$$= a_{21} \times DP_1 + a_{22} \times DP_2$$
$$FR_3 = a_{31} \times DP_1 + a_{32} \times DP_2 + a_{33} \times DP_3$$

This concept is valid during mapping between DPs and PVs. In each stage, during mapping between FRs and DPs, and mapping between DPs and PVs the principles of axiomatic design should be followed.

7.2 REDUNDANCY

One method of addressing the high failure rate of certain components is the use of redundancy, that is, the use of more than one component for the same purpose in the circuit. The philosophy behind redundancy is if one component fails, another will take its place and the operation will continue.

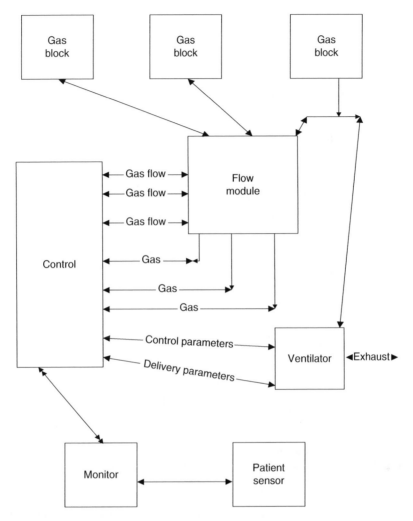

FIGURE 7.6   Block diagram.

An example would be the inclusion of two reed switches in parallel where if one fails because the reeds have stuck together, the other is available to continue the operation. (These controls, e.g., would involve the control parameters in an anesthesia machine, see Figure 7.6.) Redundancy may be of two types: active and standby.

7.2.1   ACTIVE REDUNDANCY

Active redundancy occurs when two or more components are placed in parallel, with all components being operational. Satisfactory operation occurs if at least one of the components functions. If one component fails, the remaining parts will function to sustain the operation. Active redundancy is important in improving the reliability of a device. Placing components redundantly increases the MTBF of the circuit, thus improving reliability. Consider the following example.

Figure 7.7 shows a circuit for an amplifier. Let us use the component U1 as our candidate for redundancy. The failure rate for the component in MIL-HDBK-217 gives a value for our intended use of 0.320 failures/million hours. The failure rate assumption is that the component was in its useful life period. Therefore, the reciprocal of the failure rate is the mean time between failure (MTBF). When calculating the MTBF, the failure rate must be specified in failures per hour.

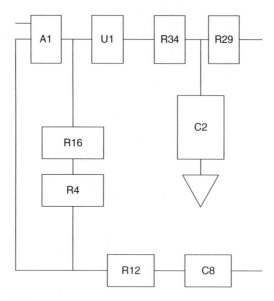

FIGURE 7.7 Circuit example.

Therefore, the failure rate, as listed in the handbook or in vendor literature must be divided by 1 million.

$$
\begin{aligned}
\text{MTBF} &= 1/\lambda \\
&= 1/0.00000032 \\
&= 3{,}125{,}000 \text{ h}
\end{aligned}
$$

Let us assume that for our particular application, this MTBF value is not acceptable. Therefore, we decide to put two components in parallel (Figure 7.8). Again, we assume the useful life period of the component. For this case:

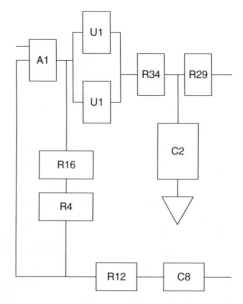

FIGURE 7.8 Active redundancy.

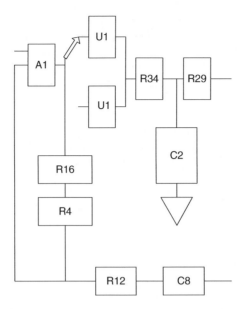

FIGURE 7.9   Standby redundancy.

$$\text{MTBF} = 3/2\lambda$$
$$= 3/2(0.00000032)$$
$$= 3/0.00000064$$
$$= 4,687,500 \text{ h}$$

By putting two components in active redundancy, the MTBF of the circuit has increased by 50%.

7.2.2   STANDBY REDUNDANCY

Standby redundancy occurs when two or more components are placed in parallel, but only one component is active. The remaining components are in standby mode.

Returning to our previous example, we have decided to use standby redundancy to increase our reliability (Figure 7.9). Again assuming the useful life period and ignoring the failure rate of the switch

$$\text{MTBF} = 2/\lambda$$
$$= 2/0.00000032$$
$$= 6,250,000 \text{ h}$$

By using standby redundancy, the MTBF has increased by 100% over the use of the single component and by 33% over active redundancy.

Obviously, the use of redundancy is dependent upon the circuit and the failure rates of the individual components in the circuit. However, the use of redundancy definitely increases the reliability of the circuit. What type of redundancy is used again depends on the individual circuit and its intended application.

7.3   COMPONENT SELECTION

As certain portions of the design become firm, the job of selecting the proper components becomes a primary concern, especially where there are long lead times for orders. How are the vendors for these components chosen? If one is honest in looking back at previous design developments and honest in listing the three main criteria for choosing a component vendor, they would be

Lowest cost
Lowest cost
Lowest cost

The only other parameter which may play a part in choosing a vendor is loyalty to a particular vendor, no matter what his incoming quality may be. Obviously, these are not the most desirable parameters to consider if the design is to be reliable. The parameters of choice include

- Fitness for use
- Criticality versus noncriticality
- Reliability
- History
- Safety

7.3.1 COMPONENT FITNESS FOR USE

Fitness for use includes analyzing a component for the purpose to which it was designed. Many vendors list common applications for their components and tolerances for those applications. Where the desired application is different than that listed, the component must be analyzed and verified in that application. This includes specifying parameters particular to its intended use, specifying tolerances, inclusion of a safety margin and a review of the history of that part in other applications.

For components being used for the first time in a particular application and for which no history or vendor data is available, testing in the desired application should be conducted.

7.3.2 COMPONENT RELIABILITY

The process of ensuring the reliability of a component is a multistep procedure, including

- Initial vendor assessment
- Vendor audit
- Vendor evaluation
- Vendor qualification

The initial vendor assessment should be a review of any past history of parts delivery, including on time deliveries, incoming rejection rate, willingness of the vendor to work with the company, and handling of rejected components. The vendor should also be questioned as to the nature of his acceptance criteria, what type of reliability tests were performed, and what the results of the tests were. It is also important to determine whether the nature of the test performed was similar to the environment the component will experience in your device.

Once the initial vendor assessment is satisfactorily completed, an audit of the vendor's facility is in order. The vendor's processes should be reviewed, the production capabilities assessed, and rejection rates and failure analysis discussed. Sometimes the appearance of the facility provides a clue as to what type of vendor you are dealing with. A facility that is unorganized or dirty may tell you about the quality of the work performed.

Once components are shipped, you need to ensure that the quality of the incoming product is what you expect. A typical approach to the evaluation would be to do 100% inspection on the first several lots to check for consistent quality. Once you have an idea of the incoming quality and you are satisfied with it, components can be randomly inspected or inspected on a skip-lot basis.

Many companies have established a system of qualified vendors to determine what components will be used and the extent of incoming inspection. Some vendors qualify through a rigorous testing scheme that determines the incoming components meet the specification. Other companies have based qualification on a certain number of deliveries with no failures at incoming. Only components from qualified vendors should be used in any medical device. This is especially important when dealing with critical components.

7.3.3 Component History

Component history is an important tool in deciding what components are to be used in a design. It is important to review the use of the component in previous products, whether similar or not. When looking at previous products, the incoming rejection history, performance of the component in field use and failure rate history need to be analyzed.

A helpful tool in looking at component history is the use of available data banks of component information. One such data bank is MIL-HDBK-217. This military standard lists component failure rates based upon the environment in which they are used. The information has been accumulated from the use of military hardware. Some environments are similar to that seen by medical devices and the data is applicable. MIL-HDBK-217 is discussed in greater detail later in this chapter.

Another component data bank is a government program named GIDEP. The only cost for joining this group is a report listing failure rates of components in your applications. You receive reports listing summaries of other reports the group has received. It is a good way to get a history on components you intend to use. More information may be obtained by contacting: http://www.gidep. org/join/revapp.pdf. A good source for both mechanical and electrical component failure rates is the books produced by the Reliability Analysis Center. They may be contacted at http://src. alionscience.com/

7.3.4 Component Safety

The safety of each component in your application must be analyzed. Do this by performing a fault tree analysis, where possible failures are traced back to the components causing them.

A failure mode analysis can be performed that looks at the results of single point failures of components. Unlike the fault tree, which works from the failure back to the component, failure mode analysis works from the component to the resultant failure. This is also discussed in more detail later in the chapter.

7.4 COMPONENT DERATING

Component failure in a given application is determined by the interaction between the strength and the stress level. When the operational stress levels of a component exceed the rated strength of the component, the failure rate increases. When the operational stress level falls below the rated strength, the failure rate decreases.

With the various ways for improving the reliability of products, derating of components is an often-used method to guarantee good performance as well as extended life of a product. Derating is the practice of limiting the stresses, which may be applied to a component, to levels below the specified maximum.

Derating enhances reliability by

- Reducing the likelihood that marginal components will fail during the life of the system
- Reducing the effects of parameter variations
- Reducing the long-term drift in parameter values
- Providing allowance for uncertainty in stress calculations
- Providing some protection against transient stresses, such as voltage spikes

An example of component derating is the use of a 2 W resistor in a 1 W application. It has been shown that derating a component to 50% of its operating value generally decreases its failure rate by a factor greater than 30%. As the failure rate is decreased, the reliability is increased.

Components are derated with respect to those stresses to which the component is most sensitive. These stresses fall into two categories, operational stresses and application stresses. Operational stresses include

- Temperature
- Humidity
- Atmospheric pressure

Application stresses include

- Voltage
- Current
- Friction
- Vibration

These latter stresses are particularly applicable to mechanical components.

Electrical stress usage rating values are expressed as ratios of maximum applied stress to the component's stress rating. The equation for table guidelines is

$$\text{Usage ratio} = \text{Maximum applied stress/component stress rating}$$

For most electronic components, the usage ratio varies between 0.5 and 0.9.

Thermal derating is expressed as a maximum temperature value allowed or as a ratio of "actual junction temperature" to "maximum allowed junction temperature" of the device. The standard expression for temperature measurement is the Celsius scale.

Derating guidelines should be considered to minimize the degradation effect on reliability. In examining the results from a derating analysis, one often finds that a design needs less than 25 components aggressively derated to greatly improve its reliability. And, depending on the design of the product, these components often relate to an increase in capacitance voltage rating, a change of propagation speed, an increase in the wattage capacity of a selected few power resistors, etc.

7.5 SAFETY MARGIN

Components or assemblies will fail when the applied load exceeds the strength at the time of application. The consideration of the load should take into account combined loads, such as voltage and temperature or humidity and friction. Combined loads can have effects that are out of proportion to their separate contributions, both in terms of instantaneous effects and strength degradation effects.

Establishing tolerances is an essential element of assuring adequate safety margins. Establishing tolerances, with appropriate controls on manufacturing provides control over the resulting strength distributions. Analysis should be based on worst-case strength or distributional analysis, rather than on an anticipated strength distribution.

Safety margin is calculated as follows:

$$\text{Safety Margin} = (\text{Mean safety factor}) - 1$$
$$= (\text{Mean strength/mean stress}) - 1$$

An example illustrates the concept:

A structure is required to withstand a pressure of 20,000 psi. A safety margin of 0.5 is to be designed into the device. What is the strength that must be designed in?

$$\text{Safety margin} = (\text{Strength/stress}) - 1$$
$$0.5 = (\text{Strength/20,000}) - 1$$
$$1.5 = \text{Strength/20,000}$$
$$(20,000 \times 1.5) = \text{Strength}$$
$$30,000 \text{ psi} = \text{Strength}$$

Most handbooks list a safety margin of 2.0 as the minimum required for high reliability devices. In some cases, this may result in an overdesign. The safety margin must be evaluated according to device function, the importance of its application and the safety requirements. For most medical applications, a minimum safety margin of 0.5 is adequate.

7.6 LOAD PROTECTION

Protection against extreme loads should be considered whenever practicable. In many cases, extreme loading situations can occur and must be protected against. When overload protection is provided, the reliability analysis should be performed on the basis of the maximum load which can be anticipated, bearing in mind the tolerances of the protection system.

7.7 ENVIRONMENTAL PROTECTION

Medical devices should be designed to withstand the worst-case environmental conditions in the product specification, with a safety margin included. Some typical environmental ranges that the device may experience include

Operating temperature	0°C to +55°C
Storage temperature	−40°C to +65°C
Humidity	95% RH at 40°C
Mechanical vibration	5–300 Hz at 2 Gs
Mechanical shock	24–48 in. drop
Mechanical impact	10 Gs at a 50 msec pulse width
Electrostatic discharge	up to 50,000 V

Electromagnetic compatibility becomes an issue in an environment, like an operating room. Each medical device should be protected from interference from other equipment, such as electrocautery and should be designed to eliminate radiation to other equipment.

7.8   PRODUCT MISUSE

An area of design concern that was briefly addressed earlier in this chapter is the subject of product misuse. Whether through failure to properly read the operation manual or through improper training, medical devices are going to be misused and even abused. There are many stories of product misuse, such as the hand held monitor that was dropped into the toilet bowl, the physician who used a hammer to pound a 9 V battery into a monitor backward or the user who spilled a can of soda on and into a device. Practically, it is impossible to make a device completely misuse-proof. But it is highly desirable to design around the ones that can be anticipated.

Some common examples of product misuse include

- Excess application of cleaning solutions
- Physical abuse
- Spills
- Excess weight applied to certain parts
- Excess torque applied to controls or screws
- Improper voltages, frequencies or pressures
- Improper or interchangeable electrical or pneumatic connections

Product misuse should be discussed with marketing to define as many possible misuse situations as can be anticipated. The designer must then design around these situations, including a safety margin, which will serve to increase the reliability of the device. Where design restrictions limit the degree of protection against misuse and abuse, the device should alarm or should malfunction in a manner that is obvious to the user.

7.9   EXTENDED TRIZ DESIGN TECHNIQUES

The initial part of the design/specify/build/test procedure is the most demanding, if the initial conception of the problem and the consequent design solution is inadequate, the product or solution may be doomed to failure. Chapter 2 covered some of the fundamental design tools used in initial attempts at a design solution. This section is meant to give an overview of another method. This section will include a partial example problem "ideation" using a software package called "Innovation Workbench," a software package that is an outgrowth of the basic TRIZ method discussed in Chapter 2. It has some properties in common with another package, TechOptimizer, which also evolved from the same roots.

An example to be discussed is fairly straightforward; it is a problem that occurs daily in our hospital environment at this time. When electrosurgical units are used to cut or cauterize, for example during the excision of a skin cancer, a need exists for a form-fitting (flexible) grounding pad to ensure that the majority of return current flow is though the grounding pad, rather than through some other small part of the body, such as a finger, that might otherwise offer a return path for the current. This smallness implies a potentially high current density and concurrent burn injury potential. After the surgery is completed, a second potential exists for injury—when the pad is removed there exists the potential for skin abrasions and tearing due to the stickiness of the pad, which obviously is what kept it on to begin with. This injury potential is more likely with elderly patients, due to the decreased elasticity of their skins. A concept map for this problem statement may be seen in Figure 7.10.

7.9.1   USE OF INNOVATION WORKBENCH

Innovation workbench is a software package that guides a designer through the initial design/solution search process by having the designer fill out material in a questionnaire. The questionnaire consists

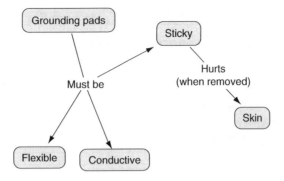

FIGURE 7.10   Concept map for grounding pad needs and problems.

of five main parts. The first part is "Innovation Situation Questionnaire." This section consists of nine main sections that serve to document the current situation and the allowable changes that the designer might be able to make. The second part, "Problem Formulation," requires that the designer build a diagram that interrelates the elements of the process/system under study, with an emphasis on the generation of good interactions versus detrimental interactions (in the chart to follow, arrows to a sharp corner rectangle versus arrows to a rounded corner rectangle). This diagram is termed a conflict diagram. The detrimental interactions are analogous to the technical contradictions discussed in Section 2.7, harmful effects in concept diagrams, and negative interactions between functions in QFD diagrams. The good and bad interactions in the conflict diagram are used to generate directions for innovation, some of which are then selected for further study in a section titled "Prioritize Directions." Finally, two sections are devoted to the "Development of Concepts" and "Evaluation of the Results," these sections will receive little development here, as the majority of the useful part of this exercise is developed in the first three sections.

In Section 7.9.2, the program headers and all numbered lines are from the program itself, the other lines and diagram are input by the user of the program. The initial sections with the "≫" delineation are the program derived suggestions based upon the user-input conflict diagram for the problem. The example is the above grounding problem.

7.9.2   IDEATION PROCESS

Innovation Situation Questionnaire

1. **Brief description of the problem**
   See PowerPoint, grounding pad injury, or above material in text
2. **Information about the system**
   2.1. **System name**
      Grounding pad
   2.2. **System structure**
      The grounding pad is placed on the subject's body in a manner so as to allow current to flow in a circuit
   2.3. **Functioning of the system**
      The primarily useful function is the removal, through the pad, of electric current
   2.4. **System environment**
      Typical situation—surgery
3. **Information about the problem situation**
   3.1. **Problem that should be resolved**
      Removal without compromise of the patient's body

    3.2. **Mechanism causing the problem**

        The adhesion of the pad to the skin pulls the skin during removal

    3.3. **Undesired consequences of unresolved problem**

        Skin gets injured, sometimes torn

    3.4. **History of the problem**

        I guess a long time, need to review the literature

    3.5. **Other systems in which a similar problem exists**

        I guess the acne/blackhead strips

    3.6. **Other problems to be solved**

        None at this time

4. **Ideal vision of solution**

    A grounding pad is designed that adheres well when needed, then can be easily removed

5. **Available resources**

    Hospital BE team, MD help, etc.

6. **Allowable changes to the system**

    Some leeway on cost, probably can change materials, etc.

7. **Criteria for selecting solution concepts**

    Easy removal with safe use

8. **Company business environment**

    BME senior design

9. **Project data**

    Gant chart, timelines, team, etc.

Problem Formulation

    1. Build the diagram

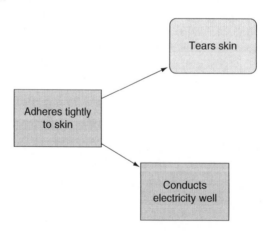

    2. Directions for Innovation

10/9/2003 9:51:20 AM Diagram 1

    ≫ 1. Find an alternative way to obtain [the] (conducts electricity well) that does not require [the] (Adheres tightly to skin).

    ≫ 2. Consider transitioning to the next generation of the system that will provide [the] (conducts electricity well) in a more effective way and/or will be free of existing problems.

    ≫ 3. Find an alternative way to obtain [the] (Adheres tightly to skin) that offers the following: provides or enhances [the] (conducts electricity well), does not cause [the] (Tears skin).

≫ 4. Try to resolve the following contradiction: The useful factor [the] (Adheres tightly to skin) should be in place in order to provide or enhance [the] (conducts electricity well), and should not exist in order to avoid [the] (Tears skin).

≫ 5. Find a way to eliminate, reduce, or prevent [the] (Tears skin) under the conditions of [the] (Adheres tightly to skin).

Prioritize Directions

1. Directions selected for further consideration
   I. Find an alternative way to obtain [the] (conducts electricity well) that does not require [the] (Adheres tightly to skin).
      A. Improve the useful factor (conducts electricity well).
      B. Obtain the useful result without the use of [the] (conducts electricity well).
      C. Increase effectiveness of the useful action of [the] (conducts electricity well).
      D. Synthesize the new system to provide [the] (conducts electricity well).
      E. Apply universal operators to provide the useful factor (conducts electricity well).
      F. Consider resources to provide the useful factor (conducts electricity well).

   II. Consider transitioning to the next generation of the system that will provide [the] (conducts electricity well) in a more effective way and/or will be free of existing problems.
      A. Improve Ideality of your system that provides [the] (conducts electricity well).
      B. Consider the possibility to transform the existing system that provides [the] (conducts electricity well) into bi- or poly-system.
      C. Consider segmentation of the existing system that provides [the] (conducts electricity well).
      D. Consider restructuring the existing system that provides [the] (conducts electricity well).
      E. Increase dynamism of the existing system that provides [the] (conducts electricity well).
      F. Increase controllability of the existing system that provides [the] (conducts electricity well).
      G. Make the existing system that provides [the] (conducts electricity well) and/or its elements more universal.

   III. Find an alternative way to obtain [the] (Adheres tightly to skin) that offers the following: provides or enhances [the] (conducts electricity well), does not cause [the] (Tears skin).
      A. Improve the useful factor (Adheres tightly to skin).
      B. Obtain the useful result without the use of [the] (Adheres tightly to skin).
      C. Increase effectiveness of the useful action of [the] (Adheres tightly to skin).
      D. Synthesize the new system to provide [the] (Adheres tightly to skin).
      E. Apply universal Operators to provide the useful factor (Adheres tightly to skin).
      F. Consider resources to provide the useful factor (Adheres tightly to skin).

   IV. Try to resolve the following contradiction: The useful factor [the] (Adheres tightly to skin) should be in place in order to provide or enhance [the] (conducts electricity well), and should not exist in order to avoid [the] (Tears skin).
      A. Apply separation principles to satisfy contradictory requirements related to [the] (Adheres tightly to skin).
      B. Apply 40 Innovation Principles to resolve contradiction between useful purpose of (Adheres tightly to skin) and its harmful result.

V. Find a way to eliminate, reduce, or prevent [the] (Tears skin) under the conditions of [the] (Adheres tightly to skin).

   A. Isolate the system or its part from the harmful effect of [the] (Tears skin).

   B. Counteract the harmful effect of [the] (Tears skin).

   C. Impact on the harmful action of [the] (Tears skin).

   D. Reduce sensitivity of the system or its part to the harmful effect of [the] (Tears skin).

   E. Eliminate the cause of the undesired action of [the] (Tears skin).

   F. Reduce the harmful results produced by [the] (Tears skin).

   G. Apply universal Operators to reduce the undesired factor (Tears skin).

   H. Consider resources to reduce the undesired factor (Tears skin).

   I. Try to benefit from the undesired factor (Tears skin).

2. List and categorize all preliminary ideas
   From the above list, read each item and place here.

Develop Concepts

1. Combine ideas into concepts.
2. Apply lines of evolution to further improve concepts.

Evaluate Results

1. Meet criteria for evaluating concepts.
2. Reveal and prevent potential failures.
3. Plan the implementation.

7.9.3 SUMMARY

The fairly extensive material in the previous several pages should leave one at least the impression that the overall process can be all-encompassing. The crux of a good solution (and solution space) lies in the good development of the system diagram, if the diagram properly captures all the relevant interactions and conflicts in a system design, the (patented) solution generation algorithm should generate a solution suggestion that will solve the design problem at hand. Other considerations that are forced by this program that are often overlooked are the requirement that one consider the environment (as this can often assist in problem solution). At the very least, the use of such a program provides a comprehensive structure for consideration of many design problems. Not shown above is the ability to reference patent databases and effects databases, the addition of this data makes this program a powerful tool.

EXERCISES

1. Use the term "reverse engineer" in a Web search. Report on the variety of firms offering this service.
2. Use the term "value engineering" in a Web search. How does it differ from reverse engineering?
3. Use the term "re-engineering" in a Web search. How does this relate to reverse engineering?
4. Perform a Web search on the term "axiomatic design." Print out and summarize an article of interest to you.
5. Pahl and Beitz do discuss idea generation techniques. Find which they stress, and why.
6. Based upon the discussion of this chapter (or via a download) perform an innovation situation questionnaire for your design project.
7. Keeping of a design notebook is considered evidence in patent litigation. How do you prove that your work, done with a computer design tool, actually took place on a given day?

BIBLIOGRAPHY

AIAG (Automotive Industry Action Group), *Potential Failure Mode and Effects Analysis*, Reference Manual Third Edition.

ANSI/AAMI/ISO–14971–2000, *Medical Devices—Application of Risk Management to Medical Devices*.

Brue, G. and Robert, G.L., *Design for Six Sigma*. New York: McGraw-Hill, 2003.

EURobust Partnership, *Use and Knowledge of Robust Design Methodology*. Sweden: Chalmers University, 2003.

Fries, R.C., *Handbook of Medical Device Design*, 2nd ed. Boca Raton, FL: CRC Press/Taylor & Francis Division, 2006.

Gaydos, W., Dynamic parameter design for an FM data demodulator, *ITT Industries Annual Taguchi Symposium*, October 1994.

Huber, C. et al., Straight talk on DFSS, *Six Sigma Forum Magazine*, 1(4), August 2002.

IEC-60812 1985, *Analysis Techniques for System Reliability—Procedure for Failure Mode and Effects Analysis (FMEA)*. www.iec.ch, last visited May 12, 2008.

Ingle, K.A., *Reverse Engineering*. New York: McGraw-Hill, 1994.

Jensen, F. and Peterson, N.E., *Burn-In*. New York: John Wiley & Sons, 1982.

Lloyd, D.K. and Lipow, M., *Reliability Management, Methods and Management*, 2nd ed. Milwaukee, WI: American Society for Quality Control, 1984.

Logothetis, N. and Wynn, H.P., *Quality through Design*. London, England: Oxford University Press, 1990.

Mason, R.L., Hunter, W.G., and Hunter, J.S., *Statistical Design and Analysis of Experiments*. New York: John Wiley & Sons, 1989.

McConnell, S., *Rapid Development*. Redmond, WA: Microsoft Press, 1996.

MIL-STD-202, *Test Methods for Electronic and Electrical Component Parts*. Washington, DC: Department of Defense, 1980.

MIL-HDBK-217, *Reliability Prediction of Electronic Equipment*. Washington, DC: Department of Defense, 1986.

MIL-STD-750, *Test Methods for Semiconductor Devices*. Washington, DC: Department of Defense, 1983.

MIL-STD-781, *Reliability Design Qualification and Production Acceptance Tests: Exponential Distribution*. Washington, DC: Department of Defense, 1977.

MIL-STD-883, *Test Methods and Procedures for Microelectronics*. Washington, DC: Department of Defense, 1983.

Montgomery, D.C., *Design and Analysis of Experiments*, 2nd ed. New York: John Wiley & Sons, 1984.

O'Connor, P.D.T., *Practical Reliability Engineering*. 3rd ed. Chichester, England: John Wiley & Sons, 1991.

Page-Jones, M., *The Practical Guide to Structured Systems Design*, 2nd ed. Englewood Cliffs, NJ: Prentice-Hall.

Phadke, M.S., *Quality Engineering Using Robust Design*. Englewood Cliff, NJ: Prentice-hall, November 1989.

Pahl, G. and Beitz, W., *Engineering Design, A Systematic Approach*. London: Springer-Verlag, 1988.

Putnam, L.H. and Ware, M., *Measures for Excellence*. Englewood Cliffs, NJ: P T R Prentice-Hall, 1992.

Reliability Analysis Center, *Nonelectronic Parts Reliability Data: 1995*. Rome, NY: Reliability Analysis Center, 1994.

Ross, P.J., *Taguchi Techniques for Quality Engineering*. New York: McGraw-Hill, 1988.

Taguchi, G., *System of Experimental Design*, Volumes 1 and 2. Clausing, D., Ed. New York: UNIPUB/Krass International Publications, 1987.

8 Software Development Methods and Tools

> The hardest single part of building a software system is deciding precisely what to build.
>
> **Frederick P. Brooks**

Software design and implementation is a multistaged process in which system and software requirements are translated into a functional program that addresses each requirement. Good software designs are based on a combination of creativity and discipline. Creativity provides resolution to new technical hurdles and the challenges of new market and user needs. Discipline provides quality and reliability to the final product.

Software design begins with the software requirements specification (SRS). The design itself is the system architecture, which addresses each of the requirements of the specification and any appropriate software standards or regulations.

The top-level design begins with the analysis of software design alternatives and their trade-offs. The overall software architecture is then established along with the design methodology to be used and the programming language to be implemented. A risk analysis that is performed and then refined to assure malfunction of any software component will not cause harm to the patient, the user, or the system. Metrics are established to check for program effectiveness and reliability. The requirements traceability matrix (RTM) is reviewed to assure all requirements have been addressed. The software design is reviewed by peers for completeness.

The detailed design begins with modularizing the software architecture; assigning specific functionality to each component and assuring both internal and external interfaces are well defined. Coding style and techniques are chosen based on their proven value and the intended function and environment of the system. Peer reviews assure the completeness and effectiveness of the design. The detailed design also establishes the basis for subsequent verification and validation activity. The use of automated tools throughout the development program is an effective method for streamlining the design and development process and assists in developing the necessary documentation.

8.1 SOFTWARE DEVELOPMENT PLANNING

Planning encompasses more than just documenting resources needed for the project and creating a schedule. It includes looking at the project from a management view including schedule and resources, as well as looking at how the work will be done. What development model should be used? What methodology? What programming language and tools? Planning also includes identifying how the software will be controlled, and how it will be tested.

A key input to planning for medical device software is safety risk. The most important factor is the greatest severity of injury that can result from a software failure. The U.S. Food and Drug Administration calls this the level of concern of the software. Since the software can pose no greater risk of injury than the medical device it is a part of, the software level of concern cannot be greater than the risk of the device. It can possibly be lower if the software is not used to control critical device functions and if a software failure could only cause a lesser severity of injury than the device itself could. Understanding the software level of concern is important, because some choices in software methods and process are not appropriate for software with a high level of concern.

Often when a software development project is planned, some of the most important considerations are not even discussed because everyone believes they are so obvious. These are the objectives for the project and the assumptions that the project team is making. Making these explicit and clear allows the project team to recognize when objectives change or assumptions are not being fulfilled. This allows the team to adjust, rather than continuing down a path that is unlikely to succeed.

Another area that good planning addresses is relationships between groups. A project may depend on the work of others. Without thinking through and getting agreement on how knowledge and information will be shared, or how deliverables are to be made available, much time and effort may be used in managing communication breakdowns that have a high level of emotion. Actions as simple as asking questions may become a problem if the person being asked is on the critical path of another project. Thinking through a process for resolving issues between groups before the project gets underway is a beneficial exercise that is too frequently overlooked or left to be thought about when it gets to be a problem.

Subcontractors are often used for developing parts of the software. This can be a very useful approach, reducing development time and adding expertise in an area where the development team may not have much experience. On the other hand, managing the contract may not be simple. The developer of the medical device is responsible for all the components used in it, and must make sure that the software being developed by the contractor is of sufficient quality for its intended purpose. Once again, careful planning before the work begins will allow for adjustments to be made when unexpected situations arise.

Project risks are another topic that planning should address. Establishing the four Cs (chronology, contingencies, consequences, and criteria) for all significant foreseeable risks will provide a great deal of help should one or more of them become a problem: chronology (when will the risk be quantified?), contingencies (what are the possible courses of action that could be followed?), consequences (what are the likely results of each action?), and criteria (how will the course of action be chosen?). Having gone through the process of analyzing potential project risks during the planning phase of the project will also provide help for dealing with unexpected problems that may come up after the project is underway.

8.2 CHOOSING THE SOFTWARE DEVELOPMENT PROCESS MODEL

A software development process model is the portion of a software life cycle model that occurs before the first release of the software. For any software development, there are some fundamental activities that always are performed, and some optional activities that are included to improve the likelihood of meeting specific project objectives such as short development schedule, high reliability, or high conformance with customer desires. The sequence in which these activities are performed, and the formality with which they are performed and documented may vary. A software development process describes the activities to be performed and defines the chronological ordering of these activities. To define the development schedule, a software development process model must be selected.

There are a number of basically different software development process models, with many variations. Whether a particular model is appropriate depends on the goals of the project and the level of concern of the software.

8.2.1 Development Process Models for High Level of Concern

If the requirements of the software are established and reliability is a major objective, such as for software with a higher level of concern, these software development process models should be considered.

8.2.1.1 Waterfall Model

This was the first documented software life cycle model. It divides the development process into steps, each of which reduces the level of abstraction of the solution. Each step includes a verification task and an exit criteria that must be met before moving on to the next step. As much as possible, iterations of a step are performed during the subsequent step.

The waterfall model's advantages are that it helps find errors early and provides a well-understood structure. Its difficulties are that the requirements must be fully specified at the beginning of the project, before any design work has started. Finding out that requirements are wrong or incomplete late in the project can lead to extensive rework.

The waterfall model works best for complex systems where the requirements and technical methodologies are well understood. Many variations have been defined such as overlapping steps, breaking implementation steps into parallel subprojects, and adding an introductory risk analysis step.

8.2.1.2 Incremental Delivery Model

The incremental delivery model is a modification of the waterfall model. It starts like the waterfall by analyzing requirements and creating an architectural design. Instead of delivering all of the software functionality at the end of the project, the functionality is divided up into increments that are delivered successively through out the project. Each increment refines the requirements and architectural design, then does detailed design, implementation, verification, and release of its functionality.

This model works well when there is a need to deliver partial functionality before all of the functionality is needed. For example, a new medical device might need some software functionality for early hardware testing, additional functionality for validating expected clinical results in animals, more functionality when a human clinical study is performed, and the complete functionality for the market release of the device.

8.2.1.3 Spiral Model

The spiral model was developed by Boehm to address some of the difficulties with the waterfall model. The spiral model iterates a set of steps, creating in effect a series of mini projects. Each of these mini projects completes a loop around the spiral. The first step in each of these mini projects is to determine the objectives, alternatives and constraints for the portion of the product being developed. The next step is to determine risks and resolve them. Then the alternatives are evaluated, the deliverables for the iteration are developed and verified and a plan for the next iteration is created.

The main advantages of the spiral model are its flexibility and that it is risk driven, and as the project progresses and costs increase, the risks decrease. The disadvantage is that the spiral model requires expertise in risk management.

8.2.1.4 Clean Room Model

The clean room model was created to develop software that has a predictable reliability. It combines a set of techniques that depend on verifying the correctness of each step in the development process model. This results in more formality in specifying requirements, performing design, and implementing the design in code. Since each step is verified as correct, clean room eliminates structural testing. It uses a statistical approach to functional testing to demonstrate that the requirements were implemented and to measure the software reliability in terms of mean time to failure.

Since the design is verified against the requirements, the requirements specification must be complete before design can begin. The requirements specification must also be written with sufficient formality that functional correctness verification of the design can be supported. This can best be achieved by using a formal specification language. The design must proceed in small steps, each being verified to be equivalent with its predecessor. Verification-based inspections, which inspect for correctness rather than defects, are used to provide independent confirmation of the design's correctness.

Software coding proceeds in a similar manner of stepwise implementation and verification using inspections. Since the code is verified for correctness to the design, no developer debugging or unit testing is needed for demonstrating that the code implements the design. Clean room eliminates this activity, and the developers do not execute their code.

Clean room relies on independent testing to ensure that the requirements were implemented correctly. It also uses statistical testing techniques, sampling inputs based on probability of usage, to determine the software's reliability. It adds a feedback loop driven by continuously measuring reliability to the incremental delivery development process model to improve the reliability of each incremental delivery. The result is a final product with very high quality and a predictable reliability.

8.3 SOFTWARE DESIGN LEVELS

Software design may be divided into two distinct stages: (1) top-level design and (2) detailed design.

1. Top-level design
   (a) Design alternatives and trade-offs
   (b) Software architecture
   (c) Choosing a methodology
   (d) Structural analysis
   (e) Object-oriented design
   (f) Choosing a language
   (g) Software risk analysis
   (h) Software RTM
   (i) Software review
2. Detailed design
   (a) Design techniques
   (b) Performance predictability and design simulation
   (c) Module specification (mspecs)
   (d) Coding
   (e) Design support tools
   (f) Design as a basis for verification and validation testing

8.4 DESIGN ALTERNATIVES AND TRADE-OFFS

The determination of the design and the allocation of requirements is a very iterative process. Alternative designs are postulated, which are candidates to satisfy the requirements. The determination of these designs is a fundamentally creative activity, a cut and try determination of what might work. The specific techniques used are numerous and call upon a broad range of skills. They include control theory, optimization, consideration of man–machine interface, use of modern control test equipment, queuing theory, communication and computer engineering, statistics, and other disciplines. These techniques are applied to factors such as performance, reliability, schedule, cost, maintainability, power consumption, weight, and life expectancy.

Some of the alternative designs will be quickly discarded, while others will require more careful analysis. The capabilities and quality of each design alternative is assessed using a set of design factors specific to each application and the methods of representing the system design.

Certain design alternatives will be superior in some aspects, while others will be superior in different aspects. These alternatives are traded-off, one against the other, in terms of the factors important for the system being designed. The design ensues from a series of technology decisions, which are documented with architecture diagrams that combine aspects of data and control flow. As an iterative component of making technology decisions, the functionality expressed by the dataflow and control flow diagrams from system requirements analysis is allocated to the various components of the system. Although the methods for selection of specific technology components are not a part of the methodology, the consequences of the decisions are documented in internal performance requirements and timing diagrams.

Finally, all factors are taken into account, including customer desires and political issues to establish the complete system design. The product of the system design is called an architecture model. The architecture includes the components of the system, allocation of requirements, and topics such as maintenance, reliability, redundancy, and self-test.

8.5 SOFTWARE ARCHITECTURE

Software architecture is the high-level part of software design, the frame that holds the more detailed parts of the design. Typically, the architecture is described in a single document referred to as the architecture specification. The architecture must be a prerequisite to the detailed design, because the quality of the architecture determines the conceptual integrity of the system. This in turn determines the ultimate quality of the system.

A system architecture first needs an overview that describes the system in broad terms. It should also contain evidence that alternatives to the final organization have been considered and the reasons the organization used was chosen over the alternatives. The architecture should also contain

- Definition of the major modules in a program. What each module does should be well defined, as well as the interface of each module.
- Description of the major files, tables, and data structures to be used. It should describe alternatives that were considered and justify the choices that were made.
- Description of specific algorithms or reference to them.
- Description of alternative algorithms that were considered and indicate the reasons that certain algorithms were chosen. In an object-oriented system, specification of the major objects to be implemented. It should identify the responsibilities of each major object and how the object interacts with other objects. It should include descriptions of the class hierarchies, of state transitions, and of object persistence. It should also describe other objects that were considered and give reasons for preferring the organization that was chosen.
- Description of a strategy for handling changes clearly. It should show that possible enhancements have been considered and that the enhancements most likely are also easiest to implement.
- Estimation of the amount of memory used for nominal and extreme cases.

Software architecture alludes to two important characteristics of a computer program: (1) the hierarchical structure of procedural components (modules) and (2) the structure of data. Software architecture is derived through a partitioning process that relates elements of a software solution to parts of a real-world problem implicitly defined during requirements analysis. The evolution of a software structure begins with a problem definition. The solution occurs when each part of the problem is solved by one or more software elements.

An architectural template may be developed, which gives a general layout for all the architectural model diagrams to follow. This template indicates the physical perspectives that had not existed in the system requirements. Areas that may be included in the template are

- User interface processing
- Maintenance, self-test, and redundancy requirements

- Input processing
- Output processing

User interface processing is the system-to-user interface, requiring some technology-based enhancements that were omitted in the requirements model. These enhancements are based on use of available technology and on various cost, operational environment, and other criteria. The architecture should be flexible so that a new user interface can be substituted without affecting the processing and output parts of the program.

Maintenance, self-test, and redundancy processing requirements are also technology dependent. These requirements cannot be identified until an implementation technology has been selected that meets the system's reliability and performance criteria.

Input/output (I/O) is another area that deserves attention in the architecture. Input processing refers to the communications across the system's boundary, which were not addressed in the system requirements and are not part of the user interface or a maintenance interface. Additional processing is added depending on technology decisions. Output processing involves the same considerations as input processing. Output processing takes the system's logical output and converts it to a physical form. The architecture should specify a look-ahead, look-behind, or just-in-time reading scheme and it should describe the level at which I/O errors are detected.

The detailed architecture may be expressed in various forms. Examples include

- Architecture context diagram: The top-level diagram for the architecture model. It contains the system's place in its environment and, in addition, the actual physical interface to the environment.
- Architecture flow diagram: The network representation of the system's physical configuration.

8.6 CHOOSING A METHODOLOGY

It seems there are about as many design methodologies as there are engineers to implement them. Typically, the methodology selection entails a prescription for the requirements analysis and design processes. Of the many popular methods, each has its own merit based on the application to which the methods are applied. The toolset and methodology selection should run hand in hand. Tools should be procured to support established or tentative design methodology and implementation plans. In some cases, tools are purchased to support a methodology already in place. In other cases, the methodology is dictated by an available toolset. Ideally, the two are selected at the same time following a thorough evaluation of need.

Selecting the right toolset and design methodology should not be based on a flashy advertisement or suggestion from an authoritative methodology guru. It is important to understand the environment in which it will be employed and the product to which it will be applied. Among other criteria, the decision should be based on the size of the project (number of requirements), type of requirements (hard or soft real-time), complexity of the end-product, number of engineers, experience and skill level of the engineers, project schedules, project budget, reliability requirements, and future enhancements to the product (maintenance concerns). Weight factors should be applied to the evaluation criteria. One way or another, whether the evaluation is done in a formal or informal way, involving one or more than one person, it should be done to assure a proper fit for the organization and product.

Regardless of the approach used, the most important factor to be considered for the successful implementation of a design methodology is software development team buy-in. The software development team must possess the confidence that the approach is appropriate for the application and be willing and excited to tackle the project. The implementation of a design methodology takes relentless discipline. Many projects have been unsuccessful as a result of lack of commitment and faith.

The two most popular formal approaches applied to the design of medical products are (1) the object-oriented analysis/design and (2) the most traditional (top-down) structured analysis/design. Each has its own advantages and disadvantages. Either approach, if done in a disciplined and systematic manner along with the electrical system design, can provide for a safe and effective product.

8.6.1 STRUCTURED ANALYSIS

Structured analysis is the process of examining the software requirements for the purposes of generating a structural model of the requirements. This activity focuses on data flowing through the system. In particular, data transformations are identified which occur in the process of delivering the required outputs from given inputs. A thorough structured analysis of the system will provide a complete and well-understood set of software requirements, which is highly conducive to the ensuing structured design process.

Structured design entails an abstraction of the analysis results into a top-down, functional decomposition of the requirements. Structured design focuses on the decomposition of the operations to be performed on the data. At the onset, a series of high-level functional blocks are identified which, in collection, address all processing expectations of the system. In a systematic manner, a hierarchy of ever smaller processing units are evolved from the high-level blocks. This iterative partitioning produces a series of small, procedural components which, when integrated together, form a system capable of satisfying all functional requirements.

Structured design is the most common approach to software design today. Designing systems from the functional decomposition perspective has been around for decades and its approach is the best understood and mature. Its weaknesses, however, lie in the emphasis on sequential thinking and the generation of solutions based on procedural connection among functional blocks. Most software developers will agree that this is not a natural representation of real-world objects and the relationships between them.

Although normally manageable in small to medium scale software systems, it is inherent that most product requirements changes result in significant design changes unless they were anticipated from the start. Certain types of changes can be very expensive to make because of their disturbance of some of the predefined high-level procedural flows. Unforeseen changes can also lead to reduced product confidence and reliability. Increased complexity often results when trying to retain harmony among existing components in the presence of new and sometimes foggy relationships.

8.6.2 OBJECT-ORIENTED DESIGN

The object-oriented design paradigm seeks to mimic the way that people form models of the real world. In contrast to procedural design methods, it de-emphasizes the underlying computer representation. Its major modeling concept is that of the object, which is used to symbolize real-world entities and their interactions. Objects are entities which have state and behavior. They can be implemented in computer systems as data and a set of operations defined over those data.

Although at its lowest level of design, object-oriented design resembles structured design and traditional code development, during the analysis and high-level design phases a different mind set surrounds the attack of the problem. Object-oriented design hinges on approaching design solutions in terms of the identification of objects, associated object attributes, and operations performed on and among the objects. This approach generates designs that map very well to real-world items and operations, thus leading to designs which can be easier to understand and maintain.

Object-oriented designs have been found to be a very successful approach to the design of some large, more complex systems. This has garnished the attention of software developers around the world. There are, however, two generally recognized blemishes currently associated with the approach. Developers often have difficulty agreeing on the definitions of objects and object classes.

This has resulted in system designs which are not as easy to understand as expectations would have. Also, an additional processing overhead is associated with the implementation of object-oriented programming languages. This inefficiency has deterred many from using the approach on embedded real-time medical systems because of the increased hardware cost incurred to deliver acceptable system performance. Still, as the price of processing power for the dollar decreases it can be expected that object-oriented programming will increase in popularity as a viable approach to the development of high performance, competitively priced medical products.

8.7   CHOOSING A LANGUAGE

Programming languages are the notational mechanisms used to implement software products. Features available in the implementation language exert a strong influence on the architectural structure and algorithmic details of the software. Choice of language has also been shown to have an influence on programmer productivity. Industry data has shown that programmers are more productive using a familiar language than an unfamiliar one. Programmers working with high-level languages achieve better productivity than those working with lower level languages. Developers working in interpreted languages tend to be more productive than those working in compiled languages. In languages that are available in both interpreted and compiled forms, programs can be productively developed in the interpreted form and then released in the better performing compiled form.

Computer languages are the malleable tools for program design and implementation alike. From one perspective, they offer representations of computer procedures that can consolidate the understanding gained from a prototyping process and then link these key requirements to machine capabilities. From another perspective, they can impose structure and clarity on the logical flow of a system with an eye toward operational efficiency and reliability. In principle, these two perspectives should converge. In actual practice, they often conflict. The problem of how to move from an initial design through the necessary revisions to implementation is the underlying issue in the choice and use of language in medical systems.

Modern programming languages provide a variety of features to support development and maintenance of software products. These features include

- Strong type checking
- Separate compilation
- User-defined data types
- Data encapsulation
- Data abstraction

The major issue in type checking is flexibility versus security. Strongly typed languages provide maximum security, while automatic type coercion provides maximum flexibility. The modern trend is to augment strong type checking with features that increase flexibility while maintaining the security of strong type checking.

Separate compilation allows retention of program modules in a library. The modules are linked into the software system, as appropriate, by the linking loader. The distinction between independent compilation and separate compilation is that type checking across compilation–unit interfaces is performed by a separate compilation facility, but not by an independent compilation facility.

User-defined data types, in conjunction with strong type checking, allow the programmer to model and segregate entities from the problem domain using a different data type for each type of problem entity.

Data encapsulation defines composite data objects in terms of the operations that can be performed on them, and the details of data representation and data manipulation are suppressed by the mechanisms. Data encapsulation differs from abstract data types in that encapsulation provides only one instance of an entity.

TABLE 8.1

Language Suitability for Programming Situations

Kind of Programs	More Effective Languages	Less Effective Languages
Structured data	Ada, C/C++, Pascal	Assembler, BASIC
Quick and dirty project	BASIC	Ada, Assembler, Pascal
Fast execution	Assembler, C	Interpreted languages
Mathematical calculations	FORTRAN	Pascal
Easy to maintain	Ada, Pascal	C, FORTRAN
Dynamic memory usage	C, Pascal	BASIC
Limited memory environments	Assembler, BASIC, C	Ada, FORTRAN
Real-time program	Ada, Assembler, C	BASIC, FORTRAN
String manipulation	BASIC, Pascal	C

Data abstraction provides a powerful mechanism for writing well-structured, easily modified programs. The internal details of data representation and data manipulation can be changed at will and, provided the interfaces of the manipulation procedures remain the same, other components of the program will be unaffected by the change, except perhaps for changes in performance characteristics and capacity limits. Using a data abstraction facility, data entities can be defined in terms of predefined types, user-defined types, and other data abstractions, thus permitting systematic development of hierarchical abstractions.

One of the most striking things about computer languages is that there are so many of them. All have struggled to keep up with the increasing individuality and complexity of modern computer systems. To be successful, a language must mediate between (1) the capabilities and limitations of the machine on which the applications run, (2) the properties of the information domain that is to be addressed, (3) the characteristics of the user, and (4) the exchange of information between machines. Ideally, every language should be a proper reflection of these four perspectives.

When choosing a language, careful evaluation is necessary for a particular program. Table 8.1 lists some languages and their suitability for various purposes. The classifications are broad, so care must be taken in their use. Among the many languages available, each has its pros and cons, depending on its specific application. The following language characteristics should be analyzed in making a choice:

- Clarity, simplicity, and unity of language concept
- Clarity of program syntax
- Naturalness of application
- Support for abstraction
- Ease of verification
- Programming environment

Table 8.2 lists some of the pros and cons for individual languages. Additions to the pros include portability of programs and cost of use.

8.8 SOFTWARE RISK ANALYSIS

Software risk analysis techniques identify software hazards and safety-critical single and multiple failure sequences, determine software safety requirements, including timing requirements, and analyze and measure software for safety. While functional requirements often focus on what the system shall do, risk requirements must also include what the system shall not do, including means of eliminating and controlling system hazards and of limiting damage in case of a mishap.

TABLE 8.2

Pros and Cons of Software Languages

Language	Pros	Cons
Ada	Some software engineering techniques are embedded in the language; portable, broad range of language constructs; built-in microprocessing	Large, overkill for many applications; development systems are expensive to purchase; life cycle costs are up front
Assembler	Very fast, low-level programming when other languages are unsuitable	High maintenance cost due to level or readability; high portability of errors, not portable, old, low-level language
BASIC	Good beginner language; straightforward commands	Slow, unstructured, difficult to maintain
C	Wide usage, portable, fast, powerful; recently became an ANSI standard language	Too powerful for the inexperienced programmer
COBOL	Good for large amounts of data, simple calculations, and business record processing	Bad for scientific applications, poor support of complex calculations, slow
FORTRAN	Well suited for scientific and engineering applications	Old technology
Modula-2	Pascal-like, yet modular	Not widely used, no language standard, several dialects
Pascal	Flexible data typing, structured, good beginner language	Monolithic, confining

An important part of the risk requirements is the specification of the ways in which the software and the system can fail safely and to what extent failure is tolerable.

Several techniques have been proposed and used for doing risk analysis, including

- Software hazard analysis
- Software fault tree analysis
- Real-time logic

Software hazard analysis, like hardware hazard analysis, is the process whereby hazards are identified and categorized with respect to criticality and probability. Potential hazards that need to be considered include normal operating modes, maintenance modes, system failure or unusual incidents in the environment, and errors in human performance. Once hazards are identified, they are assigned a severity and probability. Severity involves a qualitative measure of the worst credible mishap that could result from the hazard. Probability refers to the frequency with which the hazard occurs. Once the probability and severity are determined, a control mode is established, that is, a means of reducing the probability or severity of the associated potential hazard. Finally, a control method or methods are selected to achieve the associated control mode.

Fault tree analysis is an analytical technique used in the risk analysis of electromechanical systems. An undesired system state is specified, and the system is analyzed in the context of its environment and operation to find credible sequences of events that can lead to the undesired state. The fault tree is a graphic model of various parallel and sequential combinations of faults or system states that will result in the occurrence of the predefined undesired event. It thus depicts the logical interrelationships of basic events that lead to the hazardous event.

Real-time logic is a process wherein the system designer first specifies a model of the system in terms of events and actions. The event-action model describes the data dependency and temporal ordering of the computational actions that must be taken in response to events in a real-time application. The model can be translated into real-time logic formulas. The formulas are transformed into predicates of Presburger arithmetic with uninterpreted integer functions. Decision procedures are then used to determine whether a given risk assertion is a theorem derivable from the system specification. If so, the system is safe with respect to the timing behavior denoted by that assertion, as long as the implementation satisfies the requirements specification. If the risk assertion is unsatisfiable with respect to the specification then the system is inherently unsafe because successful implementation of the requirements will cause the risk assertion to be violated. Finally, if the negation of the risk assertion is satisfiable under certain conditions then additional constraints must be imposed on the system to assure its safety.

8.9 REQUIREMENTS TRACEABILITY MATRIX

It is becoming more and more apparent how important thorough requirements traceability is during the design and development stages of a software product, especially in large projects with requirements numbering in the thousands or tens of thousands. Regardless of the design and implementation methodology, it is important to assure the design is meeting its requirements during all phases of design.

To ensure the product is designed and developed in accordance with its requirements throughout the development cycle, individual requirements should be assigned to design components. Each software requirement, as might appear in an SRS for example, should be uniquely identifiable. Requirements resulting from design decisions (i.e., implementation requirements) should be uniquely identified and tracked along with product functional requirements.

This process not only assures that all functional and safety features are built into the product as specified but also drastically reduces the possibility of requirements slipping through the cracks. Overlooked features can be much more expensive when they become design modifications at the tail end of development.

The RTM is generally a tabular format with requirements identifiers as rows and design entities as column headings. Individual matrix cells are marked with file names or design model identifiers to denote a requirement is satisfied within a design entity.

An RTM assures completeness and consistency with the software specification. This can be accomplished by forming a table that lists the requirements from the specification versus how each is met in each phase of the software development process. Figure 8.1 is an example of an RTM.

8.10 SOFTWARE REVIEW

An integral part of all design processes include timely and well-defined reviews. Each level of design should produce design review deliverables. Software project development plans should include a list of the design phases, the expected deliverables for each phase, and a sound definition of the deliverables to be audited at each review. Reviews of all design materials have several benefits. First, knowing that their work is being reviewed, authors are more compelled to elevate the quality of their work. Second, reviews often uncover design blind spots and alternative design approaches. Finally, the documentation generated by the reviews is used to acquire agency approvals for process and product.

Requirement	Design	Code	Unit Test	Integration Test
Accept only valid input	Check_input	Check_num.c	Num_only.tc	Whole_valid.tc
		Check_char.c	Char_only.tc	Whole_inval.tc
		Check_mixed.c	Mixed.tc	
Requirement 2				

FIGURE 8.1 Requirements traceability matrix.

Software reviews may take several different forms:

- Inspections of design and code
- Code walk-throughs
- Code reading
- Dog and pony shows

An inspection is a specific kind of review that has been shown to be extremely effective in detecting defects and to be relatively economical compared to testing. Inspections differ from the usual reviews in several ways:

Checklists focus the reviewer's attention on areas that have been problems in the past:

- Emphasis is on defect detection, not correction.
- Reviewers prepare for the inspection meeting beforehand and arrive with a list of the problems they have discovered.
- Data is collected at each inspection and is fed into future inspections to improve them.

An inspection consists of several distinct stages:

Planning	The moderator, after receiving the documentation, decides who will review the material and when and where the review will take place.
Overview	The author describes the technical environment within which the design or code has been created.
Preparation	Each reviewer works alone to become familiar with the documents.
Inspection meeting	The moderator chooses someone, usually the author, to paraphrase the design or read the code. The scribe records errors as they are detected, but discussion of an error stops as soon as it is recognized as an error.
Inspection report	The moderator produces an inspection report that lists each defect, including its type and severity.
Rework	The moderator assigns defects to someone, usually the author, for repair. The assignee(s) resolve each defect on the list.
Follow-up	The moderator is responsible for seeing that all rework assigned during the inspection is carried out.

The general experience with inspections has been that the combination of design and code inspections usually removes 60%–90% of the defects in a product. Inspections identify error-prone routines early and reports indicate they result in 30% fewer defects per 1000 lines of code than walk-throughs do. The inspection process is systematic because of its standard checklists and standard roles. It is also self-optimizing because it uses a formal feedback loop to improve the checklists and to monitor preparation and inspection rates.

A walk-through usually involves two or more people discussing a design or code. It might be as informal as an impromptu bull session around a whiteboard; it might be as formal as a scheduled meeting with overhead transparencies and a formal report sent to management. Some of the characteristics of a walk-through include

- Walk-through is usually hosted and moderated by the author of the design or code under review.
- Purpose of the walk-through is to improve the technical quality of a program rather than to assess it.

TABLE 8.3

Comparison of Inspections and Walk-Throughs

Properties	Inspection	Walk-Throughs
Formal moderator training	Yes	No
Distinct participant roles	Yes	No
Who drives the inspection or walk-through	Moderator	Author
Checklists for finding errors	Yes	No
Focused review effort—looks for the most frequently found kinds of errors	Yes	No
Formal follow-up to reduce bad fixes	Yes	No
Fewer future errors because of detailed error feedback to individual programmers	Yes	Incidental
Improved inspection efficiency from analysis of results	Yes	No
Analysis of data leading to detection of problems in the process, which in turn leads to improvements in the process	Yes	No

- All participants prepare for the walk-through by reading design or code documents and looking for areas of concern.
- Emphasis is on error detection, not correction.
- Walk-through concept is flexible and can be adapted to the specific needs of the organization using it.

Used intelligently, a walk-through can produce results similar to those of an inspection, that is, it can typically find between 30% and 70% of the errors in a program. Walk-throughs have been shown to be marginally less effective than inspections, but in some circumstances, can be preferable. Table 8.3 is a comparison of inspections and walk-throughs.

Code reading is an alternative to inspections and walk-throughs. In code reading, you read source code and look for errors. You also comment on qualitative aspects of the code, such as its design, style, readability, maintainability, and efficiency.

A code reading usually involves two or more people reading code independently and then meeting with the author of the code to discuss it. To prepare for a meeting, the author hands out source listings to the code readers. Two or more people read the code independently. When the reviewers have finished reading the code, the code-reading meeting is hosted by the author of the code and focuses on problems discovered by the reviewers. Finally, the author of the code fixes the problems identified by the reviewers.

The difference between code reading on the one hand and inspections and walk-throughs on the other is that code reading focuses more on individual review of the code than on the meeting. The result is that each reviewer's time is focused on finding problems in the code. Less time is spent in meetings.

Dog and pony shows are reviews in which a software product is demonstrated to a customer. The purpose of the review is to demonstrate to the customer that the project is proceeding, so it is a management review rather than a technical review. They should not be relied on to improve the technical quality of a program. Technical improvement comes from inspections, walk-throughs and code reading. The software development process now moves into the detailed design stage.

8.11 DESIGN TECHNIQUES

Good software design practice is more than a matter of applying one of the latest design methodologies. Thorough requirement generation, requirements tracking, requirements analysis, performance predictability, system simulation, and uniform design reviewing are all activities that contribute to the development of safe and effective software designs.

8.12   PERFORMANCE PREDICTABILITY AND DESIGN SIMULATION

A key activity of design often overlooked by some software developers is the effort to predict the real-time performance of a system. During the integration phase, software designers often spend countless hours trying to finely tune a system which had bottlenecks designed in. Execution estimates for the system interfaces, response times for external devices, algorithm execution times, operating system context switch time, and I/O device access times in the forefront of the design process provide essential input into software design specifications.

For single-processor designs, mathematical modeling techniques such as rate monotonic analysis should be applied to assure all required operations of that processing unit can be performed in the expected time period. System designers often fall into the trap of selecting processors before the software design has been considered, only to experience major disappointment and finger pointing when the product is released. It is imperative to a successful project that the processor selection come after a processor loading study is complete.

In a multiprocessor application, up-front system performance analysis is equally important. System anomalies can be very difficult to diagnose and resolve in multiprocessor systems with heavy inter-processor communications and functional expectations. Performance shortcomings which appear to be the fault of one processor are often the result of a landslide of smaller inadequacies from one or more of the other processors or subsystems. Person-years of integration phase defect resolution can be eliminated by front-end system design analysis or design simulation. Commercial tools are readily available to help perform network and multiprocessor communications analysis and execution simulation. Considering the pyramid of effort needed in software design, defect correction in the forefront of design yields enormous cost savings and increased reliability in the end.

8.13   MODULE SPECIFICATIONS

The lowest level of software design is typically referred to as module specifications or mspecs. A complete set of mspecs details the actual definitions for the routine names, interfaces (inputs and outputs) of the routines, resident data structures, and pseudocode for each routine. The pseudocode for each routine should explicitly detail the flow of logic through the routine, including the lowest level algorithms, decision branches, and usage of data structures. Module specifications are generated for both structured designs and object-oriented designs and usually become part of the documentation associated with the source code as routine header information. Accurate mspecs are an essential part of all software design, regardless of toolset and methodology selection. This is especially true when the mspec designer and the coder are not the same person.

8.14   CODING

For many years the term "software development" was synonymous with coding. Today, for many software development groups, coding is now one of the shortest phases of software development. In fact, in some cases, although very rare in a world of real-time embedded software development, coding is actually done automatically from higher level design (mspecs) documentation by automated tools called code generators.

With or without automatic code generators, the effectiveness of the coding stage is dependent on the quality and completeness of the design documentation generated in the immediately preceding software development phase. The coding process should be a simple transition from the module specifications, and, in particular, the pseudocode. Complete mspecs and properly developed pseudocode leaves little to interpretation for the coding phase, thus reducing the chance of error.

The importance of coding style (how it looks) is not as great as the rules which facilitate comprehension of the logical flow (how it relates). In the same light, in-line code documentation (comments) should most often address why rather than how functionality is implemented. These two focuses help the code reader understand the context in which a given segment of code is used. With precious few exceptions (e.g., high performance device drivers) quality source code should be recognized by its readability, and not by its raw size (number of lines) or its ability to take advantage of processor features.

8.15   DESIGN SUPPORT TOOLS

Software development is very labor intensive and is therefore prone to human error. In recent years, commercial software development support packages have become increasingly more powerful, less expensive, and readily available to reduce the time spent doing things that computers do better than people. Although selection of the right tools can mean up-front dedication of some of the most talented resources in a development team, it can bring about significant long-term increase in group productivity.

Good software development houses have taken advantage of CASE tools which reduce the time spent generating clear and thorough design documentation. There are many advantages of automated software design packages. Formal documentation can be used as proof of product development procedure conformance for agency approvals. Clear and up-to-date design documents facilitate improved communications between engineers, lending to more effective and reliable designs. Standard documentation formats reduce learning curves associated with unique design depictions among software designers, leading to better and more timely design formulation. Total software life cycle costs are reduced, especially during maintenance, due to reduced ramp-up time and more efficient and reliable modifications. Finally, electronic forms of documentation can be easily backed-up and stored off-site eliminating a crisis in the event of an environmental disaster. In summary, the adaptation of computer-aided software engineering (CASE) tools have an associated up-front cost which is recovered by significant improvements in software quality and development time predictability.

8.16   DESIGN AS THE BASIS FOR VERIFICATION
AND VALIDATION ACTIVITY

Verification is the process of assuring all products of a given development phase satisfy given expectations. Before proceeding to the next/lower level or phase of design, the product (or outputs) of the current phase should be verified against the inputs of the previous stage. A design process cannot be a "good" process without the verification process ingrained. That is, they naturally go hand in hand.

Software project management plans (software quality assurance plans) should specify all design reviews. Each level of design will generate documentation to be reviewed or deliverables to verify against the demands of the previous stage. For each type of review, the software management plans should describe the purpose, materials required, scheduling rules, scope of review, attendance expectations, review responsibilities, what the minutes should look like, follow-up activities, and any other requirements that relate to company expectations.

At the code level, code reviews should assure that the functionality implemented within a routine satisfies all expectations documented in the mspecs. Code should also be inspected to satisfy all coding rules.

The output of good software designs also includes implementation requirements. At minimum, implementation requirements include the rules and expectations placed on the designers to assure design uniformity as well as constraints, controls, and expectations placed on designs to ensure upper level requirements are met. General examples of implementation requirements might include rules for accessing I/O ports, timing requirements for memory accesses, semaphore arbitration, intertask communication schemes, memory addressing assignments, and sensor or device control rules. The software verification and validation process must address implementation requirements as well as the upper level software requirements to ensure the product works according to its specifications.

8.17 SUMMARY

Software, in and as a medical device is subjected to a rigorous, multistaged design process to assure it is safe, effective, and reliable for its intended use. After taking the requirements from the SRS including reviewing current standards and regulations for appropriate requirements,

The top-level design consists of

- Establishing the software architecture
- Choosing a methodology and a language
- Estimating the potential risks the software might produce
- Defining appropriate metrics to evaluate the design
- Checking for design completeness by use of an RTM
- Conducting various types of software reviews at appropriate times throughout the process

The detailed design consists of

- Predicting real-time performance
- Conducting design simulation
- Repairing module specifications
- Coding the design
- Using support tools where appropriate

If done properly, the design will form the basis for the next phase in the development process, verification and validation. In addition to producing a safe and effective program, this process will also help reduce coding time. In the final analysis, the patient, the user, and the developer will all benefit from implementation of this structured development process.

EXERCISES

1. You are responsible for developing the software embedded in an oximeter. The software must run a self-test upon start-up, determine saturated oxygen in the blood from a finger clip, determine the heart rate, and report the results on a screen. It must also interface with an anesthesia machine and display its data on the monitor. Establish a plan for developing this software.
2. What are some of the possible design alternatives and trade-offs for this device?
3. Develop the architecture for this software. What are the most important considerations when developing the architecture?
4. Explain the pros and cons of using object-oriented design on this project versus structured analysis.
5. List some of the risk analysis activities for this project. Explain how doing a fault tree analysis early in development would help in establishing the program.
6. Check the IEEE Web site at http://standards.ieee.org/software/ for a list of standards documents that would assist in the development and testing of this software.
7. Explain how the design would assist in your verification and validation activity.

BIBLIOGRAPHY

Bass, L., Clements, P., and Kazman, R., *Software Architecture in Practice*. Reading, MA: Addison Wesley Longman, 1998.
Boehm, B.W., Industrial software metrics top 10 list, *IEEE Software*, 4(9), 84–85.
Boehm, B.W., A spiral model of software development and enhancement, *Computer*, May, 1988.

Boehm, B.W., *Software Risk Management*. Washington, DC: IEEE Computer Society Press, 1989.

Booch, G., *Object-Oriented Analysis and Design with Applications*, 2nd ed. Redwood City, CA: The Benjamin Cummings Publishing Company, 1994.

Booch, G., Jacobson, I., and Rumbaugh, J., *The Unified Modeling Language User Guide*. Reading, MA: Addison Wesley Longman, 1998.

Deutsch, M.S. and Willis, R.R., *Software Quality Engineering—A Total Technical and Management Approach*. Englewood Cliffs, NJ: Prentice-Hall, 1988.

Dyer, M., *The Cleanroom Approach to Quality Software Development*. New York: John Wiley & Sons, 1992.

Evans, M.W. and Marciniak, J.J., *Software Quality Assurance and Management*. New York: John Wiley & Sons, 1987.

Fagan, M.E., Design and code inspections to reduce errors in program development, *IBM Systems Journal*, 15(3), 1976, 182–211.

Fairley, R.E., *Software Engineering Concepts*. New York: McGraw-Hill, 1985.

Fries, R.C., *Reliable Design of Medical Devices,* 2nd ed. Boca Raton, FL: CRC Press/Taylor and Francis Group, 2006.

Fries, R.C., Pienkowski, P., and Jorgens, J. III, Safe, effective, and reliable software design and development for medical devices, *Medical Instrumentation*, 30(2), March/April 1996.

Heathfield, H., Armstrong, J., and Kirkham, N., Object-oriented design and programming in medical decision support, *Computer Methods and Programs in Biomedicine*, 36(4), December 1991.

Humphrey, W.S., *Managing the Software Process*. Reading, MA: Addison Wesley, 1989.

IEEE, *IEEE Standards Collection—Software Engineering*. New York: The Institute of Electrical and Electronic Engineers, 1993.

Jahanian, F. and Mok, A.K., Safety analysis of timing properties in real time systems, *IEEE Transactions on Software Engineering*. SE-12(9), September 1986.

Kan, S.H., *Metrics and Models in Software Quality Engineering*. Reading, MA: Addison Wesley Longman, 1995.

Keller, M. and Shumate, K., *Software Specification and Design—A Disciplined Approach for Real-Time Systems*. New York: John Wiley & Sons, 1992.

King, P.H. and Fries, R.C., *Design of Biomedical Devices and Systems*. New York: Marcel Dekker, 2002.

Leveson, N.G., *Safeware*. Reading, MA: Addison Wesley, 1995.

Leveson, N.G., Software safety: Why, what, and how, *Computing Surveys*, 18(2), June 1986.

Lincoln, T.L., Programming languages, *Clinics in Laboratory Medicine*, 11(1), March 1991.

McCabe, T., A complexity measure, *IEEE Transactions of Software Engineering*. SE-2(4), December 1976.

McConnell, S., *Code Complete*. Redmond, Washington, DC: Microsoft Press, 1993.

Möller, K.H. and Paulish, D.J., *Software Metrics—A Practitioner's Guide to Improved Product Development*. London: Chapman & Hall Computing, 1993.

Myers, G.J., *The Art of Software Testing*. New York: John Wiley & Sons, 1979.

Pressman, R., *Software Engineering—A Practitioner's Approach*. New York: McGraw-Hill, 2005.

Rakos, J.J., *Software Project Management for Small to Medium Sized Projects*. Englewood Cliffs, NJ: Prentice-Hall, 1990.

Rumbaugh, J. et al., *Object-Oriented Modeling and Design*. Englewood Cliffs, NJ: Prentice-Hall, 1991.

Rumbaugh, J., Jacobson, I., and Booch, G., *The Unified Modeling Language Reference Manual*. Reading, MA: Addison Wesley Longman, 1998.

Storey, N., *Safety-Critical Computer Systems*. Harlow, England: Addison Wesley Longman, 1996.

Thayer, R.H. and Dorfman, M., Eds., *Software Requirements Engineering,* 2nd ed. Los Alamitos, CA: IEEE Computer Society Press, 1997.

Yourdon, E., *Modern Structured Analysis*. Englewood Cliffs, NJ: Yourdon Press, 1989.

Yourdan, E., *Structured Walkthroughs*. New York: Yourdon Press, 1989.

9 Human Factors

In the sick room, ten cents' worth of human understanding equals ten dollars' worth of medical science.

Martin H. Fischer

Human factors engineering, also called ergonomics, can trace its roots to early industrial engineering studies of work efficiency and task performance using, for example, time–motion techniques. Human factors engineering emerged as a recognized discipline during World War II while focusing primarily on military system performance, including problems in signal detection, workspace constraints, and optimal task training. The widespread recognition of the importance of applying human factors engineering in the design of tools, devices, tasks, and other human activities is reflected in the increasing number of disparate professionals interested in human factors. Their work products can be found in lay and professional publications, standards, and other documents. Human factors activities have improved the quality of personal and professional life across many domains. Public and professional interest in patient safety issues has promoted increased application of human factors engineering to the medical domain.

Numerous medical device companies have established human factors engineering programs to ensure the usability and safety of their devices. These companies also believe that their human factors engineering efforts enhance the marketability of their products. National and international regulations with respect to the safety of medical devices now require that human factors engineering principles be applied to the design of medical devices, and that this process be documented.

9.1 WHAT IS HUMAN FACTORS?

Human factors is defined as the application of the scientific knowledge of human capabilities and limitations to the design of systems and equipment to produce products with the most efficient, safe, effective, and reliable operation. This definition includes several interesting concepts.

Although humans are capable of many highly technical, complex, or intricate activities, they also have limitations to these activities. Of particular interest to the medical designer are limitations due to physical size, range of motion, visual perception, auditory perception, and mental capabilities under stress. Although the user may be characterized by these limitations, the designer cannot allow them to adversely affect the safety, effectiveness, or reliability of the device. The designer should therefore identify and address all possible points of interface between the user and the equipment, characterize the operating environment, and analyze the skill level of the intended users.

Interface points are defined as those areas that the user must control or maintain to derive the desired output from the system. Interface points include control panels, displays, operating procedures, operating instructions, and user training requirements.

The environment in which the device will be used must be characterized to determine those areas that may cause problems for the user, such as lighting, noise level, temperature, criticality of the operation, and the amount of stress the user is experiencing while operating the system. The design must then be adjusted to eliminate any potential problems.

The skill level of the user is an important parameter to be analyzed during the design process and includes characteristics such as educational background, technical expertize, and computer knowledge. To assure the user's skill levels have been successfully addressed, the product should be designed to meet the capabilities of the least skilled potential user. Designing to meet this worst-case situation will assure the needs of the majority of the potential users will be satisfied.

The final and most important activity in human factors engineering is determining how these areas interact within the particular device. The points of interface are designed based on the anticipated operating environment and on the skill level of the user. The skill level may depend not only on the education and experience of the user, but on the operating environment, as well. To design for such interaction, the designer must consider the three elements that comprise human factors: human, hardware, and software.

9.2 HUMAN ELEMENT IN HUMAN FACTORS ENGINEERING

The human element addresses several user characteristics, including memory and knowledge presentation, thinking and reasoning, visual perception, dialog construction, individual skill level, and individual sophistication. Each is an important factor in the design consideration.

A human being has two types of memory. Short-term memory deals with sensory input, such as visual stimuli, sounds, and sensations of touch. Long-term memory is composed of our knowledge database. If the human–machine interface makes undue demands on either short- or long-term memory, the performance of the individual in the system will be degraded. The speed of this degradation depends on the amount of data presented, number of commands the user must remember, and stress involved in the activity.

When a human performs a problem-solving activity, they usually apply a set of guidelines or strategies based on their understanding of the situation and their experiences with similar types of problems, rather than applying formal inductive or deductive reasoning techniques. The human–machine interface must be specific in a manner enabling the user to relate to their previous experiences and develop guidelines for a particular situation.

The physical and cognitive constraints associated with visual perception must be understood when designing the human–machine interface. For example, studies have shown that since the normal line of sight is within 15° of the horizontal line of sight, the optimum position for the instrument face is within a minimum of 45° of the normal line of sight (Figure 9.1). Other physical and cognitive constraints have been categorized and are available in references located at the end of this chapter.

When people communicate with one another, they communicate best when the dialog is simple, easy to understand, direct, and to the point. The designer must assure device commands are easy to remember, error messages are simple, direct, and not cluttered with computer jargon and help messages are easy to understand and pointed. The design of dialog should be addressed to the least skilled potential user of the equipment.

The typical user of a medical device is not familiar with hardware design or computer programming. They are more concerned with the results obtained from using the device, than about how the results were obtained. They want a system that is convenient, natural, flexible, and easy to use. They do not want a system that looks imposing, is riddled with computer jargon, requires them to memorize many commands, or has unnecessary information cluttering the display areas.

In summary, the human element requires a device, which has inputs, outputs, controls, displays, and documentation that reflect an understanding of the user's education, skill, needs, experience, and the stress level when operating the equipment.

9.3 HARDWARE ELEMENT IN HUMAN FACTORS

The hardware element considers size limitations, location of controls, compatibility with other equipment, potential need for portability, and possible user training. It also addresses the height of the preferred control area and the preferred display area when the operator is standing (Figure 9.2),

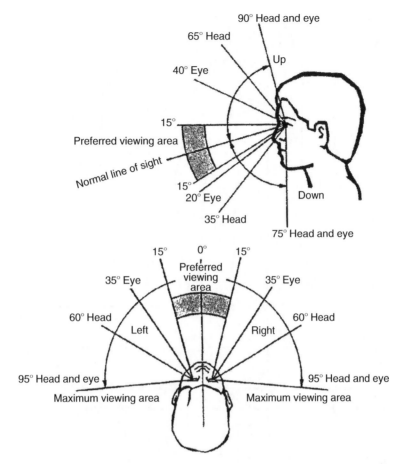

		Maximum[a]		
	Preferred	Eye Rotation	Head Rotation	Head and Eye Rotation
Up	15°	40°	65°	90°
Down	15°	20°	35°	75°
Right	15°	35°	60°	95°
Left	15°	35°	60°	95°

[a]Display area on the console defined by the angles measured from the normal line of sight.

FIGURE 9.1 Normal line of sight.

when the operator is sitting (Figure 9.3), and the size of the human hand in relation to the size of control knobs or switches (Figure 9.4).

Hardware issues are best addressed by first surveying potential customers of the device to help determine the intended use of the device, the environment in which the device will be used, and the optimum location of controls and displays. Once the survey is completed and the results are analyzed, a cardboard, foam, or wooden model of the device is built and reviewed with the potential customers. The customer can then get personal, hands-on experience with the controls, displays,

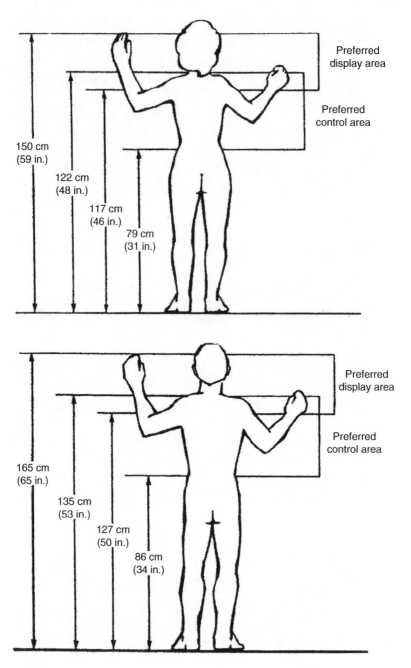

FIGURE 9.2   Display area when standing.

device framework, and offer constructive criticism on the design. Once all changes have been made, the model can be transposed into a prototype, using actual hardware.

9.4   SOFTWARE ELEMENT IN HUMAN FACTORS

The software element of the device must be easy to use and understand. It must have simple, reliable data entry, it should be menu driven if there are many commands to be learned, displays must not be

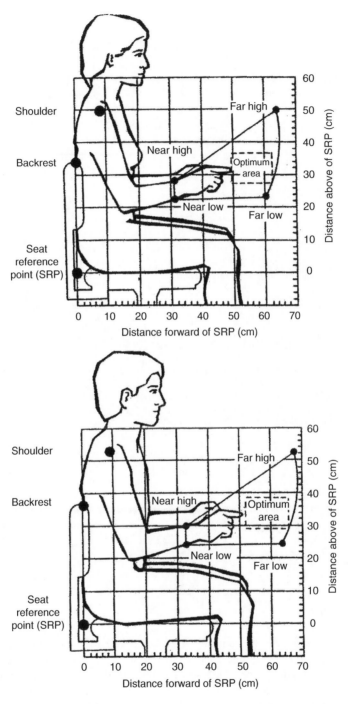

FIGURE 9.3 Display area when sitting.

overcrowded, and dialog must not be burdened with computer jargon. The software must provide feedback to the user through error messages and help messages. An indication that the process is involved in some activity is also important, as a blank screen leads to the assumption that nothing is active, and the user starts pushing keys or buttons.

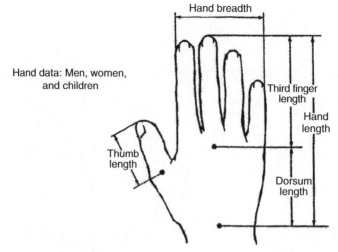

Hand data	Men			Women			Children			
	2.5% tile	50% tile	97.5% tile	2.5% tile	50% tile	97.5% tile	6 year	8 year	11 year	14 year
Hand length	173 mm (6.8 in.)	191 mm (7.5 in.)	208 mm (8.2 in.)	157 mm (6.2 in.)	175 mm (6.9 in.)	191 mm (7.5 in.)	130 mm (5.1 in.)	142 mm (5.6 in.)	160 mm (6.3 in.)	178 mm (7.0 in.)
Hand breadth	81 mm (3.2 in.)	89 mm (3.5 in.)	97 mm (3.8 in.)	66 mm (2.6 in.)	74 mm (2.9 in.)	79 mm (3.1 in.)	58 mm (2.3 in.)	64 mm (2.5 in.)	71 mm (2.8 in.)	–
Third finger length	102 mm (4.0 in.)	114 mm (4.5 in.)	127 mm (5.0 in.)	91 mm (3.6 in.)	100 mm (4.0 in.)	112 mm (4.4 in.)	74 mm (2.9 in.)	81 mm (3.2 in.)	89 mm (3.5 in.)	102 mm (4.0 in.)
Dorsum length	71 mm (2.8 in.)	75 mm (3.0 in.)	81 mm (3.2 in.)	66 mm (2.6 in.)	74 mm (2.9 in.)	79 mm (3.1 in.)	56 mm (2.2 in.)	61 mm (2.4 in.)	71 mm (2.8 in.)	75 mm (3.0 in.)
Thumb length	61 mm (2.4 in.)	69 mm (2.7 in.)	75 mm (3.0 in.)	56 mm (2.2 in.)	61 mm (2.4 in.)	66 mm (2.6 in.)	46 mm (1.8 in.)	51 mm (2.0 in.)	56 mm (2.2 in.)	61 mm (2.4 in.)

Additional data: Average man

FIGURE 9.4 Hand sizes.

Software must consider the environment in which it is to be used, especially with regard to colors of displays, type of data to be displayed, format of the data, alarm levels to be used, etc. Stress and fatigue can be reduced by consideration of color and the intensity of the displayed data. Operator effectiveness can be improved by optimizing the location of function keys, displaying more important data in the primary viewing area, and placing secondary data in the secondary display area. The inclusion of device checkout procedures and menus also improves operator effectiveness and confidence.

9.5 HUMAN FACTORS PROCESS

Human factors is the sum of several processes including the analytic process that focuses on the objectives of the proposed device and the functions that should be performed to meet those objectives; the design and development process that converts the results of the analyses into detailed equipment design features; and the test and evaluation process, which verifies that the design and development process has resolved issues identified in the analytic process.

Human factors engineering integrations begin with early planning and may continue throughout the life cycle of the device. As a minimum, human factors should continue until the device is introduced commercially. Human factors efforts following commercial introduction are important to the enhancement of the device and the development of future devices.

9.6 PLANNING

Human factors plan should be developed as an integral part of the overall plan for device development. The plan should guide human factors efforts in the interrelated processes of analysis, design and development, and test and evaluation. The plan should describe human factors tasks necessary to complete each process, the expected results of those tasks, the means of coordinating those tasks with the overall process for device development, and the schedule for that coordination. The plan should address the resources necessary for its accomplishment including levels of effort, necessary for its management and coordination as well as for accomplishment of its individual tasks.

The plan should assure that results of human factors tasks are available in time to influence the design of the proposed device as well as the conduct of the overall project. Analysis tasks should begin very early. Iterations of analysis tasks that refine earlier products may continue throughout the project. Design and development build on the products of early analysis, and iterations may also continue throughout the project. Test and evaluation should begin with the earliest products of design and development. The results of test and evaluation should influence subsequent iterations of analysis, design and development, and test and evaluation tasks.

9.7 ANALYSIS

Successful human factors is predicated on careful analyses. Early analyses should focus on the objectives of the proposed device and the functions that should be performed to meet those objectives. Later analysis should focus on the critical human performance required of specific personnel as a means of establishing the human factors parameters for design of the device and associated job aids, procedures, and training and for establishing human factors test and evaluation criteria for the device. Analyses should be updated as required to remain current with the design effort.

9.8 CONDUCT USER STUDIES

The goal of user studies is to learn as much as possible within a reasonable time frame about the customer's needs and preferences as they relate to the product under development. Several methods are available for getting to know the customer.

9.8.1 OBSERVATIONS

Observations are a productive first step toward getting to know the user. By observing people at work, a rapid sense for the nature of their jobs is developed, including the pace and nature of their interactions with the environment, coworkers, patients, equipment, and documents. Such observations may be conducted in an informal manner, possibly taking notes and photographs. Alternatively, a more formal approach may be taken that includes rigorous data collection. For example, it may be important to document a clinician's physical movements and the time they spend performing certain tasks to determine performance benchmarks. This latter approach is referred to as a time–motion analysis and may be warranted if one of the design goals is to make the customer more productive.

Enough time should be spent observing users to get a complete sense for how they perform tasks related to the product under development. A rule of thumb in usability testing is that five to eight participants provide 80%–90% of the information you seek. The same rule of thumb may be applied to observations, presuming that you are addressing a relatively homogenous user population. Significant differences in the user population (i.e., a heterogeneous user population) may warrant more extensive observations. For example, it may become necessary to observe people who have different occupational backgrounds and work in different countries.

Designers and engineers should conduct their own observations. For starters, such observations increase empathy for the customer. Also, firsthand experience is always more powerful than reading a marketing report.

9.8.2 INTERVIEWS

Similar to observations, interviews provide a wealth of information with a limited investment of time. Structured interviews based on scripted questions are generally better than unstructured interviews (i.e., a free-flowing conversation). This is because a structured interview assures that the interviewer will ask everyone the same question, enabling a comparison of answers. Structured interviews may include a few open-ended questions to produce evoke comments and suggestions that could not be anticipated. The interview script should be developed from a list of information needs. Generally, questions should progress from general to more specific design issues. Care should be taken to avoid mixing marketing and engineering related concerns with usability concerns. Interviews can be conducted after observations are completed. Conducting the interviews before the observations can be problematic as it tends to alter the way people react.

9.8.3 FOCUS GROUPS

Conducting interviews with people in their working environment (sometimes referred to as contextual interviewing) is generally best. Interviewees are likely to be more relaxed and opinionated. Interviews conducted at trade shows and medical conferences, for example, are more susceptible to bias and may be less reliable.

Conducting interviews with a group of five to ten people at a time enables easy determination of a consensus on various design issues. In preparation for such a focus group, a script should be developed from a set of information requirements. Use the script as a guide for the group interview, but feel free to let the discussion take a few tangents if they are productive ones. Also, feel at liberty to include group exercises, such as watching a video or ranking and rating existing products, as appropriate.

Conduct enough focus groups to gain confidence that an accurate consensus has been developed. Two focus groups held locally may be enough if regional differences of opinion are unlikely and the user group is relatively homogenous. Otherwise, it may be appropriate to conduct up to four groups each at domestic and international site that provides a reasonable cross section of the marketplace.

Document the results in a focus groups report. The report can be an expanded version of the script. Begin the report with a summary to pull together the results. Findings (i.e., answers to questions) may be presented after each question. The findings from various sites may be integrated or presented separately, depending on the design issue and opportunity to tailor the product under development to individual markets. Results of group exercises may be presented as attachments and discussed in the summary.

9.8.4 TASK ANALYSIS

The purpose of task analysis is to develop a detailed view of customer interactions with a product by dividing the interactions into discrete actions and decisions. Typically, a flow chart is drawn that shows the sequence and logic of customer actions and decisions. The task analysis is extended to include tables that define information and control requirements associated with each action and decision. In the course of the task analysis, characterize the frequency, urgency, and criticality of integrated tasks, such as "checking the breathing circuit."

9.8.5 BENCHMARK USABILITY TEST

The start of a new product development effort is a good time to take stock of the company's existing products. An effective way to do this is to conduct a benchmark usability test that yields, in a quantitative fashion, both objective and subjective measures of usability. Such testing will identify the strengths and weaknesses of the existing products, as well as help establish usability goals for the new product.

9.8.6 WRITE USER PROFILE

To culminate the user study effort, write the so-called user profile. A user specification (2–5 pages) summarizes the important things learned about the customers. The profile should define the user population's demographics (age, gender, education level, occupational background, and language), product-related experience, work environment, and motivation level. The user profile is a major input to the user specification that describes the product under development from the customer's point of view.

9.8.7 SET UP AN ADVISORY PANEL

To assure early and continued customer involvement, set up an advisory panel that equitably represents the user population. The panel may include three to five clinicians for limited product development efforts, or be twice as large for larger efforts. The panel participants are usually compensated for their time. Correspond with the members of the panel on a needed basis and meet them periodically to review the design in progress. Note that advisory panel reviews are not an effective replacement for usability testing.

9.9 SET USABILITY GOALS

Usability goals are comparable to other types of engineering goals in the sense that they are quantitative and provide a basis for acceptance testing. Goals may be objective or subjective. A sample objective goal might be: on average, users shall require 3 s to silence an alarm. This goal is an objective goal because the user's performance level can be determined simply by observation. For example, you can use a stop watch to determine task times. Other kinds of objective goals concentrate on the number of user errors and the rate of successful task completion.

A sample subjective goal is: on average, 75% of users shall rate the intuitiveness of the alarm system as 5 or better, where 1 = poor and 7 = excellent. This goal is subjective because it requires asking the user's opinion about their interaction with the given product. A rating sheet can be used to record their answers. Other kinds of subjective goals concentrate on mental processing and emotional response attributes, such as learning, frustration level, fear of making mistakes, etc.

Every usability goal is based on a usability attribute, for example, task, speed, or intuitiveness, include a metric such as time or scale and sets a target performance level, such as 3 s or a rating of 5 or better.

Typically, up to 50 usability goals may be written, two thirds of which are objective and one third which are subjective. The target performance level on each goal is based on findings from preceding user studies, particularly the benchmark usability testing. If there is no basis for comparison, i.e., there are no comparable products, then engineering judgment must be used to set the initial goals and adjust them as necessary to assure they are realistic.

9.10 DESIGN USER INTERFACE CONCEPTS

Concurrent design is a productive method of developing a final user interface design. It enables thorough exploration of several design concepts before converging on a final solution. In the course of exploring alternative designs, limited prototypes should be built of the most promising concepts and user feedback obtained on them. This gets users involved in the design process at its early stages and assures that the final design will be closely matched to user's expectations.

Note that the design process steps described below assume that the product includes both hardware and software elements. Some steps would be moot if the product has no software user interface.

9.10.1 DEVELOP CONCEPTUAL MODEL

When users interact with a product, they develop a mental model of how it works. This mental model may be complete and accurate or just the opposite. Enabling the user to develop a complete and accurate mental model of how a product works is a challenge. The first step is developing the so-called conceptual models of how to represent the product's functions. This exercise provides a terrific opportunity for design innovation. The conceptual model may be expressed as a bubble diagram, for example, that illustrates the major functions of the product and functional interrelationships as you would like the users to think of them. You can augment the bubble diagram with a narrative description of the conceptual model.

9.10.2 DEVELOP USER INTERFACE STRUCTURE

Develop alternative user interface structures that compliment the most promising—two to three conceptual models. These structures can be expressed in the form of screen hierarchy maps that illustrate where product functions reside and how many steps it will take users to get to them. Such maps may take the form of a single element, a linear sequence, a tree structure (cyclic or acyclic) or a network. In addition to software screens, such maps should show which functions are allocated to dedicated hardware controls.

9.10.3 DEFINE INTERACTION STYLE

In conjunction with the development of the user interface structures, alternative interaction styles should be defined. Possible styles include question and answer dialogs, command lines, menus, and direct manipulation.

9.10.4 DEVELOP SCREEN TEMPLATES

Determine an appropriate size display based on the user interface structure and interaction style, as well as other engineering considerations. Using computer-based drawing tools, draw the outline of a blank screen. Next, develop a limited number (perhaps three to five) of basic layouts for the information that will appear on the various screens. Normally, it is best to align all elements, such as titles, windows, prompts, and numerics according to a grid system.

9.10.5 DEVELOP HARDWARE LAYOUT

Apply established design principles in the development of hardware layouts that are compatible with the evolving software user interface solutions. Assure that the layouts reinforce the overall conceptual model.

9.10.6 Develop a Screenplay

Apply established design principles in the development of a detailed screenplay. Do not bother to develop every possible screen at this time. Rather, develop only those screens that would enable users to perform frequently used, critical and particularly complex functions. Base the screen designs on the templates. Create new templates or eliminate existing templates as required while continuing to limit the total number of templates. Assure that the individual screens reinforce the overall conceptual model. You may choose to get user feedback on the screenplay (what some people call a paper prototype).

9.10.7 Develop a Refined Design

Steps 5 and 6 describe prototyping and testing the user interface. These efforts will help determine the most promising design concept or suggest a hybrid of two or more concepts. The next step is to refine the preferred design. Several reiterations of the preceding steps may be necessary, including developing a refined conceptual model, developing a refined user interface structure, and developing an updated set of screen templates. Then, a refined screenplay and hardware layout may be developed.

9.10.8 Develop a Final Design

Once again, steps 5 and 6 describe prototyping and testing the user interface. These efforts will help you determine any remaining usability problems with the refined design and opportunities for further improvement. It is likely that design changes at this point will be limited in nature. Most can be made directly to the prototype.

9.11 MODEL THE USER INTERFACE

Build a prototype to evaluate the dynamics of the user interface. Early prototypes of competing concepts may be somewhat limited in terms of their visual realism and how many functions they perform. Normally, it is best to develop a prototype that (1) presents a fully functional top-level that allows users to browse their basic options, and (2) enables users to perform a few sample tasks, i.e., walkthrough a few scenarios. As much as possible, include tasks that relate to the established usability goals.

User interface prototypes may be developed using conventional programming languages or rapid prototyping languages, such as SuperCard, Altia Design, Visual Basic, Toolbook, and the like. The rapid prototyping languages are generally preferable because they allow for faster prototyping and are easier to modify based on core project team and user feedback.

Early in the screenplay development process, it may make sense to prototype a small part of the user interface to assess design alternatives or to conduct limited studies, such as how frequently to flash a warning. Once detailed screenplays of competing concepts are available, build higher fidelity prototypes that facilitate usability testing. Once a refined design is developed, build a fully functional prototype that permits a verification usability test. Such prototypes can be refined based on final test results and serve as a specification.

9.12 TEST THE USER INTERFACE

There are several appropriate times to conduct a usability test, including:

- At the start of a development effort to develop benchmarks
- When you have paper-based or computer-based prototypes of competing design concepts
- When you have a prototype of your refined design
- When you want to develop marketing claims regarding the performance of the actual product

While the rigor of the usability test may change, based on the timing of the test, the basic approach remains the same. You recruit prospective users to spend a concentrated period of time interacting with the prototype product. The users may undertake a self-exploration or perform directed tasks. During the course of such interactions, you note the test participant's comments and document their performance. At intermittent stages, you may choose to have the test participant complete a questionnaire or rating/ranking exercise. Videotaping test proceedings is one way to give those unable to attend the test a firsthand sense of user–product interactions. Sometimes it is useful to create a 10–15 min highlight tape that shows the most interesting moments of all test sessions. During testing, collect the data necessary to determine if you are meeting the established usability goals. This effort will add continuity and objectivity to the usability engineering process.

9.13   SPECIFY THE USER INTERFACE

9.13.1   STYLE GUIDE

The purpose of a style guide is to document the rules of the user interface design. By establishing such rules, you can check the evolving design to determine any inconsistencies. Also, it assures the consistency of future design changes. Style guides, usually 10–15 pages in length, normally include a description of the conceptual model, the design elements and elements of style.

9.13.2   SCREEN HIERARCHY MAP

The purpose of a screen hierarchy map is to provide an overview of the user interface structure. It places all screens that appear in the screenplay in context. It enables the flow of activity to be studied to determine if it reinforces the conceptual model. It also helps to determine how many steps users will need to take to accomplish a given task. Graphical elements of the screen hierarchy map should be cross-indexed to the screenplay.

9.13.3   SCREENPLAY

The purpose of a screenplay is to document the appearance of all major screens on paper. Typically, screen images are taken directly from the computer-based prototype. Ideally, the screenplay should present screen images in their actual scale and resolution. Each screen should be cross-indexed to the screen hierarchy map.

9.13.4   SPECIFICATION PROTOTYPE

The purpose of the specification prototype is to model accurately the majority of user interface interactions. This provides the core project team with a common basis for understanding how the final product should work. It provides a basis for writing the user documentation. It may also be used to orient those involved in marketing, sales, and training.

9.13.5   HARDWARE LAYOUTS

The hardware layout may be illustrated by the specification prototype. However, the hardware may not be located proximal to the software user interface. If this is the case, develop layout drawings to document the final hardware layout.

9.14   ADDITIONAL HUMAN FACTORS DESIGN CONSIDERATIONS

The design of medical devices should reflect human factors engineering design features that increase the potential for successful performance of tasks and for satisfaction of design objectives.

9.14.1 CONSISTENCY AND SIMPLICITY

Where common functions are involved, consistency is encouraged in controls, displays, markings, codings, and arrangement schemes for consoles and instrument panels. Simplicity in all designs is encouraged. Equipment should be designed to be operated, maintained, and repaired in its operational environment by personnel with appropriate but minimal training. Unnecessary or cumbersome operations should be avoided when simpler, more efficient alternatives are available.

9.14.2 SAFETY

Medical device design should reflect system and personnel safety factors, including the elimination or minimization of the potential for human error during operation and maintenance under both routine and nonroutine or emergency conditions. Machines should be designed to minimize consequence of human error. For example, where appropriate, a design should incorporate redundant, diverse elements arranged in a manner that increases overall reliability when failure can result in the inability to perform a critical function.

Any medical device failure should immediately be indicated to the operator and should not adversely affect safe operation of the device. Where failures can affect safe operation, simple means and procedures for averting adverse effects should be provided.

When the device failure is life-threatening or could mask a life-threatening condition, an audible alarm and a visual display should be provided to indicate device failure. Wherever possible, explicit notification of the source of failure should be provided to the user. Concise instructions on how to return to operation or how to invoke alternate backup methods should be provided.

9.14.3 ENVIRONMENTAL/ORGANIZATIONAL CONSIDERATIONS

The design of medical devices should consider the following:

- Levels of noise, vibration, humidity, and heat that will be generated by the device and the levels of noise, vibration, humidity, and heat to which the device and its operators and maintainers will be exposed in the anticipated operational environment.
- Need for protecting operators and patients from electric shock, thermal, infectious, toxicologic, radiologic, electromagnetic, visual, and explosion risks, as well as from potential design hazards, such as sharp edges and corners, and the danger of the device falling on the patient or operator.
- Adequacy of the physical, visual, auditory, and other communication links among personnel and between personnel and equipment.
- Importance of minimizing psychophysiological stress and fatigue in the clinical environment in which the medical device will be used.
- Impact on operator effectiveness of the arrangement of controls, displays, and markings on consoles and panels.
- Potential effects of natural or artificial illumination used in the operation, control, and maintenance of the device.
- Need for rapid, safe, simple, and economical maintenance and repair.
- Possible positions of the device in relation to the users as a function of the user's location and mobility.
- Electromagnetic environment(s) in which the device is intended to be used.

9.14.4 DOCUMENTATION

Documentation is a general term that includes operator manuals, instruction sheets, online help systems, and maintenance manuals. These materials may be accessed by many types of users. Therefore, the documentation should be written to meet the needs of all target populations.

Preparation of instructional documentation should begin as soon as possible during the specification phase. This assists device designers in identifying critical human factors engineering needs and in producing a consistent human interface. The device and its documentation should be developed together.

During the planning phase, a study should be made of the capabilities and information needs of the documentation users, including:

- User's mental abilities
- User's physical abilities
- User's previous experience with similar devices
- User's general understanding of the general principles of operation and potential hazards associated with the technology
- Special needs or restrictions of the environment

As a minimum, the operator's manual should include detailed procedures for setup, normal operation, emergency operation, cleaning and operator troubleshooting. The operator manual should be tested on models of the device. It is important that these test populations be truly representative of end users and that they not have advance knowledge of the device.

Maintenance documentation should be tested on devices that resemble production units. Documentation content should be presented in a language that is free of vague and ambiguous terms. Simplest words and phrases that will convey the intended meaning should be used. Terminology within the publication should be consistent. Use of abbreviations should be kept to a minimum, but defined where they are used.

Information included in warnings and cautions should be chosen carefully and with consideration of the skills and training of intended users. It is especially important to inform users about unusual hazards and hazards specific to the device.

Human factors engineering design features should assure that the device functions consistently, simply, and safely; that the environment, system organization, and documentation are analyzed and considered in the design, thus increasing the potential for successful performance of tasks and for satisfaction of design objectives.

9.14.5 ANTHROPOMETRY

Anthropometry is the science of measuring the human body and its parts and functional capacities. Generally, design limits are based on a range of values from the 5th percentile female to the 95th percentile male for critical body dimensions. The 5th percentile value indicates that 5% of the population will be equal to or smaller than that value and 95% will be larger. The 95th percentile value indicates that 95% of the population will be equal to or smaller than that value and 5% will be larger. The use of a design range from 5th to 95th percentile values will theoretically provide coverage from 90% of the user population for that dimension.

9.14.6 FUNCTIONAL DIMENSIONS

The reach capabilities of the user population play an important role in the design of the controls and displays of the medical device. The designer should take into consideration both one- and two-handed reaches in the seated and standing positions (Figures 9.5 and 9.6). Body mobility ranges should be factored into the design process. Limits of body movement should be considered relative to the age diversity and gender of the target user population.

The strength capacities of the device operators may have an impact on the design of the system controls. The lifting and carrying abilities of the personnel responsible for moving and adjusting the

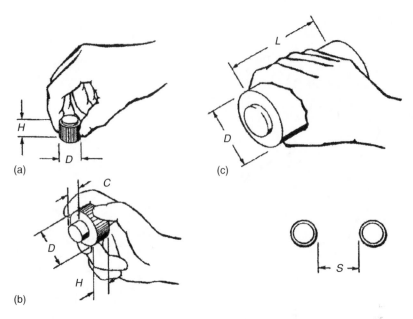

	Dimensions						
	(a) Finger grasp		(b) Thumb and fingers encircled			(c) Palm/hand grasp	
	Height, H	Diameter, D	Height, H	Diameter, D	Clearance, C	Diameter, D	Length, L
Minimum	13 mm (0.5 in.)	10 mm (0.375 in.)	13 mm (0.50 in.)	25 mm (1.0 in.)	16 mm (0.625 in.)	38 mm (1.5 in.)	75 mm (3.0 in.)
Maximum	25 mm (1.0 in.)	100 mm (4 in.)	25 mm (1.0 in.)	75 mm (3.0 in.)	—	75 mm (3 in.)	—

	Torque		Separation, S
	A	B	One hand individually
Minimum	—	—	25 mm (1.0 in.)
Preferred	—	—	50 mm (2.0 in.)
Maximum	32 mN m (4.5 in. oz.)	42 mN m (6.0 in. oz.)	—

Notes: A, To and including 25 mm (1.0 in.) diameter knobs; B, Greater than 25 mm (1.0 in.) diameter knobs.

FIGURE 9.5 Example of functional dimensions.

device need to be considered to assure the device can be transported and adjusted efficiently and safely.

9.14.7 PSYCHOLOGICAL ELEMENTS

It is crucial to consider human proficiency in perception, cognition, learning, memory, and judgment when designing medical devices to assure that operation of the system is as intuitive, effective, and safe as possible.

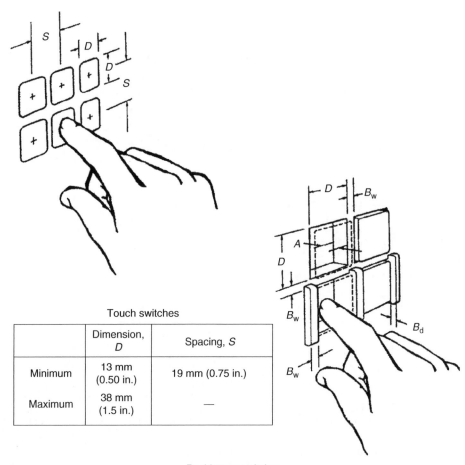

Touch switches

	Dimension, D	Spacing, S
Minimum	13 mm (0.50 in.)	19 mm (0.75 in.)
Maximum	38 mm (1.5 in.)	—

Pushbutton switches

	Dimension, D	Displacement, A	Separation/barriers[a]		Resistance
			B_w	B_d	
Minimum	19 mm (0.75 in.)	3 mm[b] (0.125 in.)	3 mm (0.125 in.)	5 mm (0.187 in.)	280 mN (10 oz)
Maximum	38 mm (1.5 in.)	6 mm (0.250 in.)	6 mm (0.250 in.)	6 mm (0.250 in.)	16.6 N (60 oz)

[a] Barriers shall have rounded edges.
[b] 5 mm (0.188 in.) for positive position switches.

FIGURE 9.6   Example of functional dimensions.

9.14.8 WORKSTATION DESIGN CONSIDERATIONS

Successful workstation design is dependent on considering the nature of the tasks to be completed, the preferred posture of the operator, and the dynamics of the surrounding environment. The design of the workstation needs to take into account the adjustability of the furniture, clearances under work surfaces, keyboard and display support surfaces, seating, footrests, and accessories.

The effectiveness with which operators perform their tasks at consoles or instrument panels depends in part on how well the equipment is designed to minimize parallax in viewing displays, allow ready manipulation of controls, and provide adequate space and support for the operator.

A horizontal or nearly horizontal work surface serves primarily as a work or writing surface or as a support for the operator's convenience items. Certain types of controls, such as joysticks or tracking controls, can also be part of the surface design.

Controls should have characteristics appropriate for their intended functions, environments, and user orientations, and their movements should be consistent with the movements of any related displays or equipment components. The shape of the control should be dictated by its specific functional requirements. In a bank of controls, those controls affecting critical or life-supporting functions should have a special shape and, if possible, a standard location.

Controls should be designed and located to avoid accidental activation. Particular attention should be given to critical controls whose accidental activation might injure patients or personnel or might compromise device performance. Feedback on control response adequacy should be provided as rapidly as possible.

9.14.9 ALARMS AND SIGNALS

The purpose of an alarm is to draw attention to the device when the operator's attention may be focused elsewhere. Alarms should not be startling but should elicit the desired action from the user. When appropriate, the alarm message should provide instructions for the corrective action that is required. In general, alarm design will be different for a device that is continuously attended by a trained operator, such as an anesthesia machine, than for a device that is unattended and operated by an untrained operator, such as a patient-controlled analgesia device. False alarms, loud and startling alarms, or alarms that recur unnecessarily can be a source of distraction for both an attendant and the patient and thus be a hindrance to good patient care.

Alarm characteristics are grouped in the following three categories:

- High priority: A combination of audible and visual signals indicating that immediate operator response is required.
- Medium priority: A combination of audible and visual signals indicating that prompt operator response is required.
- Low priority: A visual signal or a combination of audible and visual signals indicating that operator awareness is required.

A red flashing light should be used for a high priority alarm condition unless an alternative visible signal that indicates the alarm condition and its priority is employed. A red flashing light should not be used for any other purpose.

A yellow flashing light should be used for a medium priority alarm condition unless an alternative visible signal that indicates the alarm condition and its priority is employed. A yellow flashing light should not be used for any other purpose. A steady yellow light should be used for a low priority alarm condition unless an alternative visible signal that indicates the alarm condition and its priority is employed.

Audible signals should be used to alert the operator to the status of the patient or the device when the device is out of the operator's line of sight. Audible signals used in conjunction with visual displays should be supplementary to the visual signals and should be used to alert and direct the user's attention to the appropriate visual display.

Design of equipment should take into account the background noise and other audible signals and alarms that will likely be present during the intended use of the device. The lowest volume control settings of the critical life support audible alarms should provide sufficient signal strength to preclude masking by anticipated ambient noise levels. Volume control settings for other signals should similarly preclude such masking. Ambient noise levels in hospital areas can range from 50 dB in a private room to 60 dB in intensive care units and emergency rooms, with peaks as high as 65–70 dB in operating rooms due to conversations, alarms, or the activation of other devices.

The volume of monitoring signals normally should be lower than that of high priority or medium priority audible alarms provided on the same device. Audible signals should be located so as to assist the operator in identifying the device that is causing the alarm.

The use of voice alarms in medical applications should not be considered normally for the following reasons:

- Voice alarms are easily masked by ambient noise and other voice messages.
- Voice messages may interfere with communications among personnel who are attempting to address the alarm condition.
- Information conveyed by the voice alarm may reach individuals who should not be given specific information concerning the nature of the alarm.
- Types of messages transmitted by voice tend to be very specific, possibly causing complication and confusion to the user.
- In a situation where there are multiple alarms, multiple voice alarms would cause confusion.
- Different languages may be required to accommodate various markets.

The device's default alarm limits should be provided for critical alarms. These limits should be sufficiently wide to prevent nuisance alarms, and sufficiently narrow to alert the operator to a situation that would be dangerous in the average patient.

The device may retain and store one or more sets of alarm limits chosen by the user. When more than one set of user default alarm limits exists, the activation of user default alarm limits should require deliberate action by the user. When there is only one set of user default alarm limits, the device may be configured to activate this set of user default alarm limits automatically in place of the factory default alarm limits.

The setting of adjustable alarms should be indicated continuously or on user demand. It should be possible to review alarm limits quickly. During user setting of alarm limits, monitoring should continue and alarm conditions should elicit the appropriate alarms. Alarm limits may be set automatically or upon user action to reasonable ranges or percentages above and below existing values for monitored variables. Care should be taken in the design of such automatic setting systems to help prevent nuisance alarms or variables that are changing within an acceptable range.

An audible high or medium priority signal may have a manually operated, temporary override mechanism that will silence it for a period of time (e.g., 120 s). After the silencing period, the alarm should begin sounding again if the alarm condition persists or if the condition was temporarily corrected but has now returned. New alarm conditions that develop during the silencing period should initiate audible and visual signals. If momentary silencing is provided, the silencing should be visually indicated.

An audible high or medium priority signal may be equipped with a means of permanent silencing, that may be appropriate when a continuous alarm is likely to degrade user performance of associated tasks to an unacceptable extent and in cases when users would otherwise be likely to disable the device altogether. If provided, such silencing should require that the user either confirm the intent to silence a critical life support alarm or take more than one step to turn the alarm off. Permanent silencing should be visually indicated and may be signaled by a periodic audible reminder. Permanent silencing of an alarm should not affect the visual representation of the alarm and should not disable the alarm.

Life support devices and devices that monitor a life-critical variable should have an audible alarm to indicate a loss of power or failure of the device. The characteristics of this alarm should be the same as those of the highest priority alarm that becomes inoperative. It may be necessary to use battery power for such an alarm.

9.14.10 Labeling

Controls, displays, and other equipment items that need to be located, identified, or manipulated should be appropriately and clearly marked to permit rapid and accurate human performance. The characteristics of markings should be determined by such factors as the criticality of the function labeled, the distance from which the labels have to be read, the illumination level, the colors, the time available for reading, the reading accuracy required, and consistency with other markings.

Receptacles and connectors should be marked with their intended function or their intended connection to a particular cable. Convenience receptacles should be labeled with maximum allowable load in amperes or watts. The current rating of fuses should be permanently marked adjacent to the fuse holder. Fuse ratings should be indicated either in whole number, common fractions, or whole number plus common fractions. Labeling of fuses and circuit breakers should be legible in the ambient illumination range anticipated for the maintainer's location.

Operators and maintenance personnel should be warned of possible fire, radiation, explosion, shock, infection, or other hazards that may be encountered during the use, handling, storage, or repair of the device. Electromedical instruments should be labeled to show whether they may be used in the presence of flammable gases or oxygen-rich atmospheres. Hazard warnings should be prominent and understandable.

Normally, labels should be placed above panel elements that users grasp, press, or otherwise handle so the label is not obscured by the hand. However, certain panel element positions, user postures, and handling methods may dictate other label placements. Labels should be positioned to ensure visibility and readability from the position in which they should be read.

Labels should be oriented horizontally so that they may be read quickly and easily from left to right. Although not normally recommended, vertical orientation may be used, but only where its use is justified in providing a better understanding of intended function. Vertical labels should be read from top to bottom. Curved labels should be avoided except when they provide setting delimiters for rotary controls.

Labels should not cover any other information source. They should not detract from or obscure figures or scales that should be read by the operator. Labels should not be covered or obscured by other units in the equipment assembly. Labels should be visible to the operator during control activation. All markings should be permanent and should remain legible throughout the life of the equipment under anticipated use and maintenance conditions.

The words employed in the label should express exactly what action is intended. Instructions should be clear and direct. Words that have a commonly accepted meaning for all intended users should be utilized. Unusual technical terms should be avoided. Labels should be consistent within and across pieces of equipment in their use of words, acronyms, abbreviations, and part/system numbers. No mismatch should exist between the nomenclature used in documentation and that printed on the labels.

Symbols should be used only if they have a commonly accepted meaning for all intended users. Symbols should be unique and distinguishable from one another. A commonly accepted standard configuration should be used.

Human factors engineering hardware design considerations should include functional dimensions, workstation architecture considerations, alarms and signals, and labeling, and should always take the operator's psychological characteristics into account.

9.14.11 Software

Computerized systems should provide a functional interface between the system and users of that system. This interface should be optimally compatible with the intended user and should minimize conditions that can degrade human performance or contribute to human error. Thus, procedures for similar or logically related transactions should be consistent. Every input by a user should consistently produce some perceptible response or output from the computer. Sufficient online help should be provided to allow the intended but uninitiated user to operate the device effectively in its basic functional mode without reference to a user's manual or experienced operator. Users should be provided appropriate information at all times on system status either automatically or upon request. Provision of information about system dysfunction is essential.

In applications where users need to log-on to the system, log-on should be a separate procedure that should be completed before a user is required to select among any operational options. Appropriate prompts for log-on should be displayed automatically on the user's terminal with no special action required other than turning on the terminal. Users should be provided feedback relevant to the log-on procedure that indicates the status of the inputs. Log-on processes should require minimum input from the user, consistent with system access security. In the event of a partial hardware/software failure, the program should allow for orderly shutdown and establishment of a checkpoint so restoration can be accomplished without loss of data.

Where two or more users need to have simultaneous access to a computer system, under normal circumstances, operation by one person should not interfere with the operations of another person. For circumstances in which certain operators require immediate access to the system, an organized system for insuring or avoiding preemption should be provided. Provisions should be made so that preempted users are notified and can resume operations at the point of interference without data loss.

9.14.12 DATA ENTRY

Manual data entry functions should be designed to establish consistency of data entry transactions, minimize user's input actions and memory load, ensure compatibility of data entry with data display, and provide flexibility of user control of data entry. The system should provide feedback to the user about acceptance or rejection of an entry.

When a processing delay occurs, the system should acknowledge the data entry and provide the user with an indication of the delay. If possible, the system should advise the user of the time remaining for process completion.

Data entry should require an explicit completion action, such as the depression of an Enter key to post an entry into memory. Data entries should be checked by the system for correct format, acceptable value, or range of values. Where repetitive entry of data sets is required, data validation for each set should be completed before another transaction can begin.

Data should be entered in units that are familiar to the user. If several different systems of units are commonly used, the user should have the option of selecting the units either before or after data entry. Transposition of data from one system of units to another should be accomplished automatically by the device. When mnemonics or codes are used to shorten data entry, they should be distinctive and have a relationship or association to normal language or specific job-related terminology.

Data deletion or cancellation should require an explicit action, such as the depression of a Delete key. When a data delete function has been selected by a user, a means of confirming the delete action should be provided, such as a dialog box with a delete acknowledgment button or a response to a question such as Are you sure? (Y/N). In general, requiring a second press of the delete key is not preferred because of the possibility of an accidental double press. Similarly, after data have been entered, if the user fails to enter the data formally, for instance, by pressing an Enter key, the data should not be deleted or discarded without confirmation from the user.

Deleted data should be maintained in a memory buffer from which they can be salvaged, such as the undelete option. The size and accessibility of this buffer should depend on the value of the data that the user can delete from the system.

The user should always be given the opportunity to change a data entry after the data have been posted. When a user requests change or deletion of a data item that is not currently being displayed, the option of displaying the old value before confirming the change should be presented. Where a data archive is being created, the system should record both the original entry and all subsequent amendments.

9.14.13 DISPLAYS

Visual displays should provide the operator with a clear indication of equipment or system status under all conditions consistent with the intended use and maintenance of the system. The information displayed to a user should be sufficient to allow the user to perform the intended task, but should be limited to what is necessary to perform the task or to make decisions. Information necessary for performing different activities, such as equipment operation versus troubleshooting, should not appear in a single display unless the activities are related and require the same information to be used simultaneously. Information should be displayed only within the limits of precision required for the intended user activity or decision-making and within the limits of accuracy of the measure.

Graphic displays should be used for the display of information when perception of the pattern of variation is important to proper interpretation. The choice of a particular graphic display type can have significant impact on user performance. The designer should consider carefully the tasks to be supported by the display and the conditions under which the user will view the device before selecting a display type.

Numeric digital displays should be used where quantitative accuracy of individual data items is important. They should not be used as the only display of information when perception of the variation pattern is important to proper interpretation or when rapid or slow digital display rates inhibit proper perception.

Displays may be coded by various features, such as color, size, location, shape, or flashing lights. Coding techniques should be used to help discriminate among individual displays and to identify functionally related displays, the relationship among displays, and critical information within a display.

Display formats should be consistent within a system. When appropriate for users, the same format should be used for input and output. Data entry formats should match the source document formats. Essential data, text, and formats should be under computer, not user, control. When data fields have a naturally occurring order, such as chronological or sequential, such order should be reflected in the format organization of the fields. Where some displayed data items are of great significance, or require immediate user response, those items should be grouped and displayed prominently. Separation of groups of information should be accomplished through the use of blanks, spacing, lines, color coding, or other similar means consistent with the application.

The content of displays within a system should be presented in a consistent, standardized manner. Information density should be held to a minimum in displays used for critical tasks. When a display contains too much data for presentation in a single frame, the data should be partitioned into separately displayable pages. The user should not have to rely on memory to interpret new data. Each data display should provide the needed context, including the recapitulation of prior data from prior displays, as necessary.

An appropriate pointing device, such as a mouse, trackball, or touch screen, should be used in conjunction with applications that are suited to direct manipulation, such as identifying landmarks on a scanned image or selecting graphical elements from a palette of options. The suitability of a given pointing device to user tasks should be assessed.

9.14.14 INTERACTIVE CONTROL

General design objectives include consistency of control action, minimized need for control actions, and minimized memory load on the user, with flexibility of interactive control to adapt to different user needs. As a general principle, the user should decide what needs doing and when to do it. The selection of dialog formats should be based on anticipated task requirements and user skills.

System response times should be consistent with operational requirements. Required user response times should be compatible with required system response time. Required user response times should be within the limits imposed by the total user task load expected in the operational environment.

Control–display relationships should be straightforward and explicit, as well as compatible with the lowest anticipated skill levels of users. Control actions should be simple and direct, whereas potentially destructive control actions should require focused user attention and command validation/confirmation before they are performed. Steps should be taken to prevent accidental use of destructive controls, including possible erasures or memory dump.

Feedback responses to correct user input should consist of changes in the state or value of those elements of the displays that are being controlled. These responses should be provided in an expected and logical manner. An acknowledgment message should be employed in those cases where the more conventional mechanism is not appropriate. Where control input errors are detected by the system, error messages and error recovery procedures should be available.

Menu selection can be used for interactive controls. Menu selection of commands is useful for tasks that involve the selection of a limited number of options or that can be listed in a menu, or in cases when users may have relatively little training. A menu command system that involves several layers can be useful when a command set is so large that users are unable to commit all the commands to memory and a reasonable hierarchy of commands exists for the user.

Form-filling interactive control may be used when some flexibility in data to be entered is needed and when the users will have moderate training. A form-filling dialog should not be used when the computer has to handle multiple types of forms and computer response is slow.

Fixed-function key interactive control may be used for tasks requiring a limited number of control inputs or in conjunction with other dialog types.

Command language interactive control may be used for tasks involving a wide range of user inputs and when user familiarity with the system can take advantage of the flexibility and speed of the control technique.

Question and answer dialogs should be considered for routine data entry tasks when data items are known and their ordering can be constrained, when users have little or no training, and when the computer is expected to have moderate response speed.

Query language dialog should be used for tasks emphasizing unpredictable information retrieval with trained user. Query languages should reflect a data structure or organization perceived by the users to be natural.

Graphic interaction as a dialog may be used to provide graphic aids as a supplement to other types of interactive control. Graphic menus may be used that display icons to represent the control options. This may be particularly valuable when system users have different linguistic backgrounds.

9.14.15 FEEDBACK

Feedback should be provided that presents status, information, confirmation, and verification throughout the interaction. When system functioning requires the user to standby, "wait" or similar type messages should be displayed until interaction is again possible. When the standby or delay may last a significant period of time, the user should be informed. When a control process or sequence is completed or aborted by the system, a positive indication should be presented to the user about the outcome of the process and the requirements for subsequent user action. If the system rejects a user input, feedback should be provided to indicate why the input was rejected and the required corrective action.

Feedback should be self-explanatory. Users should not be made to translate feedback messages by using a reference system or code sheets. Abbreviations should not be used unless necessary.

9.14.16 Prompts

Prompts and help instructions should be used to explain commands, error messages, system capabilities, display formats, procedures, and sequences, as well as to provide data. When operating in special modes, the system should display the mode designation and the file(s) being processed. Before processing any user requests that would result in extensive or final changes to existing data, the system should require user confirmation. When missing data are detected, the system should prompt the user. When data entries or changes will be nullified by an abort action, the user should be requested to confirm the abort.

Neither humor nor admonishment should be used in structuring prompt messages. The dialog should be strictly factual and informative. Error messages should appear as close as possible in time and space to the user entry that caused the message. If a user repeats an entry error, the second error message should be revised to include a noticeable change so that the user may be certain that the computer has processed the attempted correction.

Prompting messages should be displayed in a standardized area of the display. Prompts and help instructions for system-controlled dialog should be clear and explicit. The user should not be required to memorize lengthy sequences or refer to secondary written procedural references.

9.14.17 Defaults

Manufacturer's default settings and configurations should be provided to reduce user workload. Currently defined default values should be displayed automatically in their appropriate data fields with the initiation of a data entry transaction. The user should indicate acceptance of the default values. Upon user request, manufacturers should provide a convenient means by which the user may restore factory default settings.

Users should have the option of setting their own default values for alarms and configurations on the basis of personal experience. A device may retain and store one or more sets of user default settings. Activation of these settings should require deliberate action by the user.

9.14.18 Error Management/Data Correction

When users are required to make entries into a system, an easy means of correcting erroneous entries should be provided. The system should permit correction of individual errors without requiring reentry of correctly entered commands or data elements.

9.15 FITTS' LAW

Fitts' Law is a model of human movement that predicts the time required to rapidly move to a target area, as a function of the distance to the target and the size of the target. Fitts' Law is used to model the act of pointing, both in the real world (e.g., with a hand or finger) and on computers (e.g., with a mouse). It was published in 1954 by Paul Fitts.

Theoretically, the following principles exist when applying Fitts' Law to interface designs:

- Things done more often should be assigned a larger button.
- Things done more often should be closer to the average position of the user's cursor.
- Top, bottom, and sides of the screen are infinitely targetable because of the boundary created by the edges of the screen.

9.15.1 Model

Mathematically, Fitts' Law has been formulated in several different ways. One common form is the Shannon formulation for movement in a single dimension:

$$T = A + B \log_2 (D/W + 1)$$

where
   T is the average time taken to complete the movement
   A is the start/stop time of the device
   B is the inherent speed of the device
   D is the distance from the starting point to the center of the target
   W is the width of the target measured along the axis of motion

From the equation, we can see a speed–accuracy trade-off associated with pointing, whereby targets that are smaller and further away require more time to acquire.

Fitts' Law is an unusually successful and a well-studied model. Experiment that reproduce Fitt's Law and that demonstrate the applicability of Fitts' Law in somewhat different situations are not difficult to perform. The measured data in such experiments often fit a straight line with a correlation coefficient of 0.95 or higher, a sign that the model is very accurate.

Since the advent of graphical user interfaces, Fitts' Law has been applied to tasks where the user must position the mouse cursor over an on-screen target, such as a button or other widget. Fitts' Law can model both point-and-click and drag-and-drop actions. As a result of this law, there are some consequences for user interface design, including:

- Buttons and other graphical user interface controls should be a reasonable size, as it is difficult to click on small ones.
- Edges and corners of the computer display are particularly easy to acquire because the pointer remains at the screen edge regardless of how much further the mouse is moved.
- Pop-up menus can usually be opened faster than pull-down menus, such the user avoids travel.
- Pie menu items typically are selected faster and have a lower error rate than linear menu items because (1) pie menu items are all at the same small distance from the center of the menu, and (2) their wedged-shaped target areas are very large.

Another prevalent use of Fitts' Law is to help study and compare input devices. It has been verified to be able to predict user performance in some common tasks, such as point-select and point-drag tasks, using common input devices, such as a mouse, trackball, or stylus. A study of hand and head movements in two dimensions by Jagacinski and Monk found that Fitts' Law also described head movement, and it worked for two dimensions with angular uncertainty.

EXERCISES

1. Do a Web search on the author Jeff(rey) Cooper, and isolate the papers referring to mishaps. Locate and report on one of his human factors papers relevant to anesthesia.
2. Visit the American National Standards Institute (ANSI) Web site, search for the number of standards relating to color, alarms, human factors, and labeling. Comment on your results.
3. Observe the layout of controls on your car, versus the layout of controls on a different brand. Where and why are there differences?
4. There has been a significant trend in using internationally recognized symbols rather than text to denote controls. Find and report on one example in your environment.
5. Discuss the differences in expectations for medical devices such as dialysis equipment to be used in the home versus clinic.
6. Discuss the differences in expectations for blood pressure determination in the home versus clinic.

7. Do a Web search to locate ergonomic data. Why are designs generally aimed at the 5% female to 95% male ranges?

8. Do a Web search for front panel Web simulator software, report on your results. Find an example of a car dashboard layout.

9. Prototype a front panel layout for display of pulse oximeter data for joggers.

10. How would you redesign an operating room for a deaf anesthesiologist?

11. What branches of medicine are available for a blind physician? Why?

12. Given a large 1600×1200 screen, where should the target be placed so that the user can access it the fastest, no matter where the user is originally located?

BIBLIOGRAPHY

ANSI/AAMI HE74, *Human Factors Design Process for Medical Devices*. Arlington, VA: Association for the Advancement of Medical Instrumentation, 2001.

Association for the Advancement of Medical Instrumentation (AAMI), *Human Factors Engineering Guidelines and Preferred Practices for the Design of Medical Devices*. Arlington, VA: Association for the Advancement of Medical Instrumentation, 1993.

Backinger, C. and Kingsley, P., *Write It Right: Recommendations for Developing User Instruction Manuals for Medical Devices Used in Home Health Care*. Rockville, MD: U.S. Department of Health and Human Services, 1993.

Bogner, M.S., *Human Error in Medicine*. Hillsdale, NJ: Lawrence Erlbaum Associates, 1994.

Brown, C.M., *Human–Computer Interface Design Guidelines*. Norwood, NJ: Ablex, 1989.

Card, S.K., English, W.K., and Burr, B.J., Evaluation of mouse, rate-controlled isometric joystick, step keys and text keys for text selection on a CRT, *Ergonomics*, 21, 1978: 601–613.

Fitts, P.M., The information capacity of the human motor system in controlling the amplitude of movement, *Journal of Experimental Psychology: General*, 47(6), June 1954: 381–391.

Fitts, P.M. and Peeterson, J.R., Information capacity of discrete motor responses, *Journal of Experimental Psychology*, 67(2), February 1964: 103–112.

Fries, R.C., Human factors and system reliability, *Medical Device Technology*, 3(2), March 1992.

Fries, R.C., *Handbook of Medical Device Design*. New York: Marcel Dekker, 2001.

Fries, R.C., *Reliable Design of Medical Devices*, 2nd ed. Boca Raton, FL: CRC Press/Taylor & Francis Division, 2006.

Gillian, D.J., Holden, K., Adam, S., Rudisill, M., and Magee, L., How does Fitts' law fit pointing and dragging? *Proceedings of CHI'90*. Seattle, WA: ACM Press, 1990.

Hartson, H.R., *Advances in Human–Computer Interaction*. Norwood, NJ: Ablex, 1985.

Jagacinski, R.J. and Monk, D.L., Fitts' law in two dimensions with hand and head movements, *Journal of Motor Behavior*, 17, 1985: 77–95.

King, P.H. and Richard C.F., *Design of Biomedical Devices and Systems*. New York: Marcel Dekker, 2002.

Le Cocq, A.D., Application of human factors engineering in medical product design, *Journal of Clinical Engineering*, 12(4), July–August 1987: 271–277.

MacKenzie, I.S., Sellen, A., and Buxton, W., A comparison of input devices in elemental pointing and dragging tasks, *Proceedings of the CHI'91 Conference on Human Factors in Computing Systems*. New York: ACM Press, 1991.

Mathiowetz, V. et al., Grip and pinch strength: Normative data for adults, *Archives of Physical Medicine and Rehabilitation*, 66, 1985: 69–72.

MIL-HDBK-759, *Human Factors Engineering Design for Army Material*. Washington, DC: Department of Defense, 1981.

MIL-STD-1472, *Human Engineering Design Criteria for Military Systems, Equipment and Facilities*. Washington, DC: Department of Defense, 1981.

Morgan, C.T., *Human Engineering Guide to Equipment Design*. New York: Academic Press, 1984.

Philip, J.H., Human factors design of medical devices: The current challenge, *First Symposium on Human Factors in Medical Devices*, December 13–15, 1989. Plymouth Meeting, PA: ECRI, 1990.

Weinger, M.B. and Englund, C.E., Ergonomic and human factors affecting anesthetic vigilance and monitoring performance in the operating room environment, *Anesthesiology*, 73(5), November, 1990: 995–1021.

Wiklund, M.E., How to implement usability engineering, *Medical Device and Diagnostic Industry*, 15(9), 1993.

Wiklund, M.E., *Medical Device and Equipment Design—Usability Engineering and Ergonomics*. Buffalo Grove, IL: Interpharm Press, 1995.

Woodson, W.E., *Human Factors Design Handbook*. New York: McGraw-Hill, 1981.

10 Industrial Design

The future of the aircraft industry is still the responsibility of the engineer. Money alone never did and never will create anything.

Aviation Magazine (now *Aviation Week*)

Industrial design is the professional service of creating and developing concepts and specifications that optimize the function, value, and appearance of products and systems for the mutual benefit of both user and manufacturer. Industrial designers develop these concepts and specifications through collection, analysis, and synthesis of data guided by the special requirements of the client or manufacturer. They are trained to prepare clear and concise recommendations through drawings, models, and verbal descriptions. Industrial design services are often provided within the context of cooperative working relationships with other members of a development group. Typical groups include management, marketing, engineering, and manufacturing specialists. The industrial designer expresses concepts that embody all relevant design criteria determined by the group.

The industrial designer's unique contribution places emphasis on those aspects of the product or system that relate most directly to human characteristics, needs, and interests. This contribution requires specialized understanding of visual, tactile, safety, and convenience criteria with concern for the user. Education and experience in anticipating psychological, physiological, and sociological factors that influence and are perceived by the user are essential industrial design resources. Industrial designers also maintain a practical concern for technical processes and requirements for manufacture, marketing opportunities and economic constraints, and distribution sales and servicing processes. They work to ensure that design recommendations use materials and technology effectively, and comply with all legal and regulatory requirements.

In addition to supplying concepts for products and systems, industrial designers are often retained for consultation on a variety of problems that have to do with a client's image. Such assignments include product and organization identity systems, development of communication systems, interior space planning and exhibit design, advertising devices and packaging, and other related services. Their expertise is sought in a wide variety of administrative arenas to assist in developing industrial standards, regulatory guidelines, and quality control procedures to improve manufacturing operations and products. Industrial designers, as professionals, are guided by their awareness of obligations to fulfill contractual responsibilities to clients, to protect the public safety and well-being, to respect the environment and to observe ethical business practice.

The term industrial design was coined early in the twentieth century to describe for mass-produced devices the creative role previously performed by an individual artisan. In keeping with the complexity of mass production, industrial designers work with other professions involved in conceiving, developing, and manufacturing products, including

- Marketing experts
- Design engineers
- Biomedical engineers
- Human factors specialists
- Manufacturing engineers
- Service personnel

Together with human factors specialists, industrial designers conduct usability studies to ensure that a product meets the user's needs, wants, and expectations. They often rearrange internal components to make products more efficient to manufacture and easy to assemble, service, and recycle. As about one third of all medical device incident reports involve user error, the need for improved interfaces between devices and users is evident.

10.1   SET USABILITY GOALS

Usability goals are comparable to other types of engineering goals in the sense that they are quantitative and provide a basis for acceptance testing. Goals may be objective or subjective. A sample objective goal might be—on average, users shall require 3 s to silence an alarm. This goal is an objective goal because the user's performance level can be determined simply by observation. For example, you can use a stopwatch to determine task times. Other kinds of objective goals concentrate on the number of user errors and the rate of successful task completion.

A sample subjective goal is—on average, 75% of users shall rate the intuitiveness of the alarm system as 5 or better, where $1 =$ poor and $7 =$ excellent. The range 1–7 (or other) is a Likert scale, which is a form of psychometric response scale often used to evaluate such items as usability and professors' lecture skills. This goal is subjective because it requires asking the user's opinion about their interaction with the given product. A rating sheet can be used to record their answers. Other kinds of subjective goals concentrate on mental processing and emotional response attributes, such as learning, frustration level, fear of making mistakes, etc.

Every usability goal is based on a usability attribute, for example, task, speed, or intuitiveness include a metric such as time or scale and sets a target performance level, such as 3 s or a rating of 5 or better.

Typically, up to 50 usability goals may be written for a given project, two thirds of which are objective and one third which are subjective. The target performance level on each goal is based on findings from preceding user studies, particularly the benchmark usability testing. If there is no basis for comparison, that is, there are no comparable products; then engineering judgment must be used to set the initial goals and adjust them as necessary to assure they are realistic.

10.2   DESIGN USER INTERFACE CONCEPTS

Concurrent design is a productive method of developing a final user interface design. It enables the thorough exploration of several design concepts before converging on a final solution. In the course of exploring alternative designs, limited prototypes should be built of the most promising concepts and user feedback obtained on them. This gets users involved in the design process at its early stages and assures that the final design will be closely matched to user's expectations.

Note that the design process steps described below assume that the product includes both hardware and software elements. Some steps would be moot if the product has no software user interface.

10.2.1   DEVELOP CONCEPTUAL MODEL

When users interact with a product, they develop a mental model of how it works. This mental model may be complete and accurate or just the opposite. Enabling the user to develop a complete and accurate mental model of how a product works is a challenge. The first step is developing so-called conceptual models of how to represent the product's functions. This exercise provides a terrific opportunity for design innovation. The conceptual model may be expressed as a bubble diagram that illustrates the major functions of the product and functional interrelationships as you would like the users to think of them. You can augment the bubble diagram with a narrative description of the conceptual model.

10.2.2 DEVELOP USER INTERFACE STRUCTURE

Develop alternative user interface structures that complement the most promising—two to three conceptual models. These structures can be expressed in the form of screen hierarchy maps that illustrate where product functions reside and how many steps it will take users to get to them. Such maps may take the form of a single element, a linear sequence, a tree structure (cyclic or acyclic) or a network. In addition to software screens, such maps should show which functions are allocated to dedicated hardware controls.

10.2.3 DEFINE INTERACTION STYLE

In conjunction with the development of the user interface structures, alternative interaction styles should be defined. Possible styles include question and answer dialogs, command lines, menus, and direct manipulation.

10.2.4 DEVELOP SCREEN TEMPLATES

Determine an appropriate size display based on the user interface structure and interaction style, as well as other engineering considerations. Using computer-based drawing tools, draw the outline of a blank screen. Next, develop a limited number (perhaps three to five) of basic layouts for the information that will appear on the various screens. Normally, it is best to align all elements, such as titles, windows, prompts, and numerics according to a grid system.

10.2.5 DEVELOP HARDWARE LAYOUT

Apply established design principles in the development of hardware layouts that are compatible with the evolving software user interface solutions. Assure that the layouts reinforce the overall conceptual model.

10.2.6 DEVELOP A SCREENPLAY

Apply established design principles in the development of a detailed screenplay. Do not bother to develop every possible screen at this time. Rather, develop only those screens that would enable users to perform frequently used, critical and particularly complex functions. Base the screen designs on the templates. Create new templates or eliminate existing templates as required while continuing to limit the total number of templates. Assure that the individual screens reinforce the overall conceptual model. You may choose to get user feedback on the screenplay (what some people call a paper prototype). You may show your coworker the paper concept, then ask them to control the system while giving you oral feedback on their thoughts during the process.

10.2.7 DEVELOP A REFINED DESIGN

Developing a hardware layout and developing a screenplay describe prototyping and testing the user interface. These efforts will help determine the most promising design concept or suggest a hybrid of two or more concepts. The next step is to refine the preferred design. Several reiterations of the preceding steps may be necessary, including developing a refined conceptual model, developing a refined user interface structure and developing an updated set of screen templates. Then, a refined screenplay and hardware layout may be developed.

10.2.8 DEVELOP A FINAL DESIGN

Once again, Sections 10.2.5 and 10.2.6 describe prototyping and testing the user interface. These efforts will help you determine any remaining usability problems with the refined design and opportunities for further improvement. It is likely that design changes at this point will be limited in nature. Most can be made directly to the prototype.

10.3 MODEL THE USER INTERFACE

Build a prototype to evaluate the dynamics of the user interface. Early prototypes of competing concepts may be somewhat limited in terms of their visual realism and how many functions they perform. Normally, it is best to develop a prototype that (1) presents a fully functional top level that allows users to browse their basic options, and (2) enables users to perform a few sample tasks, that is, walkthrough a few scenarios. As much as possible, include tasks that relate to the established usability goals.

User interface prototypes may be developed using conventional programming languages or rapid prototyping languages, such as SuperCard, Altia Design, Visual Basic, Toolbook, etc. The rapid prototyping languages are generally preferable because they allow for faster prototyping and they are easier to modify based on core project team and user feedback. If possible, the use of a touch screen system should be implemented at this stage, along with testing of the interfaces again by your coworkers. If neither of you are frustrated at the end of a testing session, you have likely done a reasonable job on the design so far. An additional advantage at this point is that some of the actions, such as response time to an alarm, may now be documented with your software.

Page layouts may even be mocked-up in your computer editing software, for example, with Microsoft Word. Figure 10.1 is a rendition of a computer control scheme used by author King in the development of a computerized anesthesia monitoring system.

Early in the screenplay development process, it may make sense to prototype a small part of the user interface to assess design alternatives or to conduct limited studies, such as how frequently to flash a warning. Once detailed screenplays of competing concepts are available, build higher fidelity prototypes that facilitate usability testing. Once a refined design is developed, build a fully functional prototype that permits a verification usability test. Such prototypes can be refined based on final test results and serve as a specification.

10.4 TEST THE USER INTERFACE

There are several appropriate times to conduct a usability test, including

- At the start of a development effort to develop benchmarks
- When you have paper-based or computer-based prototypes of competing design concepts
- When you have a prototype of your refined design
- When you want to develop marketing claims regarding the performance of the actual product

Main Menu			
Drug administration	View primary graph	Plotter control	
Routine events	View secondary graph	Trend select	
Comment entry	Examine variables	View/set alarm limits	
Equipment configuration	View drug dose totals	Patient/operation information entry	Exit program

FIGURE 10.1 Mock-up of a touch screen control screen.

While the rigor of the usability test may change, based on the timing of the test, the basic approach remains the same. You recruit prospective users to spend a concentrated period interacting with the prototype product. The users may undertake a self-exploration or perform directed tasks. During the course of such interactions, you note the test participants' comments and document their performance. At intermittent stages, you may choose to have the test participant complete a questionnaire or rating/ranking exercise. Videotaping test proceedings is one way to give those unable to attend the test a first-hand sense of user–product interactions. Sometimes it is useful to create a 10–15 min highlight tape that shows the most interesting moments of all test sessions. During testing, collect the data necessary to determine if you are meeting the established usability goals. This effort will add continuity and objectivity to the usability engineering process.

10.5   SPECIFY THE USER INTERFACE

10.5.1   STYLE GUIDE

The purpose of a style guide is to document the rules of the user interface design. By establishing such rules, you can check the evolving design to determine any inconsistencies. Also, it assures the consistency of future design changes. Style guides, usually 10–15 pages long, normally include a description of the conceptual model, the design elements and elements of style.

10.5.2   SCREEN HIERARCHY MAP

The purpose of a screen hierarchy map is to provide an overview of the user interface structure. It places all screens that appear in the screenplay in context. It enables the flow of activity to be studied to determine if it reinforces the conceptual model. It also helps to determine how many steps users will need to take to accomplish a given task. Graphical elements of the screen hierarchy map should be cross-indexed to the screenplay.

10.5.3   SCREENPLAY

The purpose of a screenplay is to document the appearance of all major screens on paper. Typically, screen images are taken directly from the computer-based prototype. Ideally, the screenplay should present screen images in their actual scale and resolution. Each screen should be cross-indexed to the screen hierarchy map.

10.5.4   SPECIFICATION PROTOTYPE

The purpose of the specification prototype is to model accurately the majority of user interface interactions. This provides the core project team with a common basis for understanding how the final product should work. It provides a basis for writing the user documentation. It may also be used to orient those involved in marketing, sales, and training.

10.5.5   HARDWARE LAYOUTS

The hardware layout may be illustrated by the specification prototype. However, the hardware may not be located proximal to the software user interface. If this is the case, develop layout drawings to document the final hardware layout.

10.6   ADDITIONAL INDUSTRIAL DESIGN CONSIDERATIONS

The design of medical devices should reflect industrial design features that increase the potential for successful performance of tasks and for satisfaction of design objectives.

10.6.1 CONSISTENCY AND SIMPLICITY

Where common functions are involved, consistency is encouraged in controls, displays, markings, coding, and arrangement schemes for consoles and instrument panels.

Simplicity in all designs is encouraged. Equipment should be designed to be operated, maintained, and repaired in its operational environment by personnel with appropriate but minimal training. Unnecessary or cumbersome operations should be avoided when simpler, more efficient alternatives are available.

10.6.2 SAFETY

Medical device design should reflect system and personnel safety factors, including the elimination or minimization of the potential for human error during operation and maintenance under both routine and nonroutine or emergency conditions. Machines should be designed to minimize consequence of human error. For example, where appropriate, a design should incorporate redundant, diverse elements arranged in a manner that increases overall reliability when failure can result in the inability to perform a critical function.

Any medical device failure should immediately be indicated to the operator and should not adversely affect safe operation of the device. Where failures can affect safe operation, simple means and procedures for averting adverse effects should be provided.

When the device failure is life-threatening or could mask a life-threatening condition, an audible alarm and a visual display should be provided to indicate the device failure. Wherever possible, explicit notification of the source of failure should be provided to the user. Concise instructions on how to return to operation or how to invoke alternate backup methods should be provided.

The reader should consider two other overriding factors at this point. First, if the device can be made fail-safe, it should be done. This implies that despite a failure in the device, the essential functions of the device, such as delivery of oxygen, are not compromised. Second, it is mandated that in the design of medical devices that safety considerations be considered, this interface design development will be one of the areas that must be documented.

10.6.3 ENVIRONMENTAL/ORGANIZATIONAL CONSIDERATIONS

The design of medical devices should consider the following:

- Levels of noise, vibration, humidity, and heat that will be generated by the device and the levels of noise, vibration, humidity, and heat to which the device and its operators and maintainers will be exposed in the anticipated operational environment.
- Need for protecting operators and patients from electric shock, thermal, infectious, toxicologic, radiologic, electromagnetic, visual, and explosion risks, as well as from potential design hazards, such as sharp edges and corners, and the danger of the device falling on the patient or operator.
- Adequacy of the physical, visual, auditory, and other communication links among personnel, and between personnel and equipment.
- Importance of minimizing psychophysiological stress and fatigue in the clinical environment in which the medical device will be used.
- Impact on operator effectiveness of the arrangement of controls, displays, and markings on consoles and panels the potential effects of natural or artificial illumination used in the operation, control, and maintenance of the device.
- Need for rapid, safe, simple, and economical maintenance and repair.
- Possible positions of the device in relation to the users as a function of the user's location and mobility.
- Electromagnetic environment(s) in which the device is intended to be used.

10.6.4 DOCUMENTATION

Documentation is a general term that includes operator manuals, instruction sheets, online help systems, and maintenance manuals. These materials may be accessed by many types of users. Therefore, the documentation should be written to meet the needs of all target populations.

Preparation of instructional documentation should begin as soon as possible during the specification phase. This assists device designers in identifying critical human factors engineering needs and in producing a consistent human interface. The device and its documentation should be developed together.

During the planning phase, a study should be made of the capabilities and information needs of the documentation users, including

- User's mental abilities
- User's physical abilities
- User's previous experience with similar devices
- User's general understanding of the general principles of operation and potential hazards associated with the technology
- Special needs or restrictions of the environment

As a minimum, the operator's manual should include detailed procedures for setup, normal operation, emergency operation, cleaning, and operator troubleshooting.

The operator manual should be tested on models of the device. It is important that these test populations be truly representative of end users and that they not have advance knowledge of the device. Maintenance documentation should be tested on devices that resemble production units. Documentation content should be presented in language free of vague and ambiguous terms. The simplest words and phrases that will convey the intended meaning should be used. Pictures help in understanding the document content, especially technical information. Terminology within the publication should be consistent. Use of abbreviations should be kept to a minimum, but defined where they are used. Programs exist that estimate the grade level of a particular document, some estimate that most documentation should be developed using an average of eighth grade vocabulary.

Information included in warnings and cautions should be chosen carefully and with consideration of the skills and training of intended users. It is especially important to inform users about unusual hazards and hazards specific to the device.

Human factors engineering design features should assure that the device functions consistently, simply, and safely, that the environment, system organization, and documentation are analyzed and considered in the design, thus increasing the potential for successful performance of tasks and for satisfaction of design objectives.

10.6.5 ALARMS AND SIGNALS

The purpose of an alarm is to draw attention to the device when the operator's attention may be focused elsewhere. Alarms should not be startling but should elicit the desired action from the user. When appropriate, the alarm message should provide instructions for the corrective action that is required. In general, alarm design will be different for a device that is continuously attended by a trained operator, such as an anesthesia machine, than for a device that is unattended and operated by an untrained operator, such as a patient-controlled analgesia device. False alarms, loud and startling alarms, or alarms that recur unnecessarily can be a source of distraction for both an attendant and the patient and thus be a hindrance to good patient care. Two cautions are first, the shutting off of alarms has resulted in more than one death and second, the requirement (on some systems) that a patient be admitted (or similar terminology) has also lead to patient deaths as the system was not programmed to alarm unless a condition were set.

Alarm characteristics are grouped in the following three categories

1. High priority: A combination of audible and visual signals indicating that immediate operator response is required.
2. Medium priority: A combination of audible and visual signals indicating that prompt operator response is required.
3. Low priority: A visual signal, or a combination of audible and visual signals indicating that operator awareness is required.

A red flashing light should be used for a high priority alarm condition unless an alternative visible signal that indicates the alarm condition and its priority is employed. A red flashing light should not be used for any other purpose.

A yellow flashing light should be used for a medium priority alarm condition unless an alternative visible signal that indicates the alarm condition and its priority is employed. A yellow flashing light should not be used for any other purpose.

A steady yellow light should be used for a low priority alarm condition unless an alternative visible signal that indicates the alarm condition and its priority is employed.

Audible signals should be used to alert the operator to the status of the patient or the device when the device is out of the operator's line of sight. Audible signals used in conjunction with visual displays should be supplementary to the visual signals and should be used to alert and direct the user's attention to the appropriate visual display.

Design of equipment should take into account the background noise and other audible signals and alarms that will likely be present during the intended use of the device. The lowest volume control settings of the critical life-support audible alarms should provide sufficient signal strength to preclude masking by anticipated ambient noise levels. Volume control settings for other signals should similarly preclude such a masking. Ambient noise levels in hospital areas can range from 50 dB in a private room to 60 dB in intensive care units and emergency rooms, with peaks as high as 65–70 dB in operating rooms due to conversations, alarms, or the activation of other devices. The volume of monitoring signals normally should be lower than that of high priority or medium priority audible alarms provided on the same device. Audible and visual signals should be located so as to assist the operator in identifying the device that is causing the alarm. Audible alarms also should not be able to be physically blocked from alarming (such as by a pillow, etc.)

The use of voice alarms in medical applications should normally not be considered for the following reasons:

• Voice alarms are easily masked by ambient noise and other voice messages.
• Voice messages may interfere with communications among personnel who are attempting to address the alarm condition.
• Information conveyed by the voice alarm may reach individuals who should not be given specific information concerning the nature of the alarm.
• Types of messages transmitted by voice tend to be very specific, possibly causing complication and confusion to the user.
• In the situation where there are multiple alarms, multiple voice alarms would cause confusion.
• Different languages may be required to accommodate various markets.

The device's default alarm limits should be provided for critical alarms. These limits should be sufficiently wide to prevent nuisance alarms, and sufficiently narrow to alert the operator to a situation that would be dangerous in the average patient.

The device may retain and store one or more sets of alarm limits chosen by the user. When more than one set of user default alarm limits exists, the activation of user default alarm limits should require deliberate action by the user. When there is only one set of user default alarm limits, the device may be configured to activate this set of user default alarm limits automatically in place of the factory default alarm limits.

The setting of adjustable alarms should be indicated continuously or on user demand. It should be possible to review alarm limits quickly. During user setting of alarm limits, monitoring should continue and alarm conditions should elicit the appropriate alarms. Alarm limits may be set automatically or upon user action to reasonable ranges or percentages above or below existing values for monitored variables. Care should be used in the design of such automatic setting systems to help prevent nuisance alarms or variables that are changing within an acceptable range.

An audible high or medium priority signal may have a manually operated, temporary override mechanism that will silence it for a period of time (e.g., 120 s). After the silencing period, the alarm should begin sounding again if the alarm condition persists or if the condition was temporarily corrected but has now returned. New alarm conditions that develop during the silencing period should initiate audible and visual signals. If momentary silencing is provided, the silencing should be visually indicated.

An audible high or medium priority signal may be equipped with a means of permanent silencing, that may be appropriate when a continuous alarm is likely to degrade user performance of associated tasks to an unacceptable extent and in cases when users would otherwise be likely to disable the device altogether. If provided, such silencing should require that the user either confirm the intent to silence a critical life-support alarm or take more than one step to turn the alarm off. Permanent silencing should be visually indicated and may be signaled by a periodic audible reminder. Permanent silencing of an alarm should not affect the visual representation of the alarm and should not disable the alarm.

Life-support devices and devices that monitor a life-critical variable should have an audible alarm to indicate a loss of power or failure of the device. The characteristics of this alarm should be the same as those of the highest priority alarm that becomes inoperative. It may be necessary to use battery power for such an alarm. Some consideration (generally not the industrial designer's job) should be given to the use of computer memory to document machine and patient status during and near-alarm conditions. Such a recording can be of value in debugging systems, and may be of value legally in the case of a death or injury.

10.6.6 DISPLAYS

Visual displays should provide the operator with a clear indication of equipment or system status under all conditions consistent with the intended use and maintenance of the system. The information displayed to a user should be sufficient to allow the user to perform the intended task, but should be limited to what is necessary to perform the task or to make decisions. Information necessary for performing different activities, such as equipment operation versus troubleshooting, should not appear in a single display unless the activities are related and require the same information to be used simultaneously. Information should be displayed only within the limits of precision required for the intended user activity or decision-making and within the limits of accuracy of the measure.

Graphic displays should be used for the display of information when perception of the pattern of variation is important to proper interpretation. The choice of a particular graphic display type can have significant impact on user performance. The designer should consider carefully the tasks to be supported by the display and the conditions under which the user will view the device before selecting a display type.

Numeric digital displays should be used where quantitative accuracy of individual data items is important. They should not be used as the only display of information when perception of the variation pattern is important to proper interpretation or when rapid or slow digital display rates inhibit proper perception. They should generally display only a consistent and honest number of significant figures.

Displays may be coded by various features, such as color, size, location, shape, or flashing lights. Coding techniques should be used to help discriminate among individual displays and to identify functionally related displays, the relationship among displays, and critical information within a display.

Display formats should be consistent within a system. When appropriate for users, the same format should be used for input and output. Data entry formats should match the source document formats. Essential data, text, and formats should be under computer, not user, control. When data fields have a naturally occurring order, such as chronological or sequential, such order should be reflected in the format organization of the fields. Where some displayed data items are of great significance, or require immediate user response, those items should be grouped and displayed prominently. Separation of groups of information should be accomplished through the use of blanks, spacing, lines, color coding, or other similar means consistent with the application.

The content of displays within a system should be presented in a consistent, standardized manner. Information density should be held to a minimum in displays used for critical tasks. When a display contains too much data for presentation in a single frame, the data should be partitioned into separately displayable pages. The user should not have to rely on memory to interpret new data. Each data display should provide the needed context, including the recapitulation of prior data from prior displays, as necessary.

An appropriate pointing device, such as a mouse, trackball, or touch screen, should be used in conjunction with applications that are suited to direct manipulation, such as identifying landmarks on a scanned image or selecting graphical elements from a palette of options. The suitability of a given pointing device to user tasks should be assessed. Consideration should also be given to the potential need for a backup input device.

10.6.7 INTERACTIVE CONTROL

General design objectives include consistency of control action, minimized need for control actions, and minimized memory load on the user, with flexibility of interactive control to adapt to different user needs. As a general principle, the user should decide what needs doing and when to do it. The selection of dialog formats should be based on anticipated task requirements and user skills.

System response times should be consistent with operational requirements. Required user response times should be compatible with required system response time. Required user response times should be within the limits imposed by the total user task load expected in the operational environment.

Control–display relationships should be straightforward and explicit, as well as compatible with the lowest anticipated skill levels of users. Control actions should be simple and direct, whereas potentially destructive control actions should require focused user attention and command validation/confirmation before they are performed. Steps should be taken to prevent accidental use of destructive controls, including possible erasures or memory dump.

Feedback responses to correct user input should consist of changes in the state or value of those elements of the displays that are being controlled. These responses should be provided in an expected and logical manner. An acknowledgment message should be employed in those cases where the more conventional mechanism is not appropriate. Where control input errors are detected by the system, error messages and error recovery procedures should be available.

Menu selection can be used for interactive controls. Menu selection of commands is useful for tasks that involve the selection of a limited number of options or that can be listed in a menu, or in cases when users may have relatively little training. A menu command system that involves several layers can be useful when a command set is so large that users are unable to commit all the commands to memory and a reasonable hierarchy of commands exists for the user.

Form-filling interactive control may be used when some flexibility in data to be entered is needed and when the users will have moderate training. A form-filling dialog should not be used when the computer has to handle multiple types of forms and computer response is slow.

Fixed-function key interactive control may be used for tasks requiring a limited number of control inputs or in conjunction with other dialog types.

Command language interactive control may be used for tasks involving a wide range of user inputs and when user familiarity with the system can take advantage of the flexibility and speed of the control technique.

Question and answer dialog should be considered for routine data entry tasks when data items are known and their ordering can be constrained, when users have little or no training, and when the computer is expected to have moderate response speed.

Query language dialog should be used for tasks emphasizing unpredictable information retrieval with trained user. Query languages should reflect a data structure or organization perceived by the users to be natural.

Graphic interaction as a dialog may be used to provide graphic aids as a supplement to other types of interactive control. Graphic menus may be used that display icons to represent the control options. This may be particularly valuable when system users have different linguistic backgrounds.

10.6.8 FEEDBACK

Feedback should be provided that presents status, information, confirmation, and verification throughout the interaction. When system functioning requires the user to standby, a "wait" or similar type messages should be displayed until interaction is again possible. When the standby or delay may last a significant period, the user should be informed. When a control process or sequence is completed or aborted by the system, a positive indication should be presented to the user about the outcome of the process and the requirements for subsequent user action. If the system rejects a user input, feedback should be provided to indicate why the input was rejected and the required corrective action.

Feedback should be self-explanatory. Users should not be made to translate feedback messages by using a reference system or code sheets. Abbreviations should not be used unless necessary.

10.6.9 ERROR MANAGEMENT/DATA PROTECTION

When users are required to make entries into a system, an easy means of correcting erroneous entries should be provided. The system should permit correction of individual errors without requiring re-entry of correctly entered commands or data elements.

EXERCISES

1. Compare the work done by persons concerned primarily with human factors (Chapter 9) to the work done by industrial designers (current chapter).
2. Perform a Web search using the search term "industrial design." Summarize the results from your first 10 hits.
3. Visit the Web site for the Industrial Design Society of America (idsa.org). Find and report on their definition of industrial design.
4. Visit the Web site devicelink.com; go to one of the expos listed (such as MD&M West). Search for the listing for contract manufacturers. Of the first 10 or so, how many would qualify as industrial designers (list and discuss).
5. Do a Web search for front panel Web simulator software, report on your results. Find an example of a medical device panel layout.

6. Prototype a front panel layout for display of exercise data for a weight watchers clinic.
7. Which of the 11 problems in Chapter 9 may also apply to this chapter, and why?

BIBLIOGRAPHY

Association for the Advancement of Medical Instrumentation (AAMI), *Human Factors Engineering Guidelines and Preferred Practices for the Design of Medical Devices.* Arlington, VA: Association for the Advancement of Medical Instrumentation, 1993.

Bogner, M.S., *Human Error in Medicine.* Hillsdale, NJ: Lawrence Erlbaum, 1994.

Brown, C.M., *Human–Computer Interface Design Guidelines.* Norwood, NJ: Ablex, 1998.

Fries, R.C., Human factors and system reliability, *Medical Device Technology*, 3(2), March 1992.

MIL-STD-1472, *Human Engineering Design Criteria for Military Systems, Equipment and Facilities.* Washington, DC: Department of Defense, 1981.

Morgan, C.T., *Human Engineering Guide to Equipment Design.* New York: Academic Press, 1984.

Wiklund, M.E., How to implement usability engineering, *Medical Device and Diagnostic Industry*, 15(9), 1993.

Wiklund, M.E., *Medical Device and Equipment Design—Usability Engineering and Ergonomics.* Buffalo Grove, IL: Interpharm Press, 1995.

11 Biomaterials and Material Testing

Developing products to be implanted and to function inside the highly intricate environment of the human body is among the most complex and challenging of all the academic and business pursuits in bioengineering.

Lory A. Frenkel

Biomaterials may be defined as nonviable materials used in or as a medical device with the intention of performing a medically related function. Biomaterials have been in existence for several years and their use and number of applications has exploded in the past century.

Gold was one of the first known substances to be used in dentistry. Use of gold dates back about 2000 years. Glass eyes have a shorter history. The use of wooden (George Washington had a set) and later ivory dentures date to the Middle Ages. The advent of aseptic surgery in the 1860s necessarily predated the first successful use of metal bone plates in 1900 and joint replacements in the 1930s. Accidental implantation of plastic shards from shattered airplane turrets during World War II, and the recognition that a major rejection episode did not occur, probably led to the initiation of today's market for biomaterials. Blood vessels were being replaced in the 1950s, heart valves were implanted in the 1960s, and the field has expanded radically since (Ratner et al. 1996).

Examples of uses of biomaterials include the following:

- Replacement of diseased parts: Dialysis with semipermeable membranes (cuprophane, 1960s)
- Treatment aids: Catheters
- Replacement of diseased part: Dental amalgams
- Replacement of burned or dead part: Artificial skin
- Cosmetic correction: Breast implants
- Assistive in healing: Sutures
- Diagnostic aids: Rectoscope
- Functional correction: Spinal rods (Harrington)
- Improve function: Soft contacts
- Monitor, diagnose, and treatment: Pacemaker with defibrillator

This field is sufficiently broad in scope that there exists, since 1975, the Society for Biomaterials, which coordinates the interests of students, faculty, and industry via an international organization, a searchable Web site, and national meetings.* Other Web sites contain databases for dental materials (University of Michigan)[†] and general materials (Mat Web).[‡]

Biological evaluation of biomaterial and medical devices using biomaterial is performed to determine the potential toxicity resulting from contact of the component materials of the device with

---

* www.biomaterials.org.
[†] http://www.lib.umich.edu/dentlib/Dental_tables/intro.html.
[‡] http://www.matweb.com/main.htm.

the body. The device materials should not, either directly or through the release of their material constituents:

- Produce adverse local or systemic effects
- Be carcinogenic
- Produce adverse reproductive and developmental effects

Therefore, evaluation of any new device intended for human use requires data from systematic testing to ensure that the benefits provided by the final product will exceed any potential risks produced by device materials.

When selecting the appropriate tests for biological evaluation of a medical device, one must consider the chemical characteristics of device materials and the nature, degree, frequency, and duration of its exposure to the body. In general, the tests include

- Acutes
- Subchronic and chronic toxicity
- Irritation to skin, eyes, and mucosal surfaces
- Sensitization
- Hemocompatibility
- Genotoxicity
- Carcinogenicity
- Effects on reproduction including developmental effects

However, depending on varying characteristics and intended uses of devices as well as the nature of contact, these general tests may not be sufficient to demonstrate the safety of some specialized devices. Additional tests for specific target organ toxicity, such as neurotoxicity and immunotoxicity may be necessary for some devices. For example, a neurological device with direct contact with brain parenchyma and cerebrospinal fluid (CSF) may require an animal implant test to evaluate its effects on the brain parenchyma, susceptibility to seizure, and effects on the functional mechanism of choroid plexus and arachnoid villi to secrete and absorb CSF. The specific clinical application and the materials used in the manufacture of the new device determine which tests are appropriate.

Some devices are made of materials that have been well characterized chemically and physically in the published literature and have a long history of safe use. For the purposes of demonstrating the substantial equivalence of such devices to other marketed products, it may not be necessary to conduct all the tests suggested in the Food and Drug Administration (FDA) matrix of this guidance. FDA reviewers are advised to use their scientific judgment in determining which tests are required for the demonstration of substantial equivalence under section 510(k). In such situations, the manufacturer must document the use of a particular material in a legally marketed predicate device or a legally marketed device with comparable patient exposure.

11.1 FDA AND BIOCOMPATIBILITY

In 1986, FDA, Health and Welfare (Canada), and Health and Social Services (United Kingdom) issued the Tripartite Biocompatibility Guidance for medical devices. This guidance has been used by FDA reviewers, as well as by manufacturers of medical devices, in selecting appropriate tests to evaluate the adverse biological responses to medical devices. Since then, the International Standards Organization (ISO), in an effort to harmonize biocompatibility testing, developed a standard for biological evaluation of medical devices (ISO 10993). The scope of this 12-part standard is to evaluate the effects of medical device materials on the body. The first part of this standard, Biological Evaluation of Medical Devices: Part 1: Evaluation and Testing, provides guidance for selecting the tests to evaluate the biological response to medical devices. Most of the other parts of the ISO standard deal with appropriate methods to conduct the biological tests suggested in Part 1 of the standard.

The ISO Standard, Part 1, uses an approach to test selection that is very similar to the currently used Tripartite Guidance, including the same seven principles. It also uses a tabular format (matrix) for laying out the test requirements based on the various factors discussed above (see Tables 11.1 and 11.2). In addition, FDA is in the process of preparing toxicology profiles for specific devices. These profiles will assist in determining appropriate toxicology tests for these devices.

To harmonize biological response testing with the requirements of other countries, FDA will apply the ISO standard, Part 1, in the review process in lieu of the Tripartite Biocompatibility Guidance.

FDA notes that the ISO standard acknowledges certain kinds of discrepancies. It states "due to diversity of medical devices, it is recognized that not all tests identified in a category will be necessary and practical for any given device. It is indispensable for testing that each device shall be considered on its own merits: additional tests not indicated in the table may be necessary." In keeping with this inherent flexibility of the ISO standard, FDA has made several modifications to the testing required by ISO 10993-Part 1. These modifications are required for the category of surface devices permanently contacting mucosal membranes (e.g., intra-uterine devices [IUDs]). The ISO standard would not require acute, subchronic, chronic toxicity, and implantation tests. Also, for externally communicating devices, tissue/bone/dentin with prolonged and permanent contact (e.g., dental cements, filling materials, etc.), the ISO standard does not require irritation nor systemic tests. Therefore, FDA has included these types of tests in the matrix. Although several toxicity, acute, subchronic, and chronic toxicity tests were added to the matrix, reviewers should note that some tests are commonly requested while other tests are to be considered and only asked for on a case-by-case basis. Thus, the modified matrix is only a framework for the selection of tests and not a checklist of every required test.

Reviewers should avoid proscriptive interpretation of the matrix. If a reviewer is uncertain about the applicability of a specific type of test for a specific device, the reviewer should consult toxicologists in the Office of Device Evaluation (ODE). FDA expects that manufacturers will consider performing the additional tests for certain categories of devices suggested in the FDA-modified matrix. This does not mean that all the tests suggested in the modified matrix are essential and relevant for all devices. In addition, device manufacturers are advised to consider tests to detect chemical components of device materials, which may be pyrogenic. The FDA believes that ISO 10993, Part 1, and appropriate consideration of the additional tests suggested by knowledgeable individuals will generate adequate biological data to meet its requirements.

Manufacturers are advised to initiate discussions with the appropriate review division in the Office of Device Evaluation (CDRH) before the initiation of expensive, long-term testing of any new device materials to ensure that the proper testing will be conducted. We also recognize that an ISO standard is a document that undergoes periodic review and is subject to revision. ODE will notify manufacturers of any future revisions to the ISO standard referenced here that affect this document's requirements and expectations.

11.2 INTERNATIONAL REGULATORY EFFORTS

ISO is in the process of publishing a series of standards on the biological evaluation of medical devices—ISO 10993. Many parts of this series have been accepted as international standards, while the rest are under development (see Table 11.3). The subject of the first part, ISO 10993-1, is the categorizing and performance of safety testing. Part 2 of the standard, ISO 10993-2, is concerned with animal welfare requirements; another section, ISO 10993-12, deals with sample preparation and reference materials. Most of the remaining parts of the standard treat the individual tests.

The European Union (EU) has issued a council directive—93/42/EEC, 1993—concerning medical devices. All medical devices to be sold on the EU market must comply with this directive

TABLE 11.1

Initial Evaluation Tests for Consideration

Device Categories	Body Contact	Contact Duration	Biological Effect							
			Cytotoxicity	Sensitization	Irritation	System Toxicity	Subchronic Toxicity	Geno-Toxicity	Implantation	Hemocompatibility
Surface devices	Skin	A	X	X	X	—	—	—	—	—
		B	X	X	X	—	—	—	—	—
		C	X	X	X	—	—	—	—	—
	Mucosal membrane	A	X	X	X	—	—	—	—	—
		B	X	X	X	O	O	—	O	—
		C	X	X	X	O	X	X	O	—
	Breached or compromised surfaces	A	X	X	X	O	—	—	—	—
		B	X	X	X	O	O	—	O	—
		C	X	X	X	O	X	X	O	—
External communicating devices	Blood path, indirect	A	X	X	X	X	—	—	—	X
		B	X	X	X	X	O	—	—	X
		C	X	X	O	X	X	X	O	X
	Tissue/bone/dentin communicating	A	X	X	X	O	—	—	—	—
		B	X	X	O	O	O	X	X	—
		C	X	X	O	O	O	X	X	—
	Circulating blood	A	X	X	X	X	—	O	—	X
		B	X	X	X	X	O	X	O	X
		C	X	X	X	X	X	X	O	X
Implant devices	Tissue/bone	A	X	X	X	X	—	—	—	—
		B	X	X	O	O	O	X	X	—
		C	X	X	O	O	O	X	X	—
	Blood	A	X	X	X	X	—	X	X	X
		B	X	X	X	X	O	X	X	X
		C	X	X	X	X	X	X	X	X

Notes: X is the ISO evaluation tests for consideration, O is the additional tests that may be applicable; A: 24 h, B: 24 h–30 days, C: >30 days.

TABLE 11.2

Supplementary Evaluation Tests for Consideration

Device Categories	Body Contact	Contact Duration	Biological Effects			
			Chronic Toxicity	Carcinogenicity	Reproductive Development	Biodegradable
Surface devices	Skin	A	—	—	—	—
		B	—	—	—	—
		C	—	—	—	—
	Mucosal membrane	A	—	—	—	—
		B	—	—	—	—
		C	O	—	—	—
	Breached or compromised surfaces	A	—	—	—	—
		B	—	—	—	—
		C	O	—	—	—
External communicating devices	Blood path, indirect	A	—	—	—	—
		B	—	—	—	—
		C	X	X	—	—
	Tissue/bone/dentin communicating	A	—	—	—	—
		B	—	—	—	—
		C	O	X	—	—
	Circulating blood	A	—	—	—	—
		B	—	—	—	—
		C	X	X	—	—
Implant devices	Tissue/bone	A	—	—	—	—
		B	—	—	—	—
		C	X	X	—	—
	Blood	A	—	—	—	—
		B	—	—	—	—
		C	X	X	—	—

Notes: X is the ISO evaluation tests for consideration, O is the additional tests that may be applicable; A: 24 h, B: 24 h–30 days, C: >30 days.

after June 14, 1998. The European Committee for Standardization (CEN) is currently in the process of adopting the ISO 10993 standard as the European standard. In 1986, the responsible authorities in the United Kingdom, United States, and Canada issued the Tripartite document, which was guidance on the selection of toxicological tests for medical device safety testing. This document has now been replaced by ISO 10993-1 as a first step in the process of international harmonization. In 1995, FDA chose to accept the ISO 10993-1 standard, with a modification of the matrix listing (see sidebar below). Japanese authorities have also issued a guideline for toxicological testing of medical devices. This document is available in an unofficial translation as Guidelines for Basic Biological Tests of Medical Materials and Devices. It resembles ISO 10993 in structure and content, but recommends modified tests and sample preparations.

The procedure for using the ISO 10993-1 standard is illustrated by the flowchart in Figure 11.1. The standard is applicable only for devices that are directly or indirectly in contact with the body or body fluids. If a device is to be subjected to the standard, the first step is to characterize the material. Such characterization need not always be followed by biological evaluation, because there may be sufficient historical data to verify that the device meets the requirements of the standard. If the material or the intended use of the device is different from any historical safe device, biological evaluation has to be performed. By following the standard, a suitable test program can be chosen

TABLE 11.3

Listing of Individual Parts of ISO 10993

Part	Title
1	Evaluation and testing
2	Animal welfare requirements
3	Tests for genotoxicity, carcinogenicity, and reproductive toxicity
4	Selection of tests for interactions with blood
5	Tests for cytotoxicity—in vitro methods
6	Tests for local effects after implantation
7	Ethylene oxide sterilization residuals
8	Clinical investigation of medical devices
9	Degradation of materials related to biological testing
10	Test for irritation and sensitization
11	Test for systemic toxicity
12	Sample preparation and reference material
13	Identification and quantification of degradation products from polymers
14	Identification and quantification of degradation products from ceramics
15	Identification and quantification of degradation products from coated and uncoated metals and alloys
16	Toxicokinetic study design for degradation products and leachables
17	Glutaraldehyde and formaldehyde residues in industrially sterilized medical devices

depending on the type and duration of body contact. Within the EU, all new medical devices must carry the CE mark from June 14, 1998. This should ensure the availability of relevant documentation regarding biocompatibility and the lack of health problems associated with the use of a device. It is noteworthy that the approval of such documentation is not, as it was previously, accorded by the national health authorities, but rather by the so-called notified bodies, whose experts review the products and production facilities of medical device manufacturers.

11.3   DEVICE CATEGORY AND CHOICE OF TEST PROGRAM

The need to evaluate a medical device biologically depends on the material used in the device, the intended body contact, and the duration of that contact. A device designed for surface contact for a limited time is not as likely to be bioincompatible as a permanent-exposure implant device made of the same material. The ISO 10993–1 standard divides medical devices into three main categories: surface devices, externally communicating devices, and implant devices. Each category is further divided into subcategories according to the type of contact to which the patient is exposed (see Table 11.4).

The ISO test matrix should not be considered as a checklist for the different tests that have to be performed, but rather as a guide for qualified toxicologists who also take into consideration material information and historical data from similar devices. The certifying authorities in most countries (e.g., notified bodies, FDA, Japanese authorities) are generally cooperative when a company must decide on a test program for a device. It is therefore advisable to maintain close contact with the relevant authorities during the entire process. However, testing should not be performed simply to meet regulatory requirements. This is important not only to lessen the risk of over testing and excessive use of experimental animals but also because a strict regulatory approach may mask potential negative health effects that might be identified via optional or nonroutine testing procedures.

The choice of test program for a device in a given category depends on the duration of the contact. Three different time spans are given: limited contact (<24 h), prolonged contact

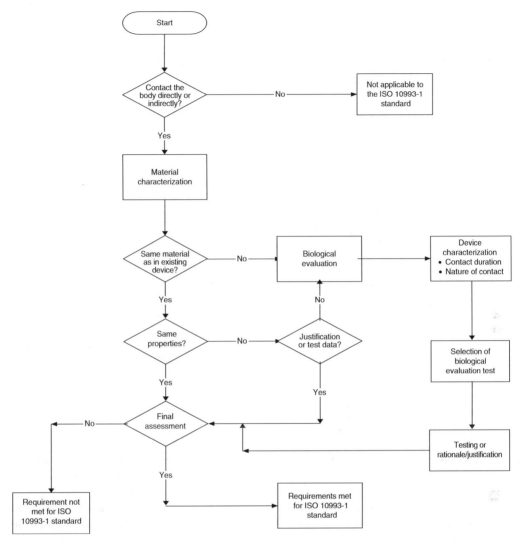

FIGURE 11.1   Steps in the biological evaluation of medical devices.

(24 h–30 days), and permanent contact (>30 days). ISO 10993-1 lists the tests that must be considered for each category.

Regarding CE marking of existing products on the market or safety evaluation of medical devices already in clinical use, appropriate historical or clinical data should be employed whenever possible to avoid unnecessary testing.

11.4   PREPARATION OF EXTRACTS

ISO 10993-12 describes how samples for biological evaluation should be selected, prepared, and extracted. Other guidelines provide similar descriptions, which differ slightly in the specifics of the extraction procedures.

The device to be tested (the test article) should be a representative specimen of the mass-produced device. It should also be finished or treated (e.g., coated or sterilized) in the same way as the mass-produced device.

TABLE 11.4

Device Categories and Examples According to ISO 10993–1

Device Categories		Examples
Surface devices	Skin	Electrodes, external prostheses, fixation tapes, compression bandages, monitors of various types
	Mucous membrane	Contact lenses, urinary catheters, intravaginal and intraintestinal devices, endotracheal tubes, bronchoscopes, dental prostheses, orthodontic devices
	Breached or compromised surfaces	Ulcer, burn, and granulation tissue dressings or healing devices, occlusive patches
Externally communicating devices	Blood path indirect	Solution administration sets, extension sets, transfer sets, blood administration sets
	Tissue/bone/dentin communicating	Laparoscopes, arthroscopes, draining systems, dental cements, dental filling materials, skin staples
	Circulating blood	Intravascular catheters, temporary pacemaker electrodes, oxygenators, extracorporeal oxygenator tubing and accessories, dialyzers, dialysis tubing and accessories, hemoadsorbents, and immunoadsorbents
	Tissue/bone implant devices	Orthopedic pins, plates, replacement joints, bone prostheses, cements and intraosseous devices, pacemakers, drug-supply devices, neuromuscular sensors and simulators, replacement tendons, breast implants, artificial larynxes, subperiosteal implants, ligation clips
	Blood	Pacemaker electrodes, artificial arteriovenous fistulae, heart valves, vascular grafts, internal drug-delivery catheters, ventricular-assist devices

Because the toxic potential of materials and devices depends to a substantial degree on the leachability and toxicity of soluble components, extracts of the device are normally used in the tests. In some tests, however, an evaluation under normal-use conditions is mimicked by using the device or a piece of the device directly. Ideally, extraction media should constitute a series of media with decreasing polarity to ensure the extraction of components of widely different solubility properties. The most commonly used extraction media are physiological saline, vegetable oil, dimethylsulfoxide, and ethanol. Other extraction media such as polyethylene glycol or aqueous dilutions of ethanol may be selected in certain cases. For in vitro cytotoxicity testing, complete cell-culture medium is most often employed. The various guidelines also differ somewhat with respect to the temperature at which the extraction is conducted. Some leachable compounds may be chemically altered at high temperatures, and it is now generally recommended that extraction be conducted at 37°C—simulating body temperature—for 72 h. This procedure will probably become increasingly accepted as the most appropriate extraction method. For in vitro cytotoxicity tests, extraction at 37°C for 24 h is usually recommended, since certain constituents of the media are relatively labile.

The amount of leachable substances released to the extraction media is related to the surface area and thickness of the product to be extracted. Recommendations vary from 1.25 to 6 cm^2 of product per milliliter of extraction medium, depending on the size and shape of the product, or from 0.1 to 0.2 g of product per milliliter of extraction medium when a surface area cannot readily be estimated (e.g., for powders or granulates). In any case, the specific properties of the product must be taken into account to make usable extracts.

For cases in which a medical device comprises several components made from different materials, the ideal procedure from a toxicological point of view would be to test extracts of the components separately. However, in some situations this is not practical, and extracts of the whole device may be used instead.

11.5   BIOLOGICAL CONTROL TESTS

Biological control tests are not described in the ISO 10993 standard for biological evaluation of medical devices because these particular tests are designed primarily for batch-control purposes. Such tests are also used during the product development phase to identify sources of contamination and to establish procedures that ensure the intended quality of the end-product.

11.5.1   MICROBIOLOGICAL CONTROL TESTS

Microbiological control tests are necessary to establish the microbiological status of an end-product—factors such as sterility, absence of pathological bacteria, or limits for microbial counts. Furthermore, it is often necessary to monitor the microbiological load of raw materials and intermediary products, or to check the efficiency of production and sterilization processes. The tests are performed by rinsing the materials or products in physiological saline and assessing the rinsing medium for microbes, or by directly incubating the products in growth media.

11.5.2   TESTS FOR ENDOTOXINS

Even sterile medical devices may contain cell wall lipopolysaccharides originating from gram-negative bacteria. Such so-called endotoxins or pyrogens can cause an abrupt fever reaction after entering directly into the body from sources such as venous catheters, syringes, or implant components. Two different biological assays can be used to measure the presence of endotoxins: the rabbit pyrogen test and the limulus test. In both the cases, an eluate is prepared—normally by rinsing the surfaces of the product with water—and then tested for endotoxins. In the rabbit pyrogen test, the eluate is injected intravenously and the rectal temperature of the animal is measured after the injection. In the limulus test, the eluate is incubated together with lysate from the blood of the horseshoe crab (*Limulus polyphemus*), which contains a substance that forms a gel in the presence of endotoxins.

11.5.3   TEST FOR NONSPECIFIC TOXICITY

This test is designed to assess any nonspecific adverse effect that occurs following intravenous injection of a device eluate in mice. The test is often performed with the same eluate used for the pyrogen test. The mice are inspected regularly for any signs of ill health, which can indicate the presence of toxic substances leaching from the product.

11.6   TESTS FOR BIOLOGICAL EVALUATION

This section provides a brief description of the individual tests included in the ISO 10993/EN 30993 standard.

11.6.1   CYTOTOXICITY

The aim of in vitro cytotoxicity tests is to detect the potential ability of a device to induce sublethal or lethal effects as observed at the cellular level. According to ISO 10993-1, the in vitro cytotoxicity assay is one of two tests; the other is the sensitization test described in Section 11.6.2, which must be considered two in the evaluation of all device categories.

Three main types of cell-culture assays have been developed:

1. Elution test
2. Direct-contact test
3. Agar diffusion test

In the elution test, an extract (eluate) of the material is prepared and added in varied concentrations to the cell cultures. Growth inhibition is a widely used parameter, but others may also be used. In the direct-contact test, pieces of test material are placed directly on top of the cell layer, which is covered only by a layer of liquid cell-culture medium. Toxic substances leaching from the test material may depress the growth rate of the cells or damage them in various ways. In the agar diffusion test, a piece of test material is placed on an agar layer covering a confluent monolayer of cells. Toxic substances leaching from the material diffuse through the thin agar layer and kill or disrupt adjacent cells in the monolayer. As always, the physical and chemical properties of the test material should be considered before the choice of the test system is made.

There is usually a good qualitative correlation between results from cell-culture tests and studies performed in vivo with respect to cytotoxicity versus primary tissue effects. It is important to recognize, however, that although cell-culture toxicity is in general a good and sensitive indicator of primary tissue compatibility, exceptions may arise in cases where leaching substances cause tissue damage in vivo through more complex mechanisms. At present, the in vitro cytotoxicity assays should be used as screening tests and considered primarily as supplements to the various in vivo tests.

11.6.2 SENSITIZATION

The sensitization test recognizes a potential sensitization reaction induced by a device, and is required by the ISO 10993-1 standard for all device categories. The sensitization reaction is also known as allergic contact dermatitis, which is an immunologically mediated cutaneous reaction. This is in contrast to irritant contact dermatitis (skin irritation)—a skin reaction caused by the primary and direct effect of a substance on the skin. In animals, the sensitization reactions manifest themselves as redness (erythema) and swelling (edema).

The preferred animal species for sensitization testing is the albino guinea pig. There is no reliable alternative in vitro test that can predict the sensitizing potential of a substance. The various available guinea pig methods have certain features in common: an induction (sensitization) phase, when the potential allergen is presented to the organism, followed by a rest period and a subsequent challenge phase to determine whether or not sensitization has occurred.

One of the most recognized and validated assays is the guinea pig maximization test (GPMT). A test design very similar to the GPMT is widely used for assessing the sensitizing potential of medical devices. After a challenge period, the skin reactions are graded on a ranking scale according to the degree of erythema and edema.

Predictive tests in guinea pigs are important tools in identifying the possible hazard to a population repeatedly exposed to a substance. Nevertheless, results from sensitization tests in guinea pigs have to be evaluated carefully. A positive test result in this assay may rate a substance as a stronger sensitizer than it appears to be during actual use. On the other hand, a negative result in such a sensitive assay ensures a considerable safety margin regarding the potential risk to humans.

11.6.3 SKIN IRRITATION

The ISO 10993-10 standard describes skin-irritation tests for both single and cumulative exposure to a device. The preferred animal species is the albino rabbit, whose highly sensitive, light skin makes it possible to detect even very slight skin irritation caused by a substance. Skin-irritation tests of medical devices are performed either with two extracts obtained with polar and nonpolar solvents or with the device itself.

In the single-exposure test, rabbits are treated for several hours only, whereas for the cumulative test the same procedure is repeated for several days. All extracts and extractants are applied to intact skin sites. Skin reaction is seen as redness or swelling and is graded according to a specified classification system. Dermal irritation is the production of reversible changes in the skin following the application of a substance, whereas dermal corrosion is the production of irreversible tissue damage (scar formation) in the skin. Materials that leak corrosive substances are not likely candidates for medical device production.

11.6.4 INTRACUTANEOUS REACTIVITY

The intracutaneous reactivity test is designed to assess the localized reaction of tissue to leachable substances. The test is required for consideration in nearly all the device categories in ISO 10993-1 (see Table 11.3). Polar and nonpolar solvent extracts are administered as intracutaneous injections to rabbits. Undesirable intracutaneous reactivity includes redness or swelling.

11.6.5 ACUTE SYSTEMIC TOXICITY

Acute systemic toxicity is the adverse effect occurring within a short time after administration of a single dose of a substance. ISO 10993-1 requires that the test for acute systemic toxicity be considered for all device categories that indicate blood contact. For this test, extracts of medical devices are usually administered intravenously or intraperitoneally in rabbits or mice.

Determining acute systemic toxicity is usually an initial step in the assessment and evaluation of the toxic characteristics of a substance. By providing information on health hazards likely to arise from short-term exposure, the acute systemic toxicity test can serve as a first step in the establishment of a dosage regimen in subchronic and other studies, and can also supply initial data on the mode of toxic action of a substance. The test is similar to the nonspecific toxicity test. Normally, only one of these two procedures is included in a test battery.

11.6.6 GENOTOXICITY

Genetic toxicology tests are used to investigate materials for possible mutagenic effects, that is, damage to the body's genes or chromosomes. The tests are performed both in vitro and in vivo. ISO 10993-1 requires the genotoxicity (mutagenicity) test to be considered for all device categories indicating permanent (>30 days) body contact (except for surface devices with skin contact only).

A mutation is a change in the formation content of the genetic material (DNA code) that is propagated through subsequent generations of cells. Mutations can be classified into two general types: (1) gene mutations and (2) chromosomal mutations.

Gene mutations are changes in nucleotide sequences at one or several coding segments within a gene; chromosomal mutations are morphological alterations or aberrations in the gross structure of the chromosomes.

The simplest and most sensitive assays for detecting induced gene mutations are those using bacteria. Gene mutations can also be detected in cultured mammalian cells. Current in vivo assays for gene mutations are cumbersome and not widely used. The simplest and most sensitive assays for investigating chromosomal aberrations are those that use cultured mammalian cells. However, two well-established in vivo procedures are also available: chromosomal aberrations can be studied in bone marrow or peripheral blood cells of rodents dosed with a suspect chemical or extract either by counting micronuclei in maturing erythrocytes (micronucleus test) or by analyzing chromosomes in metaphase cells.

In addition to these mutagenicity tests, various assays can measure the induction of an overall genotoxic response—an indirect indicator of potential damage to the genetic material.

11.6.7 IMPLANTATION

Implantation tests are designed to assess any localized effects of a device designed to be used inside the human body. Implantation testing methods essentially attempt to imitate the intended use conditions of an implanted material. Although different tests use various animal species, the rabbit has become the species of choice, with implantation performed in the paravertebral muscle. Implantation can be either surgical or nonsurgical: the surgical method involves the creation of a pouch in the muscle into which the implant is placed, while the nonsurgical method uses a cannula and stylet to insert a cylinder-shaped implant. Through a macroscopic examination (which may be supplemented with microscopic analysis), the degree of tissue reaction in the paravertebral muscle is evaluated as a measure of biocompatibility.

11.6.8 HEMOCOMPATIBILITY

The purpose of hemocompatibility testing is to look for possible undesirable changes in the blood caused directly by a medical device or by chemicals leaching from a device. Undesirable effects of device materials on the blood may include hemolysis, thrombus formation, alterations in coagulation parameters, and immunological changes. According to the ISO 10993-4 (EN 30993-4) standard, devices that only come into very brief contact with circulating blood—for example, lancets, hypodermic needles, or capillary tubes—generally do not require blood/device interaction testing.

ISO 10993–4 describes hemocompatibility tests in five categories:

1. Thrombosis
2. Coagulation
3. Platelets
4. Hematology
5. Immunology

Most of the individual tests are not discussed in detail, but they may be performed either in vivo or, preferably, in vitro. There is still some uncertainty with respect to what is actually required by the regulatory authorities for the hemocompatibility test.

11.6.9 SUBCHRONIC AND CHRONIC TOXICITY

Subchronic toxicity is the potentially adverse effect that can occur as a result of the repeated daily dosing of a substance to experimental animals over a portion of their life span. In the assessment and evaluation of the toxic characteristics of a chemical, the determination of subchronic toxicity is carried out after initial information on toxicity has been obtained by acute testing, and provides data on possible health hazards likely to arise from repeated exposures over a limited time. Such testing can furnish information on target organs and the possibilities of toxin accumulation, and provide an estimate of a no-effect exposure level that can be used to select dose levels for chronic studies and establish safety criteria for human exposure.

In subchronic or chronic toxicity studies, one or two animal species are dosed daily, usually for a period of 3–6 months; the rat is the standard animal species of choice. The animals are given the test substance in increasing doses. The dose level of the low-dose group should be at the level of human exposure. When extracts of medical devices are employed, one dose level (the highest practically applicable volume) is often sufficient, since strong toxicity is generally not expected.

11.6.10 CARCINOGENICITY

The objective of long-term carcinogenicity studies is to observe test animals over a major portion of their life span to detect any development of neoplastic lesions (tumor induction) during or after exposure to various doses of a test substance. Carcinogenicity testing is normally conducted with oral dosing. For implants and medical devices, however, only extracts can be tested and they must be administered intravenously, necessitating certain modifications of the standard procedure. There are only a very few products for which this comprehensive test can be justified.

In carcinogenicity studies, mice or rats are dosed every day for 18–24 months. For medical device extracts, one dose level (again the highest practically applicable volume) is usually sufficient. At the completion of the dosing period, all surviving animals are sacrificed and their organs and tissues examined microscopically for the presence of tumors. An increased incidence of one or more category of tumors in the dosed group would indicate that the product tested has the potential to induce tumors and could be considered a possible carcinogen in humans.

11.7 ALTERNATIVE TEST METHODS

As mentioned previously, a major goal in international toxicological testing is to reduce not only the use of in vivo studies but also the number of animals employed in these tests. A few of the in vivo procedures used today for testing medical devices may be of questionable worth for safety evaluation. However, the availability of accepted and validated in vitro assays is still limited. Substantial resources have been made available for validation of alternative in vitro assays in toxicology as replacements for animal tests, but it may take years before validated methods can be implemented and any goal of replacing all in vivo studies with in vitro assays will probably never be met.

Recently, a working group under the auspices of the European Center for Validation of Alternative Methods (ECVAM) has recommended a few alternative methods that can be used for safer testing of medical devices. These include two in vitro tests as potential substitutes for the in vivo assays for skin and eye irritation. However, the implementation of validated protocols and internationally accepted guidelines for these tests is likely to be delayed into the next century.

11.8 ENDNOTE

There have been some disasters involving biomaterials. These disasters have emphasized the need for diligence in testing of biomaterials. These include toxic shock syndrome, latex allergies, the use of talc on gloves, and perhaps reactions to silicon gel leakage from breast implants. Continuing diligence, especially when new substances are being tested, is mandated by law.

EXERCISES

1. You are charged with developing the coating material for an implantable brain stimulator for reduction of tremor due to Parkinson's disease. You may begin with a Web search to determine what materials are currently in use, if any. What materials will you consider and what tests will need to be run? Refer to Figure 11.1.
2. You are interested in building an inexpensive EKG transmitter for implantation in mice. Do a literature (or Web search) to determine a list of acceptable coatings. Which would you use if the experiment were to only last one day? Which if the work was to continue for a month? Why?
3. Do a literature search to determine the history of implant materials. What are some of the earliest signs that the human body accepted a foreign object?
4. How old is the history of implantation of materials into human teeth? Why was this done?
5. Do a Web search using the term "biocompatibility testing." Categorize the first several hits as to their relevance to this chapter. Do a similar search using the term "animal care and use form."
6. Why are rabbits so often used for pyrogen testing? What is unique about rabbits?
7. Horseshoe crab is of value in compatibility testing. What is special about this arthropod?
8. Use of earrings and other body piercing adornments has been linked to an increase in one of the hepatitis strains in the users. Do a Web or literature search to deny or defend this statement.

9. In postwar Germany an operation was performed on amputees called a cineplasty, wherein a carbon-coated rod was passed through the muscle above the amputation. With time, the tunnel often grew skin on its surface and the subject could use his or her remaining muscle to move prostheses, such as a primitive grasper hand. Research this history and speculate on what might ensue if there had been no long-term problems.

10. Research on of the four problems referenced in Section 11.8. What was involved in the problem and what was the outcome?

BIBLIOGRAPHY

Biological evaluation of medical devices, ISO 10993 Standard Series, Geneva, International Organization for Standardization, ongoing.

Bollen, L.S. and Svendsen, O., Regulatory guidelines for biocompatibility safety testing, *Medical Plastics and Biomaterials*, 4(3), May 1997, 16–43.

Council Directive 93/42/EEC of 14 June 1993 concerning medical devices, *Official Journal of the European Communities*, 36, July 1993.

Gad, S.C., *Safety Evaluation of Medical Devices*, New York: Marcel Dekker, 1997.

Guidelines for basic biological tests of medical materials and devices. Unofficial translation of Japanese guideline ISBN 4-8408-0392-7.

Hill, D., *Design Engineering of Biomaterials for Medical Devices*, John Wiley & Sons, Chichester, U.K., 1998.

Toxicology Subgroup, Tripartite Subcommittee on Medical Devices, Tripartite Biocompatibility Guidance for medical devices, Rockville, MD, FDA, Center for Devices and Radiological Health (CDRH), 1986.

Ratner, B.D. et al., *Biomaterials Science*, San Diego, CA: Academic Press, 1996, p. 11.

Svendsen, O. et al., Alternatives to the animal testing of medical devices (the report and recommendations of ECVAM workshop 17 to be published in ATLA, in press, 1996). Available at http://altweb.jhsph.edu/publications/ECVAM/ecvam17.htm, last accessed May 12, 2008.

12 Safety Engineering: Devices and Processes

Early and provident fear is the mother of safety.

Edmund Burke

The Accreditation Board for Engineering and Technology (ABET) requirements for design state that "Students must be prepared for engineering practice through the curriculum culminating in a major design experience based upon the knowledge and skills acquired in earlier coursework and incorporating engineering standards and realistic constraints that include most of the following considerations: economic; environmental; sustainability; manufacturability; ethical; health and safety; social; and political."* That biomedical engineering design work would involve health aspects is obvious. Several aspects involving safety and the potential for liability requires a discussion of the need for safety consideration in the design and redesign of medical devices, and in somewhat similar activities in process design.

Recent publication of National Academy Press, *To Err is Human: Building a Safer Health System*,[†] has provided several notable statistics, specifically "The human cost of medical errors is high. Based on the findings of one major study, medical errors kill some 44,000 people in U.S. hospitals each year. Another study puts the number much higher, at 98,000. Even using the lower estimate, more people die from medical mistakes each year than from highway accidents, breast cancer, or AIDS." This statistic permits comparing medical error deaths to the number of deaths due to other accidental causes, such as may be found from the National Safety Council (NSC) Web site (www.nsc.org), and to the number of deaths due to specific diseases. The safety council data estimates 150,445 total deaths due to injuries in 1998, a statistic on par with the above. The magnitude of the numbers above should impress one as to the need for safety considerations, not just for public activities, but also for medical activities, such as is involved in the design of medical devices and systems.

12.1 MEDICAL CASE EXAMPLE

Let us illustrate the above discussion with an example based upon an actual event. A young Down's syndrome patient with multiple heart defects died from an air embolism during a preoperative cardiac flow/oxygenation catheterization study. Evidence acquired from the hospital involved the following data and devices:

- Medical record for the patient.
- Testimony regarding the procedure.
- Complete records of blood pressure and EKG as various sites were checked for pressures and sampled for oxygen saturation levels, blood pressure was sampled periodically using the system below.

---

* From the ABET website, www.abet.org.
[†] *To Err is Human: Building a Safer Health System*, National Academy Press, 2000.

- Catheterization system, which included a three port connection system (manifold) for blood pressure determination, saline infusion, and blood sampling/injection via a syringe.
- Typical saline bag and connector assembly.
- Opaque pressurization jacket used to pressurize the saline bag to ensure saline flow when the saline port was opened.

There are two items to be determined at this point. It is given that the patient died of an air embolus. How did this occur and what could have been done to prevent this are two of the questions to be asked. Let us ask the second first, discussing some of the general procedures that must be used to analyze safe devices and procedures.

12.2 SAFETY IN DESIGN

Good design practices should consider means by which a given design may cause harm, and should—via guide sentences, structure, or checklists—assist the designer in determination of improvements to the system under study. Let us illustrate this with another example: Drink machines have been known to tip over and kill or maim persons shaking them when irate over nondelivery of drinks (a few deaths per year in the United States). How can this be prevented?

A quick checklist can be found that helps one begin the solution to this problem. This checklist can read as the following:*

- Eliminate by design
- Guard against
- Warn the user
- Train the user
- Mandate the use of personal protective equipment
- Others

As a drink machine manufacturer, what solutions are possible here? Mandating that the user wears protective equipment is not likely, nor is training the user. One might warn the user to not tilt the drink machine to get a drink, but it is not likely that you would win a case involving the death of a user due to your machine. Guarding against the machine tilting would seem to be a better approach; strapping the machine to a nearby wall should enable this outcome. A far better approach would be to eliminate the problem by design by placing the weight of the unsold drinks at the base of the machine, rather than at the top (which enables gravity feed of the drinks and thus a cheaper design).

The above checklist is simply an outline; in practice each of the subheadings can have various gradations. For example, the warning of the user can be visual or audible, color coded, flashing, etc.

Implied in the above discussion are a few other concepts that are mandatory to understand if one is to analyze unsafe designs. One is the term hazard, which may be defined as a source of potential harm or a situation with a potential for harm. Another is risk, which is a combination of the probability of occurrence of harm and the severity of that harm. If you as the manufacturer of the machine above decide not to redesign anything, you are apparently assuming that your risk of financial harm due to a lawsuit is less than your cost to prevent the problem in the first place. Good design practices include hazard analysis and risk assessment at every stage of the design, with an ultimate goal of risk and potential liability reduction. As safe design is mandated for medical devices by the Food and Drug Administration (FDA), this practice must be documented as a device is developed. As the variety of users in the medical environment varies in terms of education and responsibilities and other tasks, good design must involve a fairly comprehensive list of items.

---

* From the program "designsafe," from Designsafe Engineering.

Let us once again use the drink machine example above to look at the process of safe design. A typical approach to an analysis could include the following steps:

- Identify users (e.g., drink installer, general public).
- Identify hazards each users may be subjected to, this hazard list is associated with a checklist such as mechanical hazards, chemical hazards, and health hazards (mechanical problems would be paramount here).
- Begin the risk assessment, using a guide sentence such as "when doing the (task) the hazard may cause (harm)." One of the guide sentences here would be "When shaking the machine it could tip and crush the user."
- Identify the severity of the harm (catastrophic, serious, slight, and minimal), the exposure to harm (frequent, occasional, remote, and none), and the probability of harm (probable, possible, unlikely, and negligible) and therefore the risk level (high, moderate, or low). A high-risk level implies the outcome of severe to moderate injury or death, moderate implies moderate to low probability of harm, and low implies moderate to mild injury. The risk of death from an unsecured drink machine falling on one is high, the exposure is remote (it does happen!); the probability of it occurring with an untethered machine is high.
- Complete analysis would then involve identification of methods to reduce the risk (such as guarding), the revised exposure and risk data, and the personnel in charge of this activity.

Special attention should be given to situations where the device may be misused to "cover all bases" in an analysis. Additionally, especially in the case of devices used in a clinical environment, special consideration should be given to not only the primary users of the device, but also to casual users (cleaning crew) and special needs patients (elderly, very young, very ill, AIDS, at-risk, etc.). Human factors analyses should also be considered, especially in light of situations where there might be high stress on the part of the user (see Chapter 9).

12.3   MEDICAL CASE EXAMPLE—REVISITED

Let us revisit the case mentioned above regarding the death of a young Down's syndrome child with multiple heart defects. What considerations should there be in this case, regarding instrumentation, risk, and fault? Below are a few points to consider when looking at this case:

- Patient was described as young. A small amount of air can cause death from air embolism in young patients as compared to adults—children are more at risk for this problem compared to adults.
- Down's syndrome children often have heart problems; this particular patient was described as having multiple heart defects. Heart shunts predispose patients to risks from air emboli; this patient had shunts (the reason for the study and determination of oxygenation levels).
- Diagnosis was that the patient died of an air embolus. Where did the air come from?

The solution to this case (as a legal case) lies in determining how the air got into the patient. How can this be used to redesign the process or the mentioned devices to ensure that this does not happen again? The answer lies in the application of good safety engineering principles.

The particular hazard addressed in this discussion is one of the several that accompany this type of diagnostic workup, but let us address only one concern (the air). The hazard to be investigated is a dramatic decrease in the ability of the heart to pump blood due to the inadvertent introduction of air into the patient. How could air get into the patient? A few of the ways include

1. Air entering the patient through the catheter insertion point in the groin
2. Air entering the patient via a medication line

3. Accidental opening of the blood pressure sensor port, suction of air into the patient
4. Flushing of the blood pressure sensor port with air, rather than saline
5. Accidental infusion of air from the surgeon's sampling syringe, rather than sampling of blood at this site
6. Infusion of air from the saline drip bag

Five of these were eliminated very quickly based upon the medical records in the case. The patient was supine, thus the risk of Case 1 was minimal. Even if air had been inserted at this point, the likelihood of damage was minimal. Medication lines were patent; there was no evidence of air in them. Suction of air into the patient due to an opened pressure sensor line was unlikely, the pressures recorded in the patient at all sites were positive with respect to atmospheric, with only a few milliseconds per beat occasionally becoming subatmospheric (insufficient time to cause air to enter). Case 4 is not likely, had it occurred only a syringe full of air could have been injected (10 mL or so), it is not likely that the air could have made it to the patient. Case 5 is unlikely also; again the dose of air would have been 10 mL maximum (this has happened). Implicated air from the infusion bag, due to the pressurizing jacket entered the patient. For the particular 1 L bags in use, there is about 35 mL of air, an amount adequate to cause death in a young at-risk patient. The pressurization bag ensured that the air was indeed pumped to the patient. A simulation of the situation showed that this could occur in less than 5 s at the measured pressures involved. Too quick if one was otherwise involved with measuring data from the patient!

Why is there air in the infusion bag? To ensure that when drugs are injected into the bag, mixing can occur by shaking the bag. What would have been a safe design for this situation? Simply elect not to use a pressurization jacket.

A counter-argument may occur here, once one realizes that the air in such a bag ensures that drugs, when injected into the bag through one of the ports, cannot be properly mixed unless there is air in the bag (and the bag is shaken properly.)

12.4 PROCESS IMPROVEMENT

The prior discussion on air embolism was successful in determining the cause of the death by asking the question "How did the air get into the patient?" There are several other methods the reader will find useful in determining fault or cause of an untoward event. One such method simply uses the question "why" enough times that the questioner, assuming that there are answers to each of the more in-depth "why?" questions, finally gets to the root cause of an event.

Most hospitals have a safety process in place that looks for methods to improve the processes of health care delivery. This group may operate with the name of quality improvement, quality assurance, patient safety committee, or the like. Some of the processes involved in their work involve the types of analyses just discussed, some of their work involves flowcharting (see Chapter 2), and some involves the use of cause-and-effect diagrams. Figure 12.1 illustrates the use of this concept using an Ishikawa, or fishbone diagram. This particular diagram was generated to look at the process for "bad infusion outcomes" to assist in identifying the potential cause of this outcome.* In a major brainstorming session, the chart was first generated by considering the major items involved in the process of generating an infusion process, namely people who give the infusion, written policies regarding protocol for this infusion, the patients involved, and the equipment involved in the process. Each bone coming off of these four main "bones" relates to some attribute of the main section that might have an influence on outcomes. For example, a root cause of IV-related complications might be that the people giving the IV may be under experienced. Similarly, the antibiotic that the patient may be taking may be interfering with the IV, etc.

* Provided by Dr. Doris Quinn, Nashville, Tennesse: Vanderbilt Quality Assurance Department.

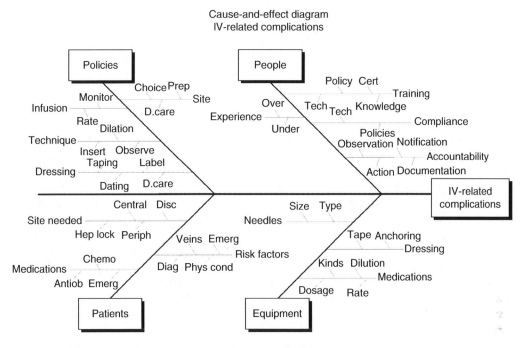

FIGURE 12.1   Causes and effect diagram, IV-related complications.

Once such a chart is completed, one can study each of the potential root causes of the outcome and try to determine which one (or ones) might have caused the outcome. Of interest is the fact that, while the administration was asking for additional in-service training for personnel doing infusions (to counter an under experienced person), this particular chart was of value in identifying the real cause of the problem, a change in the supplied concentration of the drug in question.

This technique is useful for analysis of problems in multiple areas, such as processes (here), manufacturing, management, and services.

12.5   MISCELLANEOUS ISSUES

Major design problems in some devices and drugs have resulted from drug interactions and materials failures. The need for both animal and human testing for drug interactions and possible materials testing for implanted materials serves as a beginning point for discussions of these topics. Much of this testing is mandated by the FDA in terms of required test protocols. Specifically, the drug thalidomide and some of the early experiences with heart valves deserve mention in discussions on historical problems, many of which are covered in the references cited here.[*-‡] A discussion of patent medicines and quack medical devices with their inherent risks to human safety are addressed in the following references.[*,†]

12.6   OTHER PROCESS ISSUES

Many of the mistakes made in the hospital environment are due to communication issues. Redesign of hospital systems should pay special attention to the interfaces between people, with an emphasis

---

* http://www.cyberus.ca/~sjordan/pmmain.htm.

† http://www.mtn.org/quack.

‡ *Crossing the Quality Chasm*, Institute of Medicine Report, 2001, pp. 5–6.

on correct communication. To this end, computerization of drug and medication dosing should be stressed, avoiding all oral transfer of information if at all possible. Double-checking of doses and allergies to medicines can help alleviate many medication errors. Double-checking of drug inter-actions via computer, and alerting of the health care provider to this possibility, is of value. With an average of 18 or so medications per patient in a large hospital, this is of high importance. Both private and governmental (e.g., the VA and Department of Defense) agencies are pursuing elec-tronic medical records. Medicines and most equipment will likely be bar coded in the near future to enable accurate input of this information without keystroke errors.

Governmental influences will also have an effect on improving health care. A long overdue policy on the part of the FDA is the ending of sound-alike and look-alike medication names by fiat. The U.S. government, through the Department of Health and Human Services, has established the Agency for Health care Research and Quality (AHRQ) (www.ahrq.gov) which is funding several initiatives on improving the quality of health care. A specific aim includes the use of evidence-based decision-making. The AHRQ will be the lead agency in setting up a system for reporting of medical errors and the analysis thereof. The AHRQ has set up a system for classifying and counting patient safety incidents. Such incidents include the terms failure to rescue, decubitus ulcer, and postoperative sepsis. The use of such indicators on data derived from patient safety reporting from organizations has led to a system to rank and evaluate and recommend changes in techniques among various hospitals and practitioners. The AHRQ is a research-based organization, and as such will affect medical care based upon the quality of the evidence-based research it performs and sponsors.

Figure 12.2 is an example of the type of data that might be presented in making a case for specialized studies sponsored by the AHRQ. Presented here is a histogram of patient incidents (fabricated data) versus incident type. The particular ordering, of incidents of high value to low value, allows one to study the impact (lost lives, lost incomes, etc.) of solving one or more problems. Histograms arranged in this particular order, sometimes with an overlaid percentage of total line (left to right, with 0%–100% scale on the right) are termed Pareto Charts.

The Joint Commission (once the Joint Commission on the Accreditation of Health Care Organizations) has sponsored a National Patient Safety Goals and Requirements Program since 2003. Specific goals include items such as increasing the accuracy of patient identification, recognition, and response to changes in a patient's condition, health care worker fatigue, and its potential effect on care. As this is the major accreditation agency in the United States for health care

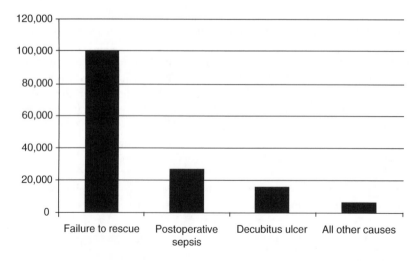

FIGURE 12.2   Histogram of safety incidents.

organizations and programs, this group has a strong voice in standards setting and potential improvements.

The National Institute for Occupational Safety and Health (NIOSH) sponsored a Prevention through Design workshop in 2007, with the aim of setting up a national group to look at the designing out of possibilities for injury via better product and process design. This group also has a health care contingent.

Private foundations also interested in improving patient safety include the National Patent Safety Foundation (www.npsf.org) and the Robert Wood Johnson Foundation. One of the interesting outcomes of the Johnson Foundation group (and others) is the LeapFrog Group, which calls for improved health care decisions and the use of incentives and rewards for providers.

12.7 SUMMARY

Designing with safety in mind is mandatory. Appropriate design for safe and effective devices and processes may cost more in the initial design, but should pay continuing benefits in the long run. Health care should be safe, equitable, effective, patient centered, timely, efficient, and equitable.* Note that the first term is safe.

EXERCISES

1. Visit a site such as www.designsafe.com; download the demo version of the software. Perform an analysis of your design project or one of your instructor's choice.
2. Visit the FDA manufacturers and users device experience (MAUDE) site, do a search for any device that caused a death in the past two months. Perform a safety analysis on this device.
3. Visit the FDA MAUDE site, do a search for any device that caused an accidental injury in the past few months. Discuss the harm caused and the possible correction of this problem.
4. There are several variations on safety analyses. Define and discuss failure modes and effects analysis (FMEA) and its applications to medicine.
5. Do a search for the term "anticipatory failure determination" and report on the value of this type of software.
6. Develop a cause-and-effect diagram for the air embolus case discussed in this chapter.
7. Search for information on thalidomide; discuss what went wrong with the use of this drug. Find and discuss at least one other drug example.
8. Some of the early heart valves had mechanical problems. Discuss how this is not a likely event today.
9. Search for information on the failure rate of implant pumps for the alleviation of male erectile dysfunction. What design problems occurred in these devices?
10. Major accidents sometimes cause a rethinking of basic procedures. Investigate the major accident at Bhopal and some of the recommendations that arose from this event.
11. Find and describe the specific wording that requires safety in medical devices, both for the FDA and for CE marking.
12. What is the value of ASA and surgical risk classification schemes?
13. Read and report on one relevant chapter from Geddes (Medical Device Accidents with Illustrative Cases or Casey (Set Phasers on Stun.)
14. Do a search and report on the term "inherently safer design."
15. Discuss the necessary components of a system to guarantee a proper patient-blood transfusion match.

---

* *Crossing the Quality Chasm*, Institute of Medicine Report, 2001, pp. 5–6.

13 Testing

Building technical systems involves a lot of hard work and specialized knowledge: languages and protocols, coding and debugging, testing and refactoring.

Jesse J. Garrett

Testing may be defined as subjecting a device to conditions that indicate its weaknesses, behavior characteristics, and modes of failure. It is a continuous operation throughout the development cycle that provides pertinent information to the development team. Testing may be performed for three basic reasons:

- Basic information
- Verification
- Validation

Basic information testing may include vendor evaluation, vendor comparison, and component limitability. Verification is the process of evaluating the products of a given phase to ensure correctness and consistency with respect to the products and standards provided as input to that phase. Validation includes proving the subsystems and the system meet the requirements of the product specification.

Testing is an essential part of any engineering development program. If the development risks are high, the test program becomes a major component of the overall development effort. To provide the basis for a properly integrated development test program, the design specification should cover all criteria to be tested including function, environment, reliability, and safety. The test program should be drawn up to cover assurance of all these design criteria.

The ultimate goal of testing is assuring that the customer is satisfied. It is the customer who pays the bills, and if we are to be successful in business, we have to solve their problems. We aim for quality, but quality is not just an abstract ideal. We are developing systems to be used, and used successfully, not to be admired on the shelf. If quality is to be a meaningful and useful goal in the real world, it must include the customer.

13.1 TESTING DEFINED

Definitions matter, although consensus as to what testing really is less important than being able to use these definitions to focus our attention on the things that should happen when we are testing. Historically, testing has been defined in several ways:

- Establishing confidence that a device does what it is supposed to do.
- Process of operating a device with the intent of finding errors.
- Detecting specification errors and deviations from the specification.
- Verifying that a system satisfies its specified requirements or identifying differences between expected and actual results.
- Process of operating a device or component under specified conditions, observing or recording the results, and making an evaluation of some aspect of the system or component.

All these definitions are useful, but in different ways. Some focus on what is done while testing, others focus on more general objectives such as assessing quality and customer satisfaction, while others focus on goals like expected results. If customer satisfaction is a goal, this satisfaction, or what would constitute it, should be expressed in the requirements. Identifying differences between expected and actual results is valuable because it focuses on the fact that when we are testing, we need to be able to anticipate what is supposed to happen. It is then possible to determine what actually does happen and compare the two.

If a test is to find every conceivable fault or weakness in the system or component, then a good test is one that has a good probability of detecting an as yet undiscovered error, and a successful test is one that detects an as yet undiscovered error. The focus on showing the presence of errors is the basic attitude of a good test.

Testing is a positive and creative effort of destruction. It takes imagination, persistence, and a strong sense of mission to systematically locate the weaknesses in a complex structure and to demonstrate its failures. This is one reason why it is so hard to test our own work. There is a natural real sense in which we do not want to find errors in our own material.

Errors are in the work product, not in the person who made the mistake. With the "test to destroy" attitude, we are not attacking an individual in an organization or team of developers, but rather are looking for errors in those developers' work products.

Everyone on the development team needs to understand that tests add value to the product by discovering errors and getting them on the table as early as possible—to save the developers from building products based on error-ridden sources, to ensure the marketing people can deliver what the customer wants, and to assure management gets the bottom line on the quality and finance they are looking for.

13.2 PARSING TEST REQUIREMENTS

No matter what type of test is conducted, there are certain requirements that must be proven as a result of the test. Before testing begins, it is helpful to place all requirements into a database where they may be sorted on a variety of attributes, such as responsible subsystem. The purpose of the database is to assure all requirements are addressed in the test protocol as well as providing a convenient tracking system for the requirements. Where the number of requirements is small, manual collation of the requirements is effective. Where the number of requirements is large, the use of a software program to parse requirements is most helpful.

Once the requirements are listed, they can be used to develop the various test protocols necessary for testing. In addition, the list of requirements can be made more useful by turning them into a checklist, as seen in Figure 13.1, by adding space for additional information, such as reference to the location of a particular requirement, location of the test protocol, location of the test results, the initials of the person performing and completing the test, and the date of test completion. This checklist is also invaluable in tracking all requirements to satisfy quality assurance and regulatory departments as well as Food and Drug Administration and International Standards Organization auditors.

13.3 TEST PROTOCOL

It has been said that testing without a plan is not testing at all, but an experiment. Therefore, it is essential that each test performed be detailed in a test protocol that includes

- Name of the device under test
- Type of the test being performed
- Purpose of the test
- Definition of potential failures during the test

Requirement	Require ment Number	Document Location	Protocol Location	Results Location	Initials of Tester	Test Date
The unit must operate according to specification after exposure to an ambient temperature of 65°C.	12	Product Specification Paragraph 10.2.1	Lab notebook # R232, p. 44	Lab notebook # R232, pp. 45–46	JR	6/12/96
The unit must operate according to specification after exposure to an ambient temperature of −40°C.	13	Product Specification Paragraph 10.2.2	Lab notebook # R232, p. 47	Lab notebook # R232, pp. 48–49	JR	6/15/96
The unit must operate according to specification after exposure to an ambient temperature of +40°C and a relative humidity of 95%.	14	Product Specification Paragraph 10.2.3	Lab notebook # R232, p. 54	Lab notebook # R232, pp. 55–57	JL	7/13/96

FIGURE 13.1 Requirements checklist.

- Any special requirements
- Number of units on test
- Length of the test in hours or cycles
- Detailed procedure for running the test or reference to a procedure in another document, such as a standard
- Parameters to be recorded

13.4 TEST METHODOLOGY

Types of testing may include time testing, event testing, stress testing, environmental testing, time-related testing, and failure-related testing.

13.4.1 TIME TESTING

Time testing is conducted primarily to determine long-term reliability parameters, such as failure rate and mean time between failure (MTBF). Time testing can also be conducted to determine what part or component fails, when it fails, the mode of failure at that particular time, the mechanism of failure, and how much more or less life the equipment has that is required for operational use. This allows priorities of criticality for reliability improvement to be established.

13.4.2 EVENT TESTING

Event testing consists of repeated testing of equipment through its cycle of operation until failure. This type of testing is analogous to time-to-failure testing. One important parameter developed from this type of test is the number of cycles to failure.

13.4.3 STRESS TESTING

Stress testing has an important place in reliability assessment, but care must be taken in its application. Too much overstress may cause the test results to be inconclusive, as overstress may precipitate a failure that the product would not normally experience during normal usage. Care should also be taken to overstress in steps, rather than getting to the maximum value immediately. If the device fails, the step method allows the determination of where in the progression the failure occurred.

TABLE 13.1
Environmental Testing Standards

Environment	Applicable Standard
Operating temperature	IEC 68-2-14
Storage temperature	IEC 68-2-1
	IEC 68-2-2
Operating humidity	IEC 68-2-30
Storage humidity	IEC 68-2-3
	IEC 68-2-30
Operating ambient pressure	IEC 68-2-13
Storage ambient pressure	IEC 68-2-13
Transportation	NSTA
Radiated electrical emissions	CISPR 11
Radiated magnetic emissions	VDE 871
Radiated electrical field	IEC 601-1-2
Electrical fast transient	IEC 601-1-2
Radiated magnetic immunity	IEC 1000-4-8
Line conducted immunity	IEC 1000-4-6
Operating vibration	IEC 68-2-6
	IEC 68-2-34
Unpackaged shock	IEC 68-2-27
Stability	UL 2601
Ingress of liquids	IEC 529
	IEC 601-1
Pneumatic supply	CEN-TC215

13.4.4 Environmental Testing

Environmental testing represents a survey of the reaction of a device to the environmental and shipping environments, it should experience in its daily usage. By investigating a broad spectrum of the environmental space, greater confidence is developed in the equipment than if it was merely subjected to ambient conditions. As with overstress testing, avoid unusually extreme or unrealistic environmental levels because of the difficulty in their interpretation. Table 13.1 lists some typical environmental tests and the standard associated with its execution.

13.4.5 Time-Related Testing

Time-related testing is conducted until a certain number of hours of operation or a certain number of cycles has been completed, for example, a switch test conducted for 100,000 on/off cycles or a monitor operated for 100,000 h. This type of test will be important in choosing the correct formula to calculate MTBF from the test data.

13.4.6 Failure-Related Testing

A test may be conducted until all test units or a certain percentage of units have failed, for example, ventilators operated until the first unit fails or power supplies power cycled until all have failed. This type of test will be important in choosing the correct formula to calculate MTBF from the test data.

13.5 PURPOSE OF THE TEST

The purposes for testing may include the feasibility of a design, comparing two or more vendors, comparing two or more configurations, testing the response to environmental stresses, developing reliability parameters, failure analysis, or validation of the device.

All testing, except the reliability demonstration, which is performed at the end of the product development cycle, is performed at a confidence level of 90%. This means one is 90% confident that the reliability parameters established in the test will be characteristic of units in the field. A 90% confidence level also yields a risk factor of (1 − confidence level) or 10%. The reliability demonstration should be conducted at a confidence level of 95%, giving a risk factor of 5%. These levels will be important in determining the number of test units and the length of test time.

13.6   FAILURE DEFINITION

For each test and for each device, a failure must be defined. This definition depends on the intended application and the anticipated environment. What is considered a failure for one component or device may not be a failure for another. The test protocol should be as detailed as possible in defining the failure.

13.7   DETERMINING SAMPLE SIZE AND TEST LENGTH

Once you determine the type of test to be performed, you need to decide on the test sample size and the length of time necessary to accomplish your testing goal. Sample size and test time are dependent upon the MTBF goal, originally defined in the product specification and on the confidence level at which the test will be conducted.

The formula for determining the sample size and test time is derived from the following equation:

$$\text{MTBF goal} = (\text{Sample size})(\text{Test time})(2)/X^2_{\alpha;2r+2} \qquad (13.1)$$

Equation 13.1 thus becomes

$$(\text{Sample size})(\text{Test time}) = (\text{MTBF goal})\left(X^2_{\alpha;2r+2}\right)/2 \qquad (13.2)$$

To complete the equation, we must first understand the Chi Square Chart, included in Appendix 1. To use this chart first find the risk factor that the chart is based upon. As mentioned earlier, the risk factor is derived from the confidence level:

$$\text{Confidence level} = 1 - \alpha$$

where α is the risk factor

Thus, a confidence level of 90% yields a risk factor of 10%, while a confidence level of 95% yields a risk factor of 5%. Using the 90% confidence level, $\alpha = 0.10$ in Equation 13.2.

The r in Equation 13.2 is the number of failures. When calculating sample size and test time, it is assumed there will be no failures. This results in the minimal test time. Thus $r = 2(0) + 2$ or 2 and Equation 13.2 becomes

$$(\text{Sample size})(\text{Test time}) = (\text{MTBF goal})\left(X^2_{\alpha;2}\right)/2$$

Looking at the Chi Square Chart in Appendix 1, go across the top row of the chart and find 0.10 for γ. Go down that column to the line for $v = 2$. There you will find the number 4.605 or 4.61. Put this into Equation 13.3:

$$(\text{Sample size})(\text{Test time}) = \text{MTBF goal}(4.61)/2 \qquad (13.3)$$

Inserting the MTBF goal into Equation 13.3 and solving it yields the unit test time or (Sample size) (Test time).

TABLE 13.2
Test Time Possibilities
from Example 1

Sample Size	Test Time (h)
3	38,417
5	23,050
10	11,525
15	7,683
20	5,763
25	4,610
50	2,305
100	1,153

13.7.1 EXAMPLE 1

We want to test some power supplies to prove an MTBF goal of 50,000 h of operation. How many units do we test and for how long, assuming no failures?

$$(\text{Sample size})(\text{Test time}) = \text{MTBF goal}(4.61)/2$$
$$(\text{Sample size})(\text{Test time}) = 50,000(4.61)/2$$
$$= 115,250 \text{ unit hours}$$

From this data, we can calculate the possibilities listed in Table 13.2. This data is based on the statistical law that states that 1 unit tested for 10,000 h is statistically equal to 10 units tested for 1000 h each and 50 units tested for 200 h each.

13.7.2 EXAMPLE 2

We want to test some power supplies to prove an MTBF goal of 50,000 h of operation. How many units do we test and for how long, assuming one failure?

$$(\text{Sample size})(\text{Test time}) = (\text{MTBF goal})\left(X_{\alpha;2}^2\right)/2$$

In this case, using the 90% confidence level, $\alpha = 0.10$ and $r = 2(1) + 2$ or 4. Looking at the Chi Square Chart in Appendix 1, go across the top row of the chart and find 0.10. Go down that column to the line for $v = 4$. There you will find the number 7.779. Put this into the equation:

$$(\text{Sample size})(\text{Test time}) = \text{MTBF goal}(7.779)/2$$
$$(\text{Sample size})(\text{Test time}) = 50,000(7.779)/2$$
$$= 194,475 \text{ unit hours}$$

From this data, we can calculate the possibilities listed in Table 13.3. Again, this data is based on the statistical law that states that 1 unit tested for 10,000 h is statistically equal to 10 units tested for 1000 h each and 50 units tested for 200 h each.

TABLE 13.3
Test Possibilities
from Example 2

Sample Size	Test Time (h)
3	64,825
5	38,895
10	19,448
15	12,965
20	9,724
25	7,779
50	3,890
100	1,945

An interesting observation is that one failure increased the test time by 69%. A second failure would yield the equation:

$$(\text{Sample size})(\text{Test time}) = 50,000(10.645)/2$$
$$= 266,125$$

This is an increase in time of 37% over the one failure example and 131% over zero failures. This proves that unreliability is costly in time and effort.

13.8 TYPES OF TESTING

13.8.1 VERIFICATION

Procedures that attempt to determine that the product of each phase of the development process is an implementation of a previous phase, that is, it satisfies it. Each verification activity is a phase of the testing life cycle. The testing objective in each verification activity is to detect as many errors as possible. The testing team should leverage its efforts by participating in any inspections and walkthroughs conducted by development and by initiating verification, especially at the early stages of development.

13.8.2 VALIDATION

Validation is the process of evaluating a system or component during or at the end of the development process to determine whether it satisfies specified requirements.

13.8.3 BLACK BOX

The easiest way to understand black box testing is to visualize a black box with a set of inputs coming into it and a set of outputs coming out of it. The black box test is performed without any knowledge of the internal structure. The black box test verifies that the end-user requirements are met from the end-user's point of view.

Black box testing is a data-driven testing scheme. The tester views the device or program as a black box, that is, the tester is not concerned about the internal behavior and structure. The tester is only interested in finding circumstances in which the device or program does not behave according to its specification. Black box testing that is used to detect errors leads to exhaustive input testing, as every possible input condition is a test case.

13.8.4 WHITE BOX

White box testing is the opposite of black box testing. It is performed by personnel who are knowledgeable of the internal structure of the device and are testing from the developer's point of view. White box testing is a logic-driven testing scheme. The tester examines the internal structure of the device or program and derives test data from an examination of the internal structure. White box testing is concerned with the degree to which test cases exercise or cover the structure of the device or program. The ultimate while box test is an exhaustive path test.

13.8.5 HARDWARE TESTING

Hardware testing includes various types of tests depending on the intended use of the device. Testing which occurs during almost every product development cycle includes

- Vendor evaluation
- Component variation
- Environmental testing
- Safety evaluation
- Shipping tests
- Standards evaluation
- Product use/misuse
- Reliability demonstration

Often, hardware testing, especially that associated with the calculation of reliability parameters is performed twice during the development process. The first occurs immediately after the design phase and evaluates the robustness and reliability of the design. The second occurs after production of customer units begins. This testing evaluates the robustness and reliability of the manufacturing process.

13.8.6 SOFTWARE TESTING

Software testing consists of several levels of evaluation. Initially, module testing occurs, where the individual modules of the software program are evaluated and stress tested. This testing consists of verifying the design and implementation of the specification at the smallest component of the program. Testing involves running each module independently to assure it works, and then inserting errors, possibly through the use of an emulator. The test is basically an interface between the programmer and the software environment.

Integration testing occurs after each module has been successfully tested. The various modules are then integrated with each other and tested to assure they work together.

System testing consists of merging the software with the hardware to assure both will work as a system. Testing involves verifying the external software interfaces, assuring the system requirements are met, and assuring the system, as a whole, is operational.

Acceptance testing is the final review of all the requirements specified for the system and assuring both hardware and software address them.

13.8.7 FUNCTIONAL TESTING

Functional testing (Table 13.4) is designed to verify that all the functional requirements have been satisfied. This type of testing verifies that given all the expected inputs then all of the expected outputs are produced. This type of testing is termed as success-oriented testing because the tests are expected to produce successful results.

Testing of the functional capabilities involves the exercising of the operational modes and the events that allow a transition between the various software operational states. These tests are performed to verify that proper mode transitions are executed and proper outputs are generated

TABLE 13.4
Examples of Functional Testing

Test Type	Examples
Functional modes	Transitions between operational modes
	Correct inputs generate correct outputs
	Inputs and outputs include switches, tones, messages, and alarms
Remote communications	Connect and disconnect tests
	Valid commands and inquiries tests
	Handling of invalid commands and inquiries
	Tests for all baud rates supported
	Corrupted frames tests
	Error handling in general and the interface to the error handler
	Control mode testing with emphasis on safety
	Monitor mode testing with emphasis on fidelity of values reported
Timing	Active failure tests are completed within the system critical time
	Passive failure tests are completed within the operational window
Battery	Ramp up and ramp down of voltages
	Test the various levels of warnings, alarms, and errors

given the correct inputs. These tests also verify that the software generates the expected output given the expected user input. A communication test tool should be utilized to test the proper operation of the remote communications protocol and functionality of the communications software located in the product under test. Timing tests should be performed for system critical functions relating to the system critical time and the operational window. Battery tests should be performed whenever a software change to the software that monitors the battery levels has been made. In addition, if new functionality is pushing the product to the absolute performance edge, then battery tests should also be performed because of its potential effect on any power down software routines.

13.8.8 ROBUSTNESS TESTING

Robustness testing (Table 13.5) is designed to determine how the software performs given unexpected inputs. Robustness testing determines whether the software recovers from an unexpected

TABLE 13.5
Examples of Robustness Testing

Test Type	Examples
Boundary	Over and under specified limits
	Numerical values which determine logic flow based on a maximum or minimum value
	Negative numerical values
Overflow and underflow	Values too large for all algorithms
	Values too small for all algorithms
User interface	Enter unexpected values
	Enter unexpected sequences
Execution time	Routines which have execution time limits are altered to introduce delays
Line processing	Tasks which have execution time limits are altered to introduce delays
	Routines with execution constraints due to parametric calculations are altered
Data transmission	Unexpected commands are transmitted to the remote communications handler
	Unexpected data is transmitted to the remote communications handler

input, locks the system in an indeterminate state or continues to operate in a manner that is unpredictable. This type of testing is termed failure oriented because the test inputs are designed to cause the product to fail given foreseeable and reasonably unforeseeable misuse of the product.

Robustness testing is performed to determine software responses at the boundary limits of the product or test and manufacturing equipment and the test cases should include negative values.

As a part of robustness testing, algorithms are tested for overflow and underflow. The user interface is tested by entering unexpected values and sequences. Routines, tasks, or processes that are time constrained are altered to introduce reasonable delays to determine the reaction of the product or equipment. Communication software is given unexpected commands and data that is then transmitted to the remote communications handler.

13.8.9 STRESS TESTING

Stress testing (Table 13.6) is designed to ascertain how the product reacts to a condition in which the amount or rate of data exceeds the amount or rate expected. Stress tests can help determine the margin of safety that exists in the product or equipment.

Stress tests are performed which exercise the equipment continuously over varying periods of time and operating parameters if latent errors exist in the software. Generally, these tests consist of overnight runs and weekend runs that gain the optimum benefit of the allotted test time. Global buffers and data structures are tested under loaded and overflow conditions to determine the response of the software. Remote communications load tests should be performed which verify the remote communications interface transfer rate at the maximum transfer rate under worst-case and maximum load conditions. Worst-case scenario tests verify the product or equipment operating capability under the projected worst-case scenario. The worst-case scenario for products generally includes highest execution rate and event overload for event-driven systems. These tests should be limited to reasonable environmental tests which do not include temperature and vibration testing.

13.8.10 SAFETY TESTING

Safety testing (Table 13.7) is designed to verify that the product performs in a safe manner and that a complete assessment of the safety design has been accomplished.

Fail-safe tests should be performed specifically to verify the fail-safe provisions of the software design. These tests cover the error conditions only and do not address warnings or alarms, which are more appropriately tested under the functional tests. Limited, nondestructive fault insertion tests

---

TABLE 13.6
Examples of Stress Testing

Test Type	Examples
Duration	Over night runs
	Weekend runs
	Others types of software burn-in tests
Buffer overload	Global buffers tested under loaded and overflow conditions
	Global data structures tested under loaded and overflow conditions
Remote communications	Verify the transfer at the maximum transfer rate
	Verify the transfer at the maximum transfer rate under maximum load conditions
Worst-case scenario	Verify the product and test and manufacturing equipments operating capability under projected worst case
	Highest execution rate
	Event overload for event-driven systems

TABLE 13.7
Examples of Safety Testing

Test Type	Examples
Fail-safe	Verify that fail-safe provisions of the software design
	Test error conditions and handling
	Test data corruption
Active failure	Tests completion within system critical time
	ROM testing via CRC computation and comparison to a stored value
	RAM testing for stuck bits in data and address paths
	RAM testing for address decoding problems
	LED indicators voltage tests
	Processor and controller tests
Passive failure	Watchdog timer test
	Watchdog disable tests
	Hardware RAM tests
	CRC generator
	Battery-test
	Audio generators and speaker tests
	EEPROM tests
Safety	Critical parameters and their duplicates
	Events that lead to a loss of audio indicators
	Events that lead to a loss of visual indicators
	Events that lead to tactile errors, such as a key press
	Error handling for corrupted vectors and structures
	Error handling for corrupted sanity checks
	Sufficiency of periodic versus aperiodic tests
From hazard analysis	Single point failures
	Normal power up, run-time and power down safety tests

should be performed by the software verification and validation engineers. Products require an analysis of the error handling routines as well as data corruption tests to ensure an acceptable level of safety. The analysis must include a review of the products active failure tests so that they are completed within the system critical time and within the product defined operational window. A number of safety aspects that must also be addressed are the protection of critical parameters and events that lead to a loss of safety critical indicators. Safety testing of the product must utilize the hazards analysis in relation to failures. In addition, validation safety tests and internal product safety self-tests that were performed on past products should be compiled, executed, and compared against the new product under test to arrive at a consistent and growing list of mandatory safety tests.

13.8.11 REGRESSION TESTING

Regression testing (Table 13.8) is performed whenever a software change or a hardware change that affects the software has occurred. Regression testing verifies that the change produces the desired effect on the altered component and that no other component that relies on the altered component is adversely affected.

Regression testing is performed on products and test and manufacturing equipment that have made a change to an established, validated baseline. Regression testing begins by comparing the new software to the existing baseline with a version difference tool and the generation of a cross-reference listing to assess the changes and to ensure that no unintended side effects are introduced. From this, an assessment of the amount of changes and their criticality is made, the level of effort

TABLE 13.8
Regression Testing Sequence

Sequence Step	Activity
1	Compare the new software to the existing baseline
2	Generate a cross-reference listing to assess changes and to ensure no unintended side effects
3	Assess the amount of changes and the criticality
4	Determine the level of effort required and assess the risk
5	Test the new functions and the debug fixes
6	Execute a predetermined set of core tests to confirm no new unintended changes
7	Devote special attention to the safety implications

that is required to perform the regression is estimated and the risk is assessed. The alterations are tested and a compiled list of core tests executed to establish that no new unintended changes have been introduced. Special attention must be made to the safety implications.

EXERCISES

1. You have the responsibility for writing the test protocol for a portable pulse oximeter that will be used in a high school science class. Detail a list of tests that you would use.
2. You must determine why your blood sugar determination kit, which worked so well when tested in Nasville, Tennessee, gives erroneous results when used in Salt Lake City. What did you not account for?
3. What common fluid spills would you plan for testing an EKG monitor in use in an operating room? How would this list differ for the same machine in a patient's room?
4. Lobby and part of the immediate exterior of the Vanderbilt University Hospital has a floor made of mortared bricks. On the inside they are shellacked, on the outside they are allowed to weather. What tests can be performed with this flooring?
5. While occupied, an electric wheelchair moved on its own accord in a hospital environment, injuring the occupant. How would you investigate this accident? What tests were probably not run properly on the wheelchair before sale?
6. You are placed in charge of specifying shipping containers for a computer-based medical device. Investigate how the ISTA can assist you in specification of tests and shipping containers. (Visit the Web site for this organization at http://www.ista.org.)
7. You are in charge of setting up the test sequence for a new heart by-pass pump system. The system has a C++ software control scheme for flow control. Maximum expected length of use of the machine will be for surgeries lasting no more than 6 h. List some of the test methods you would use for this device.
8. Estimate the types of test necessary for validation of an implanted defibrillator system. What would be the minimum number of tests be determined by?
9. Purely mechanical systems need not undergo some of the tests that software/hardware systems do. Contrast the types of tests you would perform on an artificial knee versus an insulin pump.

BIBLIOGRAPHY

Fries, R.C., *Reliability Assurance for Medical Devices, Equipment and Software*, Buffalo Grove, IL: Interpharm Press, 1991.
Fries, R.C., *Handbook of Medical Device Design*, New York: Marcel Dekker, 2001.
Fries, R.C., *Reliable Design of Medical Devices*, 2nd ed, Boca Raton, FL: CRC Press/Taylor and Francis Group, 2006.
King, P.H. and Richard, C.F., *Design of Biomedical Devices and Systems*, New York: Marcel Dekker, 2002.
Laplante, P., *Real-Time Systems Design and Analysis—An Engineer's Handbook*, New York: The Institute of Electrical and Electronic Engineers, 1993.

14 Analysis of Test Data

Testing leads to failure, and failure leads to understanding.

Burt Rutan

The heart of reliability is the analysis of data, from which desired reliability parameters can be calculated. These parameters are calculated from testing throughout the product development process. Early calculations are updated as the program progresses and the presence or lack of reliability improvement becomes apparent.

Reliability parameter calculation assumes the product is in the useful life period of the bathtub curve. During this period, the failure rate is constant and the exponential distribution is used for calculations. In standards and handbooks where failure rates and mean time between failures (MTBF) values are listed, the same assumption is made and the exponential distribution is used.

Calculations of some parameters, such as MTBF, are dependent upon the termination mode of the test. Time terminated tests, where tests are ended after a predetermined time period has elapsed, are calculated different than failure terminated tests, where tests are ended after a predetermined number of units have failed.

The calculations necessary to determine the following parameters will be reviewed:

- Failure rate
- MTBF
- Reliability
- Confidence limits

In addition, graphical analysis and its application to dealing with reliability data will be discussed.

14.1 FAILURE RATE

Failure rate is the number of failures per million hours of operation. For devices in their useful life period, the failure rate is the reciprocal of the MTBF.

$$\text{MTBF} = 1/\lambda \tag{14.1}$$

The failure rate is stated as failures per hour for Equation 14.1.

14.1.1 EXAMPLE 1

An EEG machine has an MTBF of 4380 h. What is the failure rate?

$$\lambda = 1/\text{MTBF}$$
$$= 1/4380$$
$$= 0.000228 \text{ failures per hour}$$
$$= 228 \text{ failures per million hours}$$

14.1.2 EXAMPLE 2

Nearly 10 power supplies are put on test, to be terminated after each has completed 1000 h of operation. Two power supplies fail, one at 420 h and the other at 665 h. What is the failure rate of the power supplies?

Eight units completed 1000 h

$$\text{Total test time} = 8(1000) + 420 + 665$$
$$= 9085 \text{ h}$$

$$\lambda = \text{number of failures/total test time}$$
$$= 2/9085$$
$$= 0.000220 \text{ failures per hour}$$
$$= 220 \text{ failures per million hours}$$

14.2 MEAN TIME BETWEEN FAILURES

Mean time between failures is the time at which 63% of the operational devices in the field will have failed. MTBF is the reciprocal of the failure rate. It is also calculated from test data dependent upon the type of test run, for example, time terminated or failure terminated, and upon whether the failed units were replaced or not. Five different methods of MTBF calculation are

- Time terminated, failed parts replaced
- Time terminated, no replacement
- Failure terminated, failed parts replaced
- Failure terminated, no replacement
- No failures observed during the test

14.2.1 TIME TERMINATED, FAILED PARTS REPLACED

$$\text{MTBF} = N(\text{td})/r \tag{14.2}$$

where
  N is the number of units tested
  td is the test duration
  r is the number of failures

14.2.1.1 Example 3

The performance of 10 pressure monitors is monitored while operating for a period of 1200 h. The test results are listed below. Every failed unit is replaced immediately. What is the MTBF?

Unit Number	Time of Failure (h)
1	650
2	420
3	130 and 725
4	585
5	630 and 950
6	390

(continued) Unit Number	Time of Failure (h)
7	No failure
8	880
9	No failure
10	220 and 675

$N = 10$
$r = 11$
$td = 1200$ h

$$\text{MTBF} = N(\text{td})/r$$
$$= 10(1200)/11$$
$$= 1091 \text{ h}$$

14.2.2 TIME TERMINATED, NO REPLACEMENT

$$\text{MTBF} = (\Sigma T_i) + (N - r)\text{td}/r \qquad (14.3)$$

where
   N is the number of units tested
   td is the test duration
   r is the number of failures
   T_i is the individual failure times

Using the data in Example 3

Unit Number	Time of Failure (h)
1	650
2	420
3	130
4	585
5	630
6	390
7	No failure
8	880
9	No failure
10	220

$$\text{MTBF} = ((\Sigma T_i) + (N - r)/\text{td}))/r$$
$$= ((650 + 420 + 130 + 585 + 630 + 390 + 880 + 220) + 2(1200))/8$$
$$= (3905 + 2400)/8$$
$$= 788 \text{ h}$$

14.2.3 FAILURE TERMINATED, FAILED PARTS REPLACED

$$\text{MTBF} = N(\text{td})/r$$

where

  N is the number of units tested
  td is the test duration
  r is the number of failures

14.2.3.1   Example 4

Six 10 units were placed on test until all units failed, the last occurring at 850 h. The test results are listed below. Every failed unit, except the last one, is replaced immediately. What is the MTBF?

Unit Number	Time of Failure (h)
1	130
2	850
3	120 and 655
4	440
5	725
6	580

$$\text{MTBF} = N(\text{td})/r$$
$$= 6(850)/7$$
$$= 729 \text{ h}$$

14.2.4   FAILURE TERMINATED, NO REPLACEMENT

$$\text{MTBF} = (\Sigma T_i) + (N - r)\text{td}/r$$

Using the data from Example 4

Unit Number	Time of Failure (h)
1	130
2	850
3	120
4	440
5	725
6	580

$$\text{MTBF} = (\Sigma T_i) + (N - r)\text{td}/r$$
$$= (130 + 850 + 120 + 440 + 725 + 580) + 0(850)/6$$
$$= 3945 + 0/6$$
$$= 658 \text{ h}$$

14.2.5   NO FAILURES OBSERVED

For the case where no failures are observed, an MTBF value cannot be calculated. A lower one-sided confidence limit must be calculated and the MTBF stated to be greater than that value.

$$\text{ml} = 2(\text{Ta})/\chi^2_{\alpha;2}$$

where

ml is the lower one-sided confidence limit

Ta is the total test time

$\chi^2_{\alpha;2}$ is the chi square value from the table in Appendix 1, where α is the risk level and 2 is the degrees of freedom

14.2.5.1 Example 5

Nearly 10 ventilators are tested for 1000 h without failure. What is the MTBF at a 90% confidence level?

$$N = 10$$
$$td = 1000$$
$$r = 0$$
$$1 - \alpha = 0.90$$
$$\alpha = 0.10$$
$$Ta = N(td) = 10(1000) = 10000$$
$$ml = 2(Ta)/\chi^2_{\alpha;2}$$
$$= 2(10000)/\chi^2_{0.10;2}$$
$$= 20000/4.605$$
$$= 4343 \text{ h}$$

We can then state that the MTBF > 4343 h, with 90% confidence.

14.3 RELIABILITY

Reliability has been defined as the probability that an item will perform from a required function, under specified conditions, for a specified period of time, at a desired confidence level. Reliability may be calculated from either the failure rate or the MTBF. The resultant number is the percentage of units that will survive the specified time.

Reliability can vary between 0 (no reliability) and 1.0 (perfect reliability). The closer the value is to 1.0, the better will be the reliability. To calculate the parameter "reliability," two parameters are required:

- Failure rate or MTBF
- Mission time or specified period of operation

$$\text{Reliability} = \exp(-\lambda t)$$
$$= \exp(-t/\text{MTBF})$$

14.3.1 EXAMPLE 6

Using the data in Example 2, calculate the reliability of the power supplies for an operating period of 3200 h.

λ = Failure rate = 220 failures per million hours

For the equation, λ must be in failures per hour

Thus, $220/1000000 = 0.000220$ failures per hour

$t = 3200$ h

$$\text{Reliability} = \exp(-\lambda t)$$
$$= \exp - (0.000220)(3200)$$
$$= \exp - (0.704)$$
$$= 0.495$$

This states that after 3200 h of operation, one half the power supplies in operation will not have failed.

14.3.2 EXAMPLE 7

Using the time terminated, no replacement case, calculate the reliability of the pressure monitors for 500 h of operation.

$$\text{Reliability} = \exp - (\lambda t)$$
$$= \exp - (t/\text{MTBF})$$
$$= \exp - (500/788)$$
$$= \exp - (0.635)$$
$$= 0.530$$

Thus, 53% of the pressure monitors will not fail during the 500 h of operation.

14.4   CONFIDENCE LEVEL

Confidence level is the probability that a given statement is correct. Thus, when a 90% confidence level is used, the probability that the findings are valid for the device population is 90%. Confidence level is designated as

$$\text{Confidence level} = 1 - \alpha$$

where α is the risk level.

14.4.1 EXAMPLE 8

Test sample size is determined using a confidence level of 98%. What is the risk level?

$$\text{Confidence level} = 1 - \alpha$$
$$\alpha = 1 - \text{confidence level}$$
$$= 1 - 0.98$$
$$= 0.02 \text{ or } 2\%$$

14.5   CONFIDENCE LIMITS

Confidence limits are defined as the extremes of a confidence interval within which the unknown has a probability of being included. If the identical test was repeated several times with different samples of a device, it is probable that the MTBF value calculated from each test would not be identical. However, the various values would fall within a range of values about the true MTBF value. The two values which mark the end points of the range are the lower and upper confidence limits. Confidence limits are calculated based on whether the test was time or failure terminated.

14.5.1 TIME TERMINATED CONFIDENCE LIMITS

$$mL = 2(Ta)/\chi^2_{\alpha/2;2r+2}$$

where
  mL is the lower confidence limit
  Ta is the total test time
  $\chi^2_{\alpha/2;2r+2}$ is the chi square value from Appendix 1 for α risk level and $2r+2$ degrees of freedom

$$mU = 2(Ta)/\chi^2_{1-\alpha/2;2r}$$

14.5.1.1 Example 9

Using the data from the time terminated, no replacement data from Example 3, time terminated, no replacement, at a 90% confidence level:

$$Ta = 6305 \text{ h}$$
$$\alpha = 1 - \text{confidence level} = 0.10$$
$$\alpha/2 = 0.05$$
$$r = 8$$
$$2r + 2 = 18$$
$$mL = 2(6305)/\chi^2_{0.05;18}$$
$$= 12610/28.869$$
$$= 437 \text{ h}$$
$$mU = 2(6305)/\chi^2_{0.95;16}$$
$$= 12610/7.962$$
$$= 1584 \text{ h}$$

We can thus say
  $437 < \text{MTBF} < 1584$ h
  or the true MTBF lies between 437 and 1584 h.

14.5.2 FAILURE TERMINATED CONFIDENCE LIMITS

$$mL = 2(Ta)/\chi^2_{\alpha/2;2r}$$

and

$$mU = 2(Ta)/\chi^2_{1-\alpha/2;2r}$$

Using the data from the failure terminated, no replacement data from Example 4 at a 95% confidence limit:

$$\text{Ta} = 3945 \text{ h}$$
$$\alpha = 0.05$$
$$\alpha/2 = 0.025$$
$$1 - \alpha/2 = 0.975$$
$$r = 6$$
$$2r = 12$$
$$\text{mL} = 2(3945)/\chi^2_{0.025;12}$$
$$= 7890/23.337$$
$$= 338 \text{ h}$$
$$\text{mU} = 2(3945)/\chi^2_{0.975;12}$$
$$= 7890/4.404$$
$$= 1792 \text{ h}$$

Thus
$338 < \text{MTBF} < 1792$

14.6 MINIMUM LIFE

The minimum life of a device is defined as the time of occurrence of the first failure.

14.7 GRAPHICAL ANALYSIS

Graphical analysis is a way of looking at test data or field information. It can show failure trends, determine when a manufacturing learning curve is nearly complete, indicate the severity of field problems or determine the effect of a burn-in program.

Several type of graphical analysis are advantageous in reliability analysis:

- Pareto analysis
- Graphical plotting
- Weibull analysis

14.7.1 PARETO ANALYSIS

Pareto analysis is a plot of individual failures versus the frequency of the failures. The individual failures are listed on the x-axis and the frequency of occurrence on the y-axis. The result is a histogram of problems and their severity. The problems are usually plotted with the most frequent on the left. Once the results are obtained, appropriate action can be taken. Figure 14.1 is an example of a pareto analysis based on the following data:

Problem	Frequency
Power supply problems	10
Leaks	8
Defective parts	75
Cable problems	3
Missing parts	42
Shipping damage	2

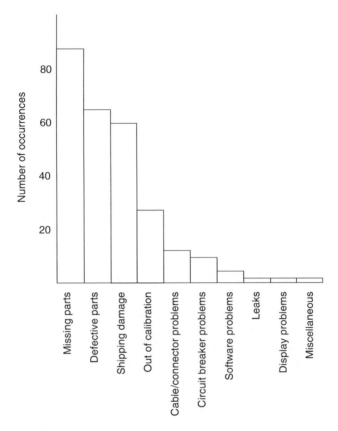

FIGURE 14.1   Pareto analysis.

14.7.2 GRAPHICAL PLOTTING

When plotting data, time is usually listed on the *x*-axis and the parameter to be analyzed on the *y*-axis.

14.7.2.1 Example 10

Nerve stimulators were subjected to 250 h of burn-in at ambient temperature before shipment to customers. Reports of early failures were grouped into 50 h intervals and showed the following pattern:

Hourly Increment	Number of Failures
0–50	12
51–100	7
101–150	4
151–200	1
201–250	1

Figure 14.2 is a plot of the data. The data indicate the number of failures begins to level off at approximately 200 h. The burn-in was changed to an accelerated burn-in, equal to 300 h of operation.

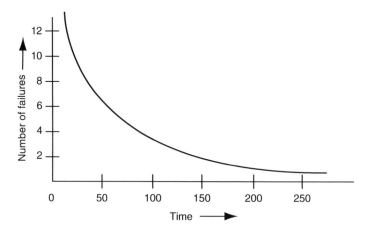

FIGURE 14.2 Plot of field data.

14.7.3 WEIBULL PLOTTING

Weibull paper is a logarithmic probability plotting paper constructed with the *y*-axis representing the cumulative probability of failure and the *x*-axis representing a time value, in hours or cycles. Data points are established from failure data, with the failure times arranged in increasing order or value of occurrence. Corresponding median ranks are assigned from a percent rank table, based on the sample size.

EXERCISES

1. Almost 50 switches are placed on test to be terminated after each switch has completed 10,000 cycles of "on" and "off." Five switches fail at the following times: 650, 925, 2000, 3500, and 7500 cycles. What is the failure rate of the switches in failures per million cycles?
2. Nearly 10 oximeters are placed on test. They are to complete 5000 h of operation. Test results are listed below. Every failed unit is replaced immediately. What is the MTBF?

Unit Number	Time of Failure (h)
1	800
2	No failure
3	1000 and 1250
4	2200
5	No failure
6	850 and 3200
7	550
8	No failure
9	4200
10	925 and 3350

3. Using the test data from Exercise 2, calculate the MTBF when the failed units are not replaced. Compare the results of Exercises 2 and 3.
4. About 50 power supplies were tested for 3000 h. No failures were observed during the test. Determine the MTBF at a 95% confidence level?

5. Resistor has a failure rate of 0.0052 failures per million hours. Calculate the reliability of the resistors for an operating period of 10,000 h.

6. Ventilator has a reliability of 99.999 for a mission time of 20,000 h. What is the MTBF?

7. Use the data in Exercise 2 to determine the upper and lower confidence limits at a 95% confidence level?

8. Repeat Exercise 7 using a 90% confidence level. Compare the results of the two confidence levels.

9. Power supplies were subjected to 168 h of burn-in with the following failures occurring. Plot the data in appropriate intervals to show the failure pattern. Discuss the possible actions that would result from the plot.

   Failures (hours):

   10, 22, 35, 38, 42, 45, 48, 52, 63, 77, 88, 94, 122, 135, 148, 165

10. Using the results from Exercise 1, determine the MTBF for the switches. The metric for the MTBF will be cycles.

BIBLIOGRAPHY

Frankel, E.G., *System Reliability and Risk Analysis*, The Hague: Martinus Nijhoff, 1984.

Fries, R.C., *Handbook of Medical Device Design*, New York: Marcel Dekker, 2001.

Fries, R.C., *Reliable Design of Medical Devices*, 2nd ed. Boca Raton, FL: CRC Press/Taylor & Francis Group, 2006.

Ireson, W.G. and Clyde, F.C. Jr., *Handbook of Reliability Engineering and Management*, New York: McGraw-Hill, 1988.

King, J.R., *Probability Charts for Decision Making*, New Hampshire: Team, 1971.

King, P.H. and Richard C.F., *Design of Biomedical Devices and Systems*, New York: Marcel Dekker, 2002.

Lloyd, D.K. and Lipow, M., *Reliability Management, Methods and Mathematics*, 2nd ed. Milwaukee, WI: The American Society for Quality Control, 1984.

Mann, N.R. et al., *Methods for Statistical Analysis of Reliability and Life Data*, New York: John Wiley & Sons, 1974.

Nelson, W., *Applied Life Data Analysis*, New York: John Wiley & Sons, 1982.

O'Connor, P.D.T., *Practical Reliability Engineering*, 3rd ed. Chichester, England: John Wiley & Sons, 1991.

15 Reliability and Liability

Why is there never enough time to develop a product correctly, but always enough time to do it over?

Anonymous

The law can be defined as the collection of rules and regulations by which society is governed. The law regulates social conduct in a formal binding way while it reflects society's needs, attitudes, and principles. Law is a dynamic concept that lives, grows, and changes. It can be described as a composite of court decisions, regulations, and sanctioned procedures, by which laws are applied and disputes adjudicated.

The three most common theories of liability for which a manufacturer may be held liable for personal injury caused by its product are negligence, strict liability, and breach of warranty. These are referred to as common-law causes of action, which are distinct from causes of action based on federal or state statutory law. Although within the last decade federal legislative action that would create a uniform federal product liability law has been proposed and debated, no such law exists today. Thus, such litigation is governed by the laws of each state.

These three doctrines are called theories of recovery because an injured person cannot recover damages against a defendant unless he alleges and proves, through use of one or more of these theories, that the defendant owed him a legal duty and that the defendant breached that duty, thereby causing the plaintiff's injuries. Although each is conceptually distinct, similarities exist between them. Indeed, two or more theories are asserted in many product defect suits.

15.1 NEGLIGENCE

Since much of medical malpractice litigation relies on negligence theory, it is important to clearly establish the elements of the cause of action. Negligence may be defined as conduct, which falls below the standard established by law for the protection of others against unreasonable risk of harm. There are four major elements of the negligence action:

1. Person or business owes a duty of care to another.
2. Applicable standard for carrying out the duty be breached.
3. Proximate cause of the breach of duty, a compensable injury results.
4. Compensable damages or injury to the plaintiff.

The burden is on the plaintiff to establish each and every element of the negligence action.

The basic idea of negligence law is that one should have to pay for injuries that he or she causes when acting below the standard of care of a reasonable, prudent person participating in the activity in question. This standard of conduct relates to a belief that centers on potential victims that people have a right to be protected from unreasonable risks of harm. A fundamental aspect of the negligence standard of care resides in the concept of foreseeability.

A plaintiff in a product liability action grounded in negligence, then, must establish a breach of the manufacturer's or seller's duty to exercise reasonable care in the manufacture and preparation of a product. The manufacturer in particular must be certain that the product is free of any potentially dangerous defect that might become dangerous upon the happening of a reasonably anticipated emergency. The obligation to exercise reasonable care has been expanded to include reasonable care in the inspection or testing of the product, the design of the product, or the giving of warnings concerning the use of the product.

A manufacturer must exercise reasonable care even though he or she is but a link in the production chain that results in a finished product. For example, a manufacturer of a product which is designed to be a component part of another manufactured product is bound by the standard of reasonable care. Similarly, a manufacturer of a finished product which incorporates component parts fabricated elsewhere has the same legal obligation.

A seller of a product, on the other hand, is normally held to a less stringent standard of care than a manufacturer. The lesser standard is also applied to distributors, wholesalers, or other intermediary in the marketing chain. This rule pertains because a seller or intermediary is viewed as simply a channel through which the product reaches the consumer.

In general, the duty owed at any particular time varies with the degree of risk involved in a product. The concept of reasonable care is not static, but changes with the circumstances of the individual case. The care must be commensurate with the risk of harm involved. Thus, manufacturers or sellers of certain hazardous products must exercise a greater degree of care in their operations than manufacturers or sellers of other less dangerous products.

15.2   STRICT LIABILITY

Unlike the negligence suit, in which the focus is on the defendant's conduct, in a strict liability suit, the focus is on the product itself. The formulation of strict liability states that one who sells any product in a defective condition unreasonably dangerous to the user or consumer or to his property is subject to liability for physical harm thereby caused to the ultimate user or consumer or to his property if the seller is engaged in the business of selling such a product, and it is expected to and does reach the user or consumer without substantial change to the condition in which it is sold. Therefore, the critical focus in a strict liability case is on whether the product is defective and unreasonably dangerous. A common standard applied in medical device cases to reach that determination is the risk/benefit analysis, that is, whether the benefits of the device outweigh the risks attendant with its use.

The result of strict liability is that manufacturers, distributors, and retailers are liable for the injuries caused by defects in their products, even though the defect may not be shown to be the result of any negligence in the design or manufacture of the product. Moreover, under strict liability, the manufacturer cannot assert any of the various defenses available to him in a warranty action.

Strict liability means that a manufacturer may be held liable even though he has exercised all possible care in the preparation and sale of this product. The sole necessity for manufacturer liability is the existence of a defect in the product and a causal connection between this defect and the injury which resulted from the use of the product.

15.3   BREACH OF WARRANTY

A warranty action is contractual rather than tortuous in nature. Its basis lies in the representations, either express or implied, which a manufacturer or a seller makes about its product.

A third cause of action that may be asserted by a plaintiff is breach of warranty. There are three types of breaches of warranty that may be alleged:

1. Breach of the implied warranty of merchantability
2. Breach of the implied warranty of fitness for a particular purpose
3. Breach of an express warranty

15.3.1 IMPLIED WARRANTIES

Some warranties accompany the sale of an article without any express conduct on the part of the seller. These implied warranties are labeled the warranties of merchantability and of fitness for a particular purpose.

A warranty that goods shall be merchantable is implied in a contract for their sale, if the seller is a merchant who commonly deals with such goods. At a minimum, merchantable goods must

- Pass without objection in the trade under the contract description
- Fit the ordinary purposes for which they are used
- Be within the variations permitted by the sales agreement, of even kind, quality, and quantity within each unit and among all units involved
- Be adequately contained, packaged, and labeled as the sales agreement may require
- Conform to the promises or affirmations of fact made on the container or label

The implied warranty of fitness for a particular purpose arises when a buyer makes known to the seller the particular purpose for which the goods are to be used, and the buyer, relying on the seller's skill or judgment, receives goods which are warranted to be sufficient for that purpose.

15.3.2 EXCLUSION OF WARRANTIES

The law has always recognized that sellers may explicitly limit their liability upon a contract of sale by including disclaimers of any warranties under the contract. The Uniform Commercial Code embodies this principle and provides that any disclaimer, exclusion, or modification is permissible under certain guidelines. However, a disclaimer is not valid if it deceives the buyer.

These warranty causes of action do not offer any advantages for the injured plaintiff that cannot be obtained by resorting to negligence and strict liability claims and, in fact, pose greater hurdles to recovery. Thus, although a breach of warranty claim is often pled in the plaintiff's complaint, it is seldom relied on at trial as the basis for recovery.

15.4 DEFECTS

The term "defect" is used to describe generically the kinds and definitions of things that courts find to be actionably wrong with products when they leave the seller's hands. In the decisions, however, the courts sometimes distinguish between defectiveness and unreasonable danger. Other considerations in determining defectiveness are

- Consumer expectations
- Presumed seller knowledge
- Risk–benefit balancing
- State of the art
- Unavoidably unsafe products

A common and perhaps the prevailing definition of product unsatisfactoriness is that of unreasonable danger. This has been defined as the article sold must be dangerous to an extent beyond that which would be contemplated by the ordinary consumer who purchases it, with the ordinary knowledge common to the community as to its characteristics.

Another test of defectiveness sometimes used is that of presumed seller knowledge: would the seller be negligent in placing a product on the market if he had knowledge of its harmful or dangerous condition? This definition contains a standard of strict liability, as well as one of defectiveness, since it assumes the seller's knowledge of a product's condition even though there may be no such knowledge or reason to know.

Sometimes a risk–benefit analysis is used to determine defectiveness, particularly in design cases. The issue is phrased in terms of whether the cost of making a safer product is greater or less than the risk or danger from the product in its present condition. If the cost of making the change is greater than the risk created by not making the change, then the benefit or utility of keeping the product as is outweighs the risk and the product is not defective. If on the other hand, the cost is less than the risk then the benefit or utility of not making the change is outweighed by the risk and the product in its unchanged condition is defective.

Risk–benefit or risk–burden balancing involves questions concerning state of the art, since the burden of eliminating a danger may be greater than the risk of that danger if the danger cannot be eliminated. State of the art is similar to the unavoidably unsafe defense where absence of the knowledge or ability to eliminate a danger is assumed for purposes of determining if a product is unavoidably unsafe. State of the art is defined as the state of scientific and technological knowledge available to the manufacturer at the time the product was placed in the market.

Determining defectiveness is one of the more difficult problems in products liability, particularly in design litigation. There are three types of product defects:

1. Manufacturing or production defects
2. Design defects
3. Defective warnings or instructions

The issue implicates questions of the proper scope of the strict liability doctrine, and the overlapping definitions of physical and conceptual views of defectiveness.

Manufacturing defects can rarely be established on the basis of direct evidence. Rather, a plaintiff who alleges the existence of a manufacturing defect in the product must usually resort to the use of circumstantial evidence to prove that the product was defective. Such evidence may take the form of occurrence of other similar injuries resulting from the use of the product, complaints received about the performance of the product, defectiveness of other units of the product, faulty methods of production, testing or analysis of the product, elimination of other causes of the accident, and comparison with similar products.

A manufacturer has a duty to design his product so as to prevent any foreseeable risk of harm to the user or patient. A product that is defectively designed can be distinguished from a product containing a manufacturing defect. While the latter involves some aberration or negligence in the manufacturing process, the former encompasses improper planning in connection with the preparation of the product. Failure to exercise reasonable care in the design of a product is negligence. A product that is designed in a way which makes it unreasonably dangerous will subject the manufacturer to strict liability. A design defect, in contrast to a manufacturing defect, is the result of the manufacturer's conscious decision to design the product in a certain manner.

Product liability cases alleging unsafe design may be divided into three basic categories:

1. Cases involving concealed dangers
2. Cases involving a failure to provide appropriate safety features
3. Cases involving construction materials of inadequate strength

A product has a concealed danger when its design fails to disclose a danger inherent in the product which is not obvious to the ordinary user.

Some writers treat warning defects as a type of design defect. One reason for doing this is that a warning inadequacy, like a design inadequacy, is usually a characteristic of a whole line of products, while a production or manufacturing flaw is usually random and atypical of the product.

15.5   FAILURE TO WARN OF DANGERS

An increasingly large portion of product liability litigation concerns the manufacturer's or seller's duty to warn of actual or potential dangers involved in the use of the product. Although the duty to warn may arise under all three theories of product liability, as mentioned above, most warnings cases rely on negligence principles as the basis for the decision. The general rule is that a manufacturer or seller who has knowledge of the dangerous character of the product has a duty to warn users of this danger. Thus, failure to warn where a reasonable person would do so is negligence.

15.6   PLAINTIFF'S CONDUCT

A manufacturer or seller may defend a product liability action by demonstrating that the plaintiff either engaged in negligent conduct that was a contributing factor to his injury or used a product when it was obvious that a danger existed and thereby assumed the risk of his injury. Another type of misconduct which may defeat recovery is when the plaintiff misuses the product by utilizing it in a manner not anticipated by the manufacturer. The applicability of these defenses in any given product suit is dependent upon the theory or theories of recovery which are asserted by the plaintiff.

15.7   DEFENDANT'S CONDUCT

Compliance with certain standards by a manufacturer may provide that party with a complete defense if the product leaves the manufacturer's or seller's possession or control and when it is a substantial or proximate cause of the plaintiff's injury. Exceptions to this rule include alterations or modifications made with the manufacturer's or seller's consent, or according to manufacturer's/seller's instructions.

15.8   DEFENDANT-RELATED ISSUES

When a medical device proves to be defective, potential liability is created for many parties who may have been associated with the device. Of all the parties involved, the injured patient is least able to bear the financial consequences. To place the financial obligation upon the proper parties, the courts must consider the entire history of the product involved, often from the time the design concept was spawned until the instant the injury occurred.

The first parties encountered in this process are the designers, manufacturers, distributors, and sellers of the product. Physicians and hospitals are subject to liability through medical malpractice actions for their negligence, whether or not a defective product is involved. Where such a product is involved, the doctor or hospital may be liable for

- Negligent misuse of the product
- Negligent selection of the product
- Failure to inspect or test the product
- Using the product with knowledge of its defect

15.9   MANUFACTURER'S AND PHYSICIAN'S RESPONSIBILITIES

Manufacturers of medical devices have a duty with regard to manufacture, design, warnings, and labeling. A manufacturer is required to exercise that degree of care, which a reasonable, prudent manufacturer would use under the same or similar conditions. A manufacturer's failure to comply with the standard in the industry, including failing to warn or give adequate instructions, may result in a finding of liability against the manufacturer.

With regard to medical devices, a manufacturer must take reasonable steps to warn physicians of dangers of which it is aware or reasonably should be aware where the danger would not be obvious to the ordinary competent physician dispensing a particular device. The responsibility for the prudent use of the medical device is with a physician. A surgeon who undertakes to perform a surgical procedure has the responsibility to act reasonably.

It is therefore required of the manufacturer to make a full disclosure of all known side effects and problems with a particular medical device by use of appropriate warnings given to physicians. The physician is to act as the learned intermediary between the manufacturer and the patient and transmit appropriate information to the patient. The manufacturer, however, must provide the physician with the information in order that he can pass it on to the patient.

In addition, the manufacturer's warnings must indicate the scope of potential danger from the use of a medical device and the risks of its use. This is particularly important where there is off-label use (the practice of using a product approved for one application in a different application) by a physician.

The manufacturer's warnings must detail the scope of potential danger from the use of a medical device, including the risks of misuse. The warnings must alert a reasonably competent physician to the dangers of not using a product as instructed. It would seem then the manufacturer may be held liable for failing to disclose the range of possible consequences of the use of a medical device if it has knowledge that the particular device is being used off label.

The duty of a manufacturer and physician for use of a medical device will be based upon the state of knowledge at the time of the use. The physician therefore has a responsibility to be aware of the manufacturer's warnings as he considers the patient's condition. This dual responsibility is especially relevant in deciding what particular medical device to use. Physician judgment and an analysis of the standard of care in the community should predominate the court's analysis in determining liability for possible misuses of the device.

A concern arises if the surgeon has received instruction as to the specific device from a manufacturer outside an investigative device exemption clinical trail approved by the Food and Drug Administration. In such circumstances, plaintiffs will maintain that the manufacturer and physician conspired to promote a product that is unsafe for off-label use.

15.10   ACCIDENT RECONSTRUCTION AND FORENSICS

Biomedical engineers, due to their generally broad-based education, may sometimes be called upon to analyze accidents. Analysis of medical device accidents is discussed first, followed by a brief discussion on biomechanics and accident (physical injury due to car, etc. impact) investigation. Both of these have implications for improved designs of devices and processes that biomedical engineers may be involved in.

15.10.1   Medical Device Accidents

Medical device accident investigation follows a fairly typical chain of events, of which most are in common for accidents in general. The overall process for a medical device accident investigation takes roughly the following outline:

- Incident occurs, someone is injured, and a cause for action is established.
- You are contacted by the wrong person, by his/her lawyer, or by one of the parties or their representative needing an investigation.
- After an initial familiarization with the problem, you may opt to work on the problem or opt out.
- You need to research the device or process in question. This means that you will use manufacturers and users device experience (MAUDE), if necessary. You will need to

access the operators' manual for the device, if necessary. You will need to investigate maintenance manuals, if necessary. You will need to run simulations on the device, if necessary. You likely will need to use the Web, other than just MAUDE for key word searches. You will need to obtain agreement for the use of specialists, such as personnel who perform calibration or maintenance on the devices as necessary. You may need to do some basic research and mock-ups of the device as necessary.

- As a result of your work, you will need to estimate causes and their likelihood. If you can demonstrate the error, so much the better.
- Branching point is reached here. A report (oral or written, this should be preagreed to) should be submitted to the person who contacted and contracted you. You must be prepared to continue the investigation, await further court action, or be released from further work. The latter is generally the case when you find for someone other than those who hired you.
- You must be prepared to answer questioning from opposition lawyers, if necessary. This can take the form of both oral and written testimony as to the current status of the investigation.
- Most of the time, a final formal report designating the fault will end your work. On a small number of occasions, expect to go to court, get sworn in, and testify regarding your work.

Two brief cases serve to illustrate the range of efforts that may come of a medical device accident.

- Scene: A patient was sent home from a nursing facility with an enteral feeding pump (direct to stomach tube feeding), a supply of feeding compound, and a supply of enteral feeding pump tubes. On the first use of the pump, the patient wound up with too high a flow of food such that food filled the stomach and entered the lungs. The patient expired due to pneumonia induced by the flow within a few days.
- Resolution: After a very brief overview of the material, a panel was convened, comprised of representatives of the nursing facility, a biomedical engineer from academia, a representative from the company that manufactured the pump, and the opposition lawyers. Within 5 min the determination was made that the pump had been sent home with the wrong pump tubing installed; the tubing that was in place allowed for gravity feed of the feeding fluid independent of the pump speed. Thus a direct cause of the accident was found in a timely manner.
- Scene: A pressure-limited pump was used to ventilate a baby who had a very small plastic airway directly in place in the throat. The baby was found asphyxiated after the airway had withdrawn from it. The unit, though the nursing service has presumably properly set the upper and lower pressure controls, was not alarming.
- Resolution: The unit was retested and still did not alarm. Various settings were tried. It was determined that the extremely small airway element enabled sufficient backpressure to the system that the recommended pressure settings were meaningless.

15.10.2   BIOMECHANICS AND TRAFFIC ACCIDENT INVESTIGATIONS

A very basic understanding of biomechanics is necessary before any undertaking in which a bioengineer may be involved in design or accident investigation. Some of the concepts that need to be understood are the following:

Data collection: Data for analyses involving traffic accidents involves data collected from reported and analyzed accidents. Large data sets are collected by individual states; some of which are recollected and analyzed by the National Highway Transportation Safety Administration (NHTSA). The agency also specifically maintains data on fatal accidents, including information on vehicle type, rollover, ejection, alcohol use, etc. A smaller data set includes data for cases specifically investigated by the agency, data that includes medical information as well as more specific conclusions as to the cause of the accident, and many other details. Other related data sets have been obtained from cadaver studies, anthropometric dummy studies, animal studies, and mathematical modeling analyses. Much of the data obtained and related issues may be found on the NHTSA Web site (http://www.nhtsa.dot.gov/) and in the *Proceedings of the Annual STAPP Car Crash Conferences*.

Injury estimation: In studies of human survival following trauma, an early scheme involved the development of an abbreviated injury scale (AIS); this scale ranges from 0 (minor sprain) to 6 (unsurvivable injury.) This scale is developed for each of the six body regions of interest in survivability: head, face, chest, abdomen, extremities (including pelvis), and external. The highest squared scores from the three most injured areas are added together to generate a new score, the injury severity score (ISS). With the exception that any 6 AIS rating automatically yields the maximum ISS of 75; this score relates linearly to rates of mortality, morbidity, and length of hospital stay.*

Impact analyses: Often, an engineer must estimate the relative speeds of the vehicles and personnel involved. This means that an engineer must, from the data involved in the accident report, crush patterns on the vehicles involved, vehicle data sheets, weather conditions reported, etc, estimate the relative speeds, angles of impact, and probable outcome of an accident. For example, working backward from skid length data, one can find that a vehicle's initial velocity before the skid is directly related to the square root of twice the product of skid length, skid friction coefficient, and the value of gravity (*g*). The skid friction coefficient is a function of the type of surface (e.g., pavement vs. dirt road), the weather conditions (dry, wet, or icy), and the type of braking system the vehicle has (two or four wheel, antilock, etc.). If a subject has been thrown or ejected from a vehicle, simple trajectory analysis can be used to determine the initial velocity, if sufficient information exists. If there is little body damage on the two vehicles, a combination of conservation of momentum analysis and elastic collision analysis might apply, along with skid analysis. Alternatively, damage analysis combined with inelastic collision analysis must be used.

The biomedical engineer doing design or doing forensic analysis after the fact in on matters involving vehicular accidents must understand the above, and be able to apply background material learned in a biomechanics class to real-life problems. A few examples follow:

- Occupant-restraint systems may be designed to absorb energy during an impact. Consider the alternatives for air bags, especially for situations with low body weight passengers.
- During a motorcycle–truck accident, the helmet of the motorcyclist came off, resulting in death of the motorcyclist due to blunt head trauma. Where was the error in the design of the helmet system?
- Current seat belts are a trade-off between convenience and safety. Determine the ideal design.

15.11 CONCLUSION

Products liability will undoubtedly continue to be a controversial field of law, because it cuts across so many fundamental issues of our society. It will also remain a stimulating field of study and practice, since it combines a healthy mixture of the practical and theoretical. The subject will certainly continue to change, both by statutory and by common-law modification.

Products liability implicates many of the basic values of our society. It is a test of the ability of private industry to accommodate competitiveness and safety. It tests the fairness and the workability of the tort system of recovery, and of the jury system as a method of resolving disputes.

---

* See www.trauma.org for more information.

EXERCISES

1. Do a MAUDE search for deaths caused by enteral feeders printout and discuss at least one case.
2. Do a MAUDE search for deaths caused in one week of the year. Comment on your results.
3. Visit the new car assessment pages at the NHTSA Web site (http://www.nhtsa.dot.gov/ NCAP/Info.html). Copy the frequently asked questions list, and comment on five of the particular items.
4. Visit the Stapp Conference Web site (www.stapp.org). Determine the history of the conferences.
5. One of the authors of this book owns a 1996 Chrysler Voyager van and a 1995 Volvo 950. Visit the NHTSA site to determine which is the safest.
6. Find data for the chance of survival for a patient with a major liver laceration and a closed tibial fracture as a result of a vehicular injury.
7. A lawyer asks you to testify about an injury that was received during a low-speed (10 mph or less) two-vehicle collision. Specifically he asks that you testify that no data exists that can prove the correct speeds of the vehicles and the likelihood of injury. Is this correct?
8. A 3 year old girl sustained neck injuries on a child roller coaster at a theme park. What would you do to prove or disprove this claim? By the way, the father has a videotape of the injury occurring, and the girl seemed to have a long neck. This particular ride had been in use for 10 years.
9. Find and report on the use of the Apgar score. Compare this to the AIS score in this section.
10. How might tissue engineering change the field of trauma care?
11. Brain poses a special case when studying injury patterns. Research the term "contrecoup." Report on its significance.

BIBLIOGRAPHY

Boardman, T.A. and Thomas, D., Product liability implications of regulatory compliance or non-compliance, in *The Medical Device Industry-Science, Technology, and Regulation in a Competitive Environment*. New York: Marcel Dekker, 1990.

Boumil, M.M. and Clifford, E.E., *The Law of Medical Liability in a Nutshell*. St. Paul, MN: West Publishing Co., 1995.

Buchholz, S.D., Defending pedicle screw litigation, *For the Defense*, 38(3), March, 1996.

Food and Drug Administration, *Device Recalls: A Study of Quality Problems*. Rockville, MD: Food and Drug Administration, 1990.

Food and Drug Administration, *Evaluation of Software Related Recalls for the Period FY83-FY89*. Rockville, MD: Food and Drug Administration, Center for Devices and Radiological Health, 1990.

Fries, R.C., *Handbook of Medical Device Design*. New York: Marcel Dekker, 2001.

Fries, R.C., *Reliable Design of Medical Devices*, 2nd ed. Boca Raton, FL: CRC Press, Taylor and Francis Group, 2006.

Gingerich, D., *Medical Product Liability: A Comprehensive Guide and Sourcebook*. New York: F&S Press, 1981.

Kanoti, G.A., Ethics, medicine, and the law, in *Legal Aspects of Medicine*. New York: Springer-Verlag, 1989.

King, P.H. and Fries, R.C., *Design of Biomedical Devices and Systems*. New York: Marcel Dekker, 2002.

Nahum, A.M. and Melvin, J.W., *Accidental Injury, Biomechanics and Prevention*. New York: Springer, 2002.

Olivier, D.P., Software safety: Historical problems and proposed solutions, *Medical Device and Diagnostic Industry*, 17(7), July, 1995.

Phillips, J.J., *Products Liability in a Nutshell*, 4th ed. St. Paul, MN: West Publishing Co., 1993.

Shapo, M.S., *Products Liability and the Search for Justice*. Durham, NC: Carolina Academic Press, 1993.

16 Food and Drug Administration

There is a greater law than the FDA and that is an obligation of a doctor to try to do anything he can to save a life when he thinks there is a chance.

Dr. Cecil Vaughn

When designing any device that will be used medically, it is important to consider all safety aspects, including the repercussions of design flaws and misuse of the device. Regulation of medical devices is intended to protect consumer's health and safety by attempting to ensure that marketed products are effective and safe. Before 1976, the Food and Drug Administration (FDA) had limited authority over medical devices under the Food, Drug, and Cosmetic (FD&C) Act of 1938. Since 1968, Congress established a radiation control program to authorize the establishment of standards for electronic products, including medical and dental radiology equipment. From the early 1960s to 1975, concern over devices increased and six U.S. Presidential messages were given to encourage medical device legislation.

In 1969, the Department of Health, Education, and Welfare appointed a special committee (the Cooper Committee) to review the scientific literature associated with medical devices. The Committee estimated that over a 10 year period, 10,000 injuries were associated with medical devices, of which 731 resulted in death. The majority of problems were associated with three device types: artificial heart valves, cardiac pacemakers, and intrauterine contraceptive devices. There activities culminated in passage of the Medical Devices Amendments of 1976.

Devices marketed after 1976 are subject to full regulation unless they are found substantially equivalent to a device already on the market in 1976. By the end of 1981, only about 300 of the 17,000 products submitted for clearance to the FDA after 1976 had been found not substantially equivalent.

16.1 HISTORY OF DEVICE REGULATION

In 1906, the FDA enacted its first regulations addressing public health. While these regulations did not address medical devices per se, they did establish a foundation for future regulations. It was not until 1938, with the passage of the Federal Food, Drug and Cosmetic (FFD&C) Act that the FDA was authorized, for the first time, to regulate medical devices. This act provided for regulation of adulterated or misbranded drugs, cosmetics and devices that were entered into interstate commerce. A medical device could be marketed without being federally reviewed and approved.

In the years following World War II, the FDA focused much of the attention on drugs and cosmetics. Over-the-counter drugs became regulated in 1961. In 1962, the FDA began requesting safety and efficacy data on new drugs and cosmetics.

By the mid 1960s, it became clear that the provisions of the FFD&C Act were not adequate to regulate the complex medical devices of the times to assure both patient and user safety. Thus, in 1969, the Cooper Committee was formed to examine the problems associated with medical devices and to develop concepts for new regulations.

In 1976, with input from the Cooper Committee, the FDA created the Medical Device Amendments to the FFD&C Act, which were subsequently signed into law. The purpose of the amendments was to assure that medical devices were safe, effective, and properly labeled for their intended use. To accomplish this mandate, the amendments provided the FDA with the authority to regulate devices during most phases of their development, testing, production, distribution, and use. This marked the first time the FDA clearly distinguished between devices and drugs. Regulatory requirements were derived from this 1976 law.

In 1978, with the authority granted the FDA by the amendments, the good manufacturing practices (GMPs) were promulgated. The GMP represents a quality assurance program intended to control the manufacturing, packaging, storage, distribution, and installation of medical devices. This regulation was intended to allow only safe and effective devices to reach the market place. It is this regulation that has had the greatest effect on the medical device industry. It allows the FDA to inspect a company's operations and take action on any noted deficiencies, including prohibition of device shipment.

In 1990, the Safe Medical Devices Act (SMDA) was passed by Congress. It gave the FDA authority to add preproduction design validation controls to the GMP regulations. The act also encouraged the FDA to work with foreign countries toward mutual recognition of GMP inspections.

On July 31, 1996, the new medical device reporting (MDR) regulation became effective for user facilities and device manufacturers. The MDR regulation provides a mechanism for the FDA and manufacturers to identify and monitor significant adverse events involving medical devices. The goals are to detect and correct problems in a timely manner. Although the requirements of the regulation can be enforced through legal sanctions authorized by the FFD&C Act, the FDA relies on the goodwill and cooperation of all affected groups to accomplish the objectives of the regulation. The statutory authority for the MDR regulation is section 519 of the FD&C Act, as amended by the SMDA. The SMDA requires user facilities to

- Report device-related deaths to the FDA and the device manufacturer
- Report device-related serious injuries and serious illnesses to the manufacturer, or to the FDA, if the manufacturer is not known
- Submit to the FDA on a semiannual basis a summary of all reports submitted during that period

In 1990, the FDA proposed revised GMP regulations. Almost 7 years of debate and revision followed, but finally, on October 7, 1996, the FDA issued its final rules. The new quality system (QS) regulations, incorporating the required design controls, went into effect June 1, 1997. The design control provisions were not enforced until June 14, 1998.

16.2 DEVICE CLASSIFICATION

A medical device is any article or health care product intended for use in the diagnosis of disease or other condition or for use in the care, treatment, or prevention of disease that does not achieve any of its primary intended purposes by chemical action or by being metabolized.

From 1962, when Congress passed the last major drug law revision and first attempted to include devices, until 1976 when device laws were finally written, there were almost constant congressional hearings. Testimony was presented by medical and surgical specialty groups, industry, basic biomedical sciences, and various government agencies, including the FDA. Nearly two dozen bills were rejected as either inadequate or inappropriate.

The Cooper Committee concluded that many inherent and important differences between drugs and devices necessitated a regulatory plan specifically adapted to devices. They recognized that some degree of risk is inherent in the development of many devices, so that all hazards cannot be eliminated, that there is often little or no prior experience on which to base judgments about safety and effectiveness that devices undergo performance improvement modifications during the course of clinical trials, and that results also depend upon the skill of the user.

They therefore rejected the drug-based approach and created a new and different system for evaluating devices. All devices were placed into classes based upon the degree of risk posed by each individual device and its use. The premarket notification process or 510(k) and the premarket approval application (PMAA) became the regulatory pathways for device approval. The investigational device exemption (IDE) became the mechanism to establish safety and efficacy in clinical studies for PMAAs.

16.2.1 CLASS I DEVICES

Class I devices are defined as nonlife sustaining. Their failure poses no risk to life, and there is no need for performance standards. Basic standards, however, such as premarket notification or 510(k) process, registration, device listing, GMPs, and proper record keeping are all required. Nonetheless, the FDA has exempted many of the simpler Class I devices from some or all of these requirements. For example, tongue depressors and stethoscopes are both Class I devices. Both are exempt from GMP; tongue depressors are exempt from 510(k) filing, whereas stethoscopes are not.

16.2.2 CLASS II DEVICES

Class II devices were also defined in 1976 as not life sustaining. However, they must not only comply with the basic standards for Class I devices, but also must meet specific controls or performance standards. For example, sphygmomanometers, although not essential for life, must meet standards of accuracy and reproducibility.

Premarket notification is a documentation submitted by a manufacturer who notifies the FDA that a device is about to be marketed. It assists the agency in making a determination about whether a device is substantially equivalent to a previously marketed predecessor device. As provided for in section 510(k) of the FD&C Act, the FDA can clear a device for marketing on the basis of premarket notification that the device is substantially equivalent to a pre-1976 predecessor device. The decision is based on premarket notification information that is provided by the manufacturer and includes the intended use, physical composition, and specifications of the device. Additional data usually submitted include in vitro and in vivo toxicity studies.

The premarket notification or 510(k) process was designed to give manufacturers the opportunity to obtain rapid market approval of these noncritical devices by providing evidence that their device is substantially equivalent to a device that is already marketed. The device must have the same intended use and the same or equally safe and effective technological characteristics as a predicate device.

Class II devices are usually exempt from the need to prove safety and efficacy. The FDA, however, may require additional clinical or laboratory studies. On occasion these may be as rigorous as for an IDE in support of a premarket approval (PMA), although this is rare. The FDA responds with an order of concurrence or nonconcurrence with the manufacturer's equivalency claims.

The SMDA of 1990 and the amendments of 1992 attempted to take advantage of what had been learned since 1976 to give both the FDA and manufacturers greater leeway by permitting reduction in the classification of many devices, including some life-supporting and life-sustaining devices previously in Class III, provided that reasonable assurance of safety and effectiveness can be obtained by application of special controls such as performance standards, postmarket surveillance, guidelines, and patient and device registries.

16.2.3 CLASS III DEVICES

Class III devices were defined in 1976 as either sustaining or supporting life so that their failure is life threatening. For example, heart valves, pacemakers, and PCTA balloon catheters are all Class III devices. Class III devices almost always require a PMAA, a long and complicated task fraught with many pitfalls that have caused the greatest confusion and dissatisfaction for both industry and the FDA.

The new regulations permit the FDA to use data contained in four prior PMAs for a specific device that demonstrate safety and effectiveness, to approve future PMA applications by establishing performance standards or actual reclassification. Composition and manufacturing methods that companies wish to keep as proprietary secrets are excluded. Advisory medical panel review is now elective.

However, for PMAAs that continue to be required, all of the basic requirements for Class I and II devices must be provided, plus failure mode analysis, animal tests, toxicology studies and human clinical studies, directed to establish safety and efficacy under an IDE.

It is necessary that preparation of the PMA must actually begin years before it will be submitted. It is only after the company has the results of all of the laboratory testing, preclinical animal testing, failure mode analysis and manufacturing standards on their final design that their proof of safety and efficacy can begin, in the form of a clinical study under an IDE.

At this point the manufacturer must not only have settled on a specific, fixed design for his device, but with his marketing and clinical consultants must also have decided on what the indications, contraindications, and warnings for use will be. The clinical study must be carefully designed to support these claims.

Section 520(g) of the FFD&C Act, as amended, authorizes the FDA to grant an IDE to a researcher using a device in studies undertaken to develop safety and effectiveness data for that device when such studies involve human subjects. An approved IDE application permits a device that would otherwise be subject to marketing clearance to be shipped lawfully for the purpose of conducting a clinical study. An approved IDE also exempts a device from certain sections of the act. All new significant risk devices not granted substantial equivalence under section 510(k) of the Act must pursue clinical testing under an IDE. An institutional review board (IRB) is a group of physicians and laypeople at a hospital who must approve clinical research projects before their initiation.

16.3 REGISTRATION AND LISTING

Under section 510 of the act, every person engaged in the manufacture, preparation, propagation, compounding or processing of a device shall register their name, place of business, and such establishment. This includes manufacturers of devices and components, repackers, relabelers, and initial distributors of imported devices. Those not required to register include manufacturers of raw materials, licensed practitioners, manufacturers of devices for use solely in research or teaching, warehousers, manufacturers of veterinary devices, and those who only dispense devices, such as pharmacies.

Upon registration, the FDA issues a device registration number. A change in the ownership or corporate structure of the firm, the location, or person designated as the official correspondent must be communicated to the FDA device registration and listing branch within 30 days. Registration must be done when first beginning to manufacture medical devices and must be updated yearly.

Section 510 of the act also requires all manufacturers to list the medical devices they market. Listing must be done when beginning to manufacture a product and must be updated every 6 months. Listing includes not only informing the FDA of products manufactured but also providing the agency with copies of labeling and advertising.

Foreign firms that market products in the United States are permitted but not required to register, and are required to list. Foreign devices that are not listed are not permitted to enter the country.

Registration and listing provides the FDA with information about the identity of manufacturers and the products they make. This information enables the agency to schedule inspections of facilities and also to follow up on problems. When the FDA learns about a safety defect in a particular type of device, it can use the listing information to notify all manufacturers of those devices about that defect.

16.4 510(k) PROCESS

16.4.1 DETERMINING SUBSTANTIAL EQUIVALENCY

A new device is substantially equivalent if, in comparison to a legally marketed predicate device, it has the same intended use and (1) has the same technological characteristics as the predicate device or (2) has different technological characteristics and submitted information that does not raise different questions of safety and efficacy and demonstrates that the device is as safe and effective as the legally marketed predicate device. Figure 16.1 is an overview of the substantial equivalence decision-making process.

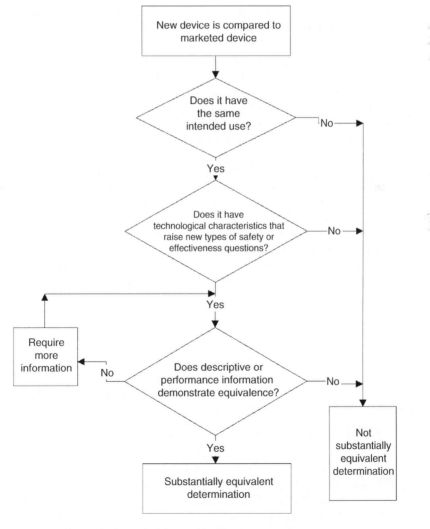

FIGURE 16.1 Substantial equivalence decision-making process.

16.4.2 Regular 510(k)

16.4.2.1 Types of 510(k)s

There are several types of 510(k) submissions that require different formats for addressing the requirements. These include

- Submissions for identical devices
- Submissions for equivalent but not identical devices
- Submissions for complex devices or for major differences in technological characteristics
- Submissions for software-controlled devices

The 510(k) for simple changes or for identical devices should be kept simple and straightforward. The submission should refer to one or more predicate devices. It should contain samples of labeling, it should have a brief statement of equivalence, and it may be useful to include a chart listing similarities and differences.

The group of equivalent but not identical devices includes combination devices where the characteristics or functions of more than one predicate device are relied on to support a substantially equivalent determination. This type of 510(k) should contain all of the information listed above as well as sufficient data to demonstrate why the differing characteristics or functions do not affect safety or effectiveness. Submission of some functional data may be necessary. It should not be necessary, however, to include clinical data—bench or preclinical testing results should be sufficient. Preparing a comparative chart showing differences and similarities with predicate devices can be particularly helpful to the success of this type of application.

Submission for complex devices or for major differences in technological characteristics is the most difficult type of submission, since it begins to approach the point at which the FDA will need to consider whether a 510(k) is sufficient of whether a PMAA must be submitted. The key is to demonstrate that the new features or the new uses do not diminish safety or effectiveness and that there are no significant new risks posed by the device. In addition to the types of information described above, this type of submission will almost always require submission of some data, possibly including clinical data.

As a general rule, it is often a good idea to meet with FDA to explain why the product is substantially equivalent, to discuss the data that will be submitted in support of a claim of substantial equivalence, and to learn the FDA's concerns and questions so that these may be addressed in the submission. The FDA's guidance documents can be of greatest use in preparing this type of submission.

The term "software" includes programs and or data that pertain to the operation of a computer-controlled system, whether they are contained on floppy disks, hard disks, magnetic tapes, laser disks, or embedded in the hardware of a device. The depth of review by the FDA is determined by the "level of concern" for the device and the role that the software plays in the functioning of the device. Levels of concern are listed as minor, moderate, and major and are tied very closely with risk analysis.

In reviewing such submissions, the FDA maintains that end-product testing may not be sufficient to establish that the device is substantially equivalent to the predicate devices. Therefore, a firm's software development process or documentation should be examined for reasonable assurance of safety and effectiveness of the software-controlled functions, including incorporated safeguards. Types of 510(k) submissions that are heavily software dependent will receive greater FDA scrutiny, and the questions posed must be satisfactorily addressed.

16.4.2.2 510(k) Format

The actual 510(k) submission will vary in complexity and length according to the type of device or product change for which substantial equivalency is sought. A submission shall be in sufficient detail to provide an understanding of the basis for a determination of substantial equivalence. All submissions shall contain the following information:

- Submitter's name, address, telephone number, a contact person, and the date the submission was prepared.
- Name of the device, including the trade or proprietary name, if applicable, the common or usual name, and the classification name.
- Identification of the predicate or legally marketed device or devices to which substantial equivalence is being claimed.
- Description of the device that is the subject of the submission, including an explanation of how the device functions, the basic scientific concepts that form the basis for the device, and the significant physical and performance characteristics of the device such as device design, materials used, and physical properties.
- Statement of the intended use of the device, including a general description of the diseases or conditions the device will diagnose, treat, prevent, cure, or mitigate, including a description, where appropriate, of the patient population for which the device is intended. If the indication statements are different from those of the predicate or legally marketed device identified above, the submission shall contain an explanation as to why the differences are not critical to the intended therapeutic, diagnostic, prosthetic, or surgical use of the device and why the differences do not affect the safety or effectiveness of the device when used as labeled.
- Statement of how the technological characteristics (design, material, chemical composition, or energy source) of the device compare to those of the predicate or legally marketed device identified above.

The 510(k) summaries for those premarket notification submissions in which a determination of substantial equivalence is based on an assessment of performance data shall contain the following information in addition to that listed above:

- Brief discussion of the nonclinical tests and their results submitted in the premarket notification.
- Brief discussion of the clinical tests submitted, referenced, or relied on in the premarket notification submission for a determination of substantial equivalence. This discussion shall include, where applicable, a description of the subjects upon whom the device was tested, a discussion of the safety and effectiveness data obtained with specific reference to adverse effects and complications, and any other information from the clinical testing relevant to a determination of substantial equivalence.
- Conclusions drawn from the nonclinical and clinical tests that demonstrate that the device is safe, effective, and performs as well as or better than the legally marketed device identified above.

The summary should be in a separate section of the submission beginning on a new page and ending on a page not shared with any other section of the premarket notification submission, and should be clearly identified as a 501(k) summary.

A 510(k) statement submitted as part of a premarket notification shall state as follows:

I certify that (name of person required to submit the pre-market notification) will make available all information included in this pre-market notification on safety and effectiveness that supports a finding of substantial equivalence within 30 days of request by any person. The information I agree to make available does not include confidential patient identifiers.

The above statement should be made in a separate section of the premarket notification submission and should be clearly identified as a 510(k) statement.

A Class III certification submitted as part of a premarket notification shall state as follows:

I certify that a reasonable search of all information known or otherwise available to (name of pre-market notification submitter) about the types and causes of reported safety and/or effectiveness problems for the (type of device) has been conducted. I further certify that the types of problems to which the (type of device) is susceptible and their potential causes are listed in the attached class III summary, and that this class III summary is complete and accurate.

The above statement should be clearly identified as a Class III certification and should be made in the section of the premarket notification submission that includes the Class III summary.

A 510(k) should be accompanied by a brief cover letter that clearly identifies the submission as a 510(k) premarket notification. To facilitate prompt routing of the submission to the correct reviewing division within FDA, the letter can mention the generic category of the product and its intended use.

When the FDA receives a 510(k) premarket notification, it is reviewed according to a checklist to assure its completeness. A sample 510(k) checklist is shown in Figure 16.2.

	Critical Elements	Yes	No
1	Is the product a device?		
2	Is the device exempt from 510(k) by regulation or policy?		
3	Is the device subject to review by CDRH?		
4	Are you aware that this device has been the subject of a previous not substantially equivalent (NSE) decision? If yes, does this new 510(k) address the NSE issues?		
5	Are you aware of the submitter being the subject of an integrity investigation? If yes, consult the Office of Device Evaluation (ODE) Integrity Officer.		
6	Has the ODE Integrity Officer given permission to proceed with the review? (Blue Book Memo #191–2 and *Federal Register* 90N-0332, September 10, 1990.)		
7	Does the submission contain the information required under Sections 510 (k), 513(f), and 513(I) of the FFD&C Act and Subpart E of Part 807 in Title 21 of the *Code of Federal Regulations*?		
8	Device trade or proprietary name?		
9	Device common or usual name or classification name?		
10	Establishment registration number? (Only applies if the establishment is registered.)		
11	Class into which the device is classified under 21 CFR Parts 862–892?		
12	Classification panel?		
13	Action taken to comply with section 514 of the act?		
14	Proposed labels, labeling, and advertisements (if available) that describe the device, its intended use, and directions for use? (Blue Book Memo #G91–1)		
15	A 510(k) summary of safety and effectiveness or a 510(k) statement that safety and effectiveness information will be made available to any person upon request?		
16	For Class III devices only, a Class III certification and a Class III summary?		
17	Photographs of the device?		
18	Engineering drawings for the device with dimensions and tolerances?		
19	The marketed device(s) to which equivalence is being claimed including labeling and description of the device?		
20	Statement of similarities and/or differences with marketed devices?		

FIGURE 16.2 Sample 510(k) checklist.

	Critical Elements	Yes	No
21	Data to show consequences and effects of a modified device(s)?		
22	Additional information that is necessary under 21 CFR 807.87 (h)?		
23	Submitter's name and address?		
24	Contact person, telephone number, and fax number?		
25	Representative/consultant, if applicable?		
26	Table of Contents, with pagination?		
27	Address of manufacturing facility/facilities and, if appropriate, sterilization site(s)?		
28	Additional information that may be necessary under 21 CFR 807.87 (h)?		
29	Comparison table of the new device to the marketed device?		
30	Action taken to comply with voluntary standards?		
31	Performance data		
	Marketed device?		
	Bench testing?		
	Animal testing?		
	Clinical data?		
	New device?		
	Bench testing?		
	Animal testing?		
	Clinical data?		
32	Sterilization information?		
33	Software information?		
34	Hardware information?		
35	Is this 510(k) is for a kit, has the kit certification statement been provided?		
36	Is this device subject to issues that have been addressed in specific guidance document(s)? If yes, continue review with checklist from any appropriate guidance document. If no, is 510(k) sufficiently complete to allow substantive review?		
37	Truthfulness certification?		
38	Other as required?		

FIGURE 16.2   (continued)

16.4.3   SPECIAL 510(K)

Under this option, a manufacturer who is intending to modify their own legally marketed device will conduct the risk analysis and the necessary verification and validation activities to demonstrate that the design outputs of the modified device meet the design input requirements. Once the manufacturer has ensured the satisfactory completion of this process, a Special 510(k): Device Modification may be submitted. While the basic content requirements of the 510(k) will remain the same, this type of submission should also reference the cleared 510(k) number and contain a declaration of conformity with design control requirements.

Under the quality system regulation, manufacturers are responsible for performing internal audits to assess their conformance with design controls. A manufacturer could, however, use a third party to provide a supporting assessment of the conformance. In this case, the third party will perform a conformance assessment for the device manufacturer and provide the manufacturer with a statement to this effect. The marketing application should then include a declaration of conformity signed by the manufacturer, while the statement from the third party should be maintained in the device master record (DMR). As always, responsibility for conformance with design control requirements rests with the manufacturer.

To provide an incentive for manufacturers to choose this option, the ODE intends to process Special 510(k)s within 30 days of receipt by the Document Mail Center. The Special 510(k) option will allow the agency to review modifications that do not affect the device's intended use or alter the device's fundamental scientific technology within this abbreviated time frame. The agency does not believe that modifications that affect the intended use or alter the fundamental scientific technology of the device are appropriate for review under this type of application, but rather should continue to be subject to the traditional 510(k) procedures.

To ensure the success of the Special 510(k) option, there must be a common understanding of the types of device modifications that may gain marketing clearance by this path. Therefore, it is critical that industry and agency staff can easily determine whether a modification is appropriate for submission by this option. To optimize, the chance that this option will be accepted and promptly cleared, manufacturers should evaluate each modification against the considerations described below to ensure that the particular change does not

- Affect the intended use
- Alter the fundamental scientific technology of the device

16.4.3.1   Special 510(k) Content

A Special 510(k) should include the following:

- Coversheet clearly identifying the application as a "Special 510(k): Device Modification."
- Name of the legally marketed (unmodified) device and the 510(k) number under which it was cleared.
- Item required under paragraph 807.87, including a description of the modified device and a comparison to the cleared device, the intended use of the device, and the proposed labeling for the device.

Concise summary of the design control activities, including (1) an identification of the risk analysis method(s) used to assess the impact of the modification on the device and its components as well as the results of the analysis, (2) based on the risk analysis, an identification of the verification or validation activities required, including methods or tests used and the acceptance criteria applied, (3) a declaration of conformity with design controls. The declaration of conformity should include

- Statement that, as required by the risk analysis, all verification and validation activities were performed by the designated individual(s) and the results demonstrated that the predetermined acceptance criteria were met.
- Statement that the manufacturing facility is in conformance with the design control procedure requirements as specified in 21 CFR 820.30 and the records are available for review.

The above two statements should be signed by the designated individual(s) responsible for those particular activities.

16.4.4   Abbreviated 510(k)

Device manufacturers may choose to submit an Abbreviated 510(k) when

- Guidance document exists.
- Special control has been established.
- FDA has recognized a relevant consensus standard.

An Abbreviated 510(k) submission must include the required elements identified in 21 CFR 807.87. In addition, manufacturers submitting an Abbreviated 510(k) that relies on a guidance document or special control(s) should include a summary report that describes how the guidance document or special control(s) were used during device development and testing. The summary report should include information regarding the manufacturer efforts to conform with the guidance document or special control(s) and should outline any deviations. Persons submitting an Abbreviated 510(k) that relies on a recognized standard should provide the information described below (except for the summary report) and a declaration of conformity to the recognized standard.

In an Abbreviated 510(k), a manufacturer will also have the option of using a third party to assess conformance with the recognized standard. Under this scenario, the third party will perform a conformance assessment to the standard for the device manufacturer and should provide the manufacturer with a statement to this effect. Like a Special 510(k), the marketing application should include a declaration of conformity signed by the manufacturer, while the statement from the third party should be maintained in the DMR pursuant to the quality system regulation. Responsibility for conformance with the recognized standard, however, rests with the manufacturer, not the third party.

The incentive for manufacturers to elect summarizes reports on the use of guidance documents or special controls or declarations of conformity to a recognized standard will be an expedited review of their submissions. While abbreviated submissions will compete with traditional 510(k) submissions, it is anticipated that their review will be more efficient than that of the traditional 510(k) submissions, which tend to be data intensive. In addition, by allowing ODE reviewers to rely on a manufacturer's summary report on the use of a guidance document or special controls and declarations of conformity with recognized standards, review resources can be directed at more complicated issues and thus should expedite the process.

16.4.4.1 Abbreviated 510(k) Content

An Abbreviated 510(k) should include

- Coversheet clearly identifying the application as an Abbreviated 510(k)
- Items required under paragraph 807.87, including a description of the device, the intended use of the device, and the proposed labeling for the device
- For a submission that relies on a guidance document or special control(s) were used to address the risks associated with the particular device type
- For a submission that relies on a recognized standard, a declaration of conformity to the standard
- Data/information to address issues not covered by guidance documents, special controls, or recognized standards
- Indications for use enclosure

16.5 DECLARATION OF CONFORMANCE TO A RECOGNIZED STANDARD

Declarations of conformity to recognized standards should include the following information:

- Identification of the applicable recognized consensus standards that were met
- Specification, for each consensus standard, that all requirements were met, except for inapplicable requirements or deviations noted below
- Identification for each consensus standard, of any manner(s) in which the standard may have been adopted for application to the device under review (e.g., an identification of an alternative series of tests that were performed)
- Identification for each consensus standard of any requirements that were not applicable to the device

- Specification of any deviations from each applicable standard that were applied
- Specification of the differences that may exist, if any, between the tested device and the device to be marketed and a justification of the test results in these areas of difference
- Name and address of any test laboratory or certification body involved in determining the conformance of the device with the applicable consensus standards and a reference to any accreditation of those organizations

16.6 PMA APPLICATION

PMA is an approval application for a Class III medical device, including all information submitted with or incorporated by reference. The purpose of the regulation is to establish an efficient and thorough device review process to facilitate the approval of PMAs for devices that have been shown to be safe and effective for their intended use and that otherwise meet the statutory criteria for approval, while ensuring the disapproval of PMAs for devices that have not been shown to be safe and effective or that do not otherwise meet the statutory criteria for approval.

16.6.1 PMA PROCESS

The first step in the PMAA process is the filing of the IDE application for significant risk devices. The IDE is reviewed by the FDA and once accepted, the sponsor can proceed with clinical trials.

16.6.2 CONTENTS OF A PMAA

Section 814.20 of 21 CFR defines what must be included in an application, including

- Name and address
- Application procedures and table of contents
- Summary
- Complete device description
- Reference to performance standards
- Nonclinical and clinical investigations
- Justification for single investigator
- Bibliography
- Sample of device
- Proposed labeling
- Environmental assessment
- Other information

The summary should include indications for use, a device description, a description of alternative practices and procedures, a brief description of the marketing history, and a summary of studies. This summary should be of sufficient detail to enable the reader to gain a general understanding of the application. The PMAA must also include the applicant's foreign and domestic marketing history as well as any marketing history of a third party marketing the same product.

The description of the device should include a complete description of the device, including pictorial presentations. Each of the functional components or ingredients should be described, as well as the properties of the device relevant to the diagnosis, treatment, prevention, cure, or mitigation of a disease or condition. The principles of the device's operation should also be explained. Information regarding the methods used in, and the facilities and controls used for the manufacture, processing, packing, storage, and installation of the device should be explained in sufficient detail so that a person generally familiar with current GMP can make a knowledgeable judgment about the quality control used in the manufacture of the device.

To clarify which performance standards must be addressed, applicants may ask members of the appropriate reviewing division of the ODE or consult FDA's list of relevant voluntary standards or the Medical Device Standards Activities Report.

16.7 INVESTIGATIONAL DEVICE EXEMPTIONS

The purpose of the IDE regulation is to encourage the discovery and development of useful devices intended for human use while protecting the public health. It provides the procedures for the conduct of clinical investigations of devices. An approved IDE permits a device to be shipped lawfully for the purpose of conducting investigations of the device without complying with a performance standard or having marketing clearance.

16.7.1 INSTITUTIONAL REVIEW BOARDS

Any human research is covered by federal regulation will not be funded unless it has been reviewed by an IRB. The fundamental purpose of an IRB is to ensure that research activities are conducted in an ethical and legal manner. Specifically, IRBs are expected to ensure that each of the basic elements of informed consent, as defined by regulation, are included in the document presented to the research participant for signature or verbal approval.

The deliberations of the IRB must determine that

- Risks to subjects are equitable.
- Selection of subjects is equitable.
- Informed consent will be sought from each prospective subject or their legally authorized representative.
- Informed consent will be appropriately documented.
- Where appropriate, the research plan makes adequate provision for monitoring the data collected to assure the safety of the subjects.
- Where appropriate, there are adequate provisions to protect the privacy of subjects and to maintain the confidentiality of data.

It is axiomatic that the IRB should ensure that the risks of participation in a research study should be minimized. The IRB must determine that this objective is to be achieved by ensuring that investigators use procedures that are consistent with sound research design and that do not necessarily expose subjects to excessive risk. In addition, the IRB needs to assure that the investigators, whenever appropriate, minimize risk and discomfort to the research participants by using, where possible, procedures already performed on the subjects as part of routine diagnosis or treatment.

The IRB is any board, committee, or other group formally designated by an institution to review, to approve the initiation of, and to conduct periodic review of biomedical research involving human subjects. The primary purpose of such review is to assure the protection of the rights and welfare of human subjects.

An IRB must comply with all applicable requirements of the IRB regulation and the IDE regulation in reviewing and approving device investigations involving human testing. An IRB has the authority to review and approve, require modification, or disapprove an investigation. If no IRB exists or if FDA finds an IRB's review to be inadequate, a sponsor may submit an application directly to FDA.

An investigator is responsible for

- Ensuring that the investigation is conducted according to the signed agreement, the investigational plan, and applicable FDA regulations
- Protecting the rights, safety, and welfare of subjects
- Control of the devices under investigation

An investigator is also responsible for obtaining informed consent and maintaining and making reports.

16.7.2 IDE FORMAT

There is no preprinted form for an IDE application, but the following information must be included in an IDE application for a significant risk device investigation. Generally, an IDE application should contain the following:

- Name and address of sponsor
- Complete report of prior investigations
- Description of the methods, facilities, and controls used for the manufacture, processing, packing, storage, and installation of the device
- Example of the agreements to be signed by the investigators and a list of the names and addresses of all investigators
- Certification that all investigators have signed the agreement, that the list of investigators includes all investigators participating in the study, and that new investigators will sign the agreement before being added to the study
- List of the names, addresses, and chairpersons of all IRBs that have or will be asked to review the investigation and a certification of IRB action concerning the investigation
- Name and address of any institution (other than those above) where a part of the investigation may be conducted
- Amount, if any, charged for the device and an explanation of why sale does not constitute commercialization
- Claim for categorical exclusion or an environmental assessment
- Copies of all labeling for device
- Copies of all informed consent forms and all related information materials provided to subjects
- Any other relevant information that FDA requests for review of the IDE application

16.8 GOOD LABORATORY PRACTICES

In 1978, FDA adopted good laboratory practices (GLPs) rules and implemented a laboratory audit and inspection procedure covering every regulated entity that conducts nonclinical laboratory studies for product safety and effectiveness. The GLPs were amended in 1984.

The GLP standard addresses all areas of laboratory operations including requirements for a quality assurance unit to conduct periodic internal inspections and keep records for audit and reporting purposes, standard operating procedures for all aspects of each study and for all phases of laboratory maintenance, a formal mechanism for evaluation and approval of study protocols and their amendments, and reports of data in sufficient detail to support conclusions drawn from them. The FDA inspection program includes GLP compliance, and a data audit to verify that information submitted to the agency accurately reflects the raw data.

16.9 GOOD MANUFACTURING PRACTICES

FDA is authorized, under section 520(f) of the act, to promulgate regulations detailing compliance with current GMPs. GMPs include the methods used in, and the facilities and controls used for, the manufacture, packing, storage, and installation of a device. The GMP regulations were established as manufacturing safeguards to ensure the production of a safe and effective device and include all of the essential elements of a quality assurance program. Because manufacturers cannot test every device, the GMPs were established as a minimum standard of manufacturing to ensure that each device produced would be safe. If a product is not manufactured according to GMPs, even if it is later shown not to be a health risk, it is in violation of the act and subject to FDA enforcement action.

The general objectives of the GMPs, not specific manufacturing methods, are found in Part 820 of the *Code of Federal Regulations*. The GMPs apply to the manufacture of every medical device. The newest GMP regulations were released in 1996 and gave the FDA the authority to examine the design area of the product development cycle for the first time. The regulation also parallels very closely the ISO 9000 set of standards.

16.10 HUMAN FACTORS

In April 1996, the FDA issued a draft primer on the use of human factors in medical device design, entitled *Do It By Design*. The purpose of the document was to improve the safety of medical devices by minimizing the likelihood of user error by systematic, careful design of the user interface, that is, the hardware and software features that define the interaction between the users and the equipment. The document contains background information about human factors as a discipline, descriptions and illustrations of device problems, and a discussion of human factors methods. It also contains recommendations for manufacturers and health facilities.

As the source for this document, the FDA extensively used the guideline *Human Factors Engineering Guidelines and Preferred Practices for the Design of Medical Devices* published by the Association for the Advancement of Medical Instrumentation as well as interfacing with human factors consultants. It is expected that human factors requirements will become part of the product submission as well as the GMP inspection.

16.11 DESIGN CONTROL

With the publication of the new GMP regulations, the FDA will have the authority to cover design controls in their inspections. The FDA issued a draft guidance document in March 1996 entitled *Design Control Guidance for Medical Device Manufacturers*. The purpose of the document was to provide readers with an understanding of what is meant by control in the context of the requirements. By providing an understanding of what constitutes control of a design process, readers could determine how to apply the concepts in a way that was both consistent with the requirements and best suited for their particular situation.

Three underlying concepts served as a foundation for the development of this guidance:

- Nature of the application of design controls for any device should be proportional to both the complexity of and the risks associated with that device.
- Design process is a multifunctional one that involves other departments beside design and development if it is to work properly, thus involving senior management as an active participant in the process.
- Product life cycle concept serves throughout the document as the framework for introducing and describing the design control activities and techniques.

Design control concepts are applicable to process development as well as product development. The extent is dependent upon the nature of the product and processes used to manufacture the product. The safety and performance of a new product is also dependent on an intimate relationship between product design robustness and process capability.

The document covers the areas of

- Risk management
- Design and development planning
- Organizational and technical interfaces
- Design input

- Design output
- Design review
- Design verification
- Design validation
- Design changes
- Design transfer

These topics are covered in detail in Sections 3 through 6 of the document.

16.12   FDA AND SOFTWARE

The subject of software in and as a medical device has become an important topic for the FDA. This interest began in 1985 when software in a radiation treatment therapy device is alleged to have resulted in a lethal overdose. The FDA then analyzed recalls by fiscal year (FY) to determine how many were caused by software problems. In FY 1985, for example, 20% of all neurology device recalls were attributable to software problems, while 8% of cardiovascular problems had the same cause. This type of analysis, along with the results of various corporate inspections, led the FDA to conclude that some type of regulation was required.

Since there are many type of software in use in the medical arena, the problem of the best way to regulate it has become an issue for the FDA. Discussions have centered on what type of software is a medical device, the type of regulation required for such software, and what could be inspected under current regulations. Agency concerns fall into three major categories: medical device software, software used in manufacturing, and software information systems used for clinical decision-making.

For medical device software, FDA is responsible for assuring that the device utilizing the software is safe and effective. It only takes a few alleged serious injuries or deaths to sensitize the agency to a particular product or generic component that deserves attention. The agency's review of MDR incidents and analysis of product recalls has convinced the agency that software is a factor contributing to practical problems within devices.

When software is used during manufacturing, FDA is concerned with whether or not the software controlling a tool or automatic tester is performing as expected. The FDA's perceptions are rooted in experiences with GMP inspections of pharmaceutical manufacturers, where computers are heavily depended upon for control of manufacturing processes. Although there are few incidents of device or manufacturing problems traceable to flaws in manufacturing software, GMP inspections have focused intensively on validation of software programs used in industry for control of manufacturing operations.

With regard to stand-alone software used to aid clinical decision-making, the FDA is concerned with hypothetical problems rather than extensive records of adverse incidents. While most commercially available health care information systems replace manual systems that had a far higher potential for errors, FDA believes that regulations should apply to the kinds of systems that may influence clinical treatment or diagnoses. FDA has observed academic work of expert systems used by medical professionals and is concerned that such systems may be commercialized without sufficient controls.

The FDA has published guidelines for developing quality software, off-the-shelf software, the requirements for product approval submissions (510k) and the inspection of software-controlled test fixtures as a part of GMP inspections. They have also conducted training courses for their inspectors and submission reviewers on the subject of software and computer basics.

16.13   SOFTWARE CLASSIFICATION

When a computer product is a component, part, or accessory of a product recognized as a medical device in its own right, the computer component is regulated according to the requirements for the parent device unless the component of the device is separately classified. Computer products that are medical devices and not components, parts or accessories of other products that are themselves medical devices are subject to one of the three degrees of regulatory control depending on their characteristics. These products are regulated with the least degree of control necessary to provide reasonable assurance of safety and effectiveness. Computer products that are substantially equivalent to a device previously classified will be regulated to the same degree as the equivalent device. Those devices that are not substantially equivalent to a preamendment device or that are substantially equivalent to a Class III device are regulated as Class III devices.

Medical software is divided into three classes with regulatory requirements specific to each:

- Class I software is subject to the act's general controls relating to such matters as misbranding, registration of manufacturers, record keeping, and GMPs. An example of Class I software would be a program that calculates the composition of infant formula.
- Class II software is that for which general controls are insufficient to provide reasonable assurance of safety and effectiveness and for which performance standards can provide assurance. This is exemplified by a computer program designed to produce radiation therapy treatment plans.
- Class III software is that for which insufficient information exists to assure that general controls and performance standards will provide reasonable assurance of safety and effectiveness. Generally, these devices are represented to be life sustaining or life supporting and may be intended for a use that is of substantial importance in preventing impairment to health. They may be implanted in the body or present a potential unreasonable risk of illness or injury. A program that measures glucose levels and calculates and dispenses insulin based upon those calculations without physician intervention would be a Class III device.

16.14 FDA INSPECTION

The FDA's power to inspect originates in section 704 of the FFD&C Act. This provision allows FDA officials to inspect any factory, warehouse, or establishment in which devices are manufactured, processed, packed, or held, for introduction into interstate commerce of after such introduction. In addition to the establishment specification, FDA is permitted to enter any vehicle used to transport or hold regulated products for export or in interstate commerce. The inspection power is specifically extended to medical device manufacturers by Sections 519 and 520 of the FFD&C Act.

Every FDA inspector is authorized by law to inspect all equipment that is used in the manufacturing process. Furthermore, investigators may examine finished and unfinished devices and device components, containers, labeling for regulated products, and all documents that are required to be kept by the regulations, such as DMRs and device history records.

Despite the broad inspectional authority over restricted devices, the statute provides that regardless of the device's unrestricted status, certain information is excluded from FDA's inspectional gambit. The kind of information to which FDA does not have access includes financial data, sales data, and pricing data. The new GMPs give the FDA authority to inspect the design area and the qualifications of personnel in all aspects of the product development process.

16.15 ADVICE ON DEALING WITH THE FDA

Several recommendations can be made regarding how to deal with the FDA and its regulatory process. None of these bits of advice are dramatic or new, but in the course of observing a firm's interaction with the agency, it is amazing how many times the failure to think of these steps can result in significant difficulties.

Know your district office. This may not be an easy thing to accomplish, since, understandably, there is a great reluctance to walk into a regulatory agency and indicate you are there to get acquainted. As opportunities arise, however, they should not be overlooked. Situations such as responding to a notice of an investigator's observations at the conclusion of an inspection or a notice of adverse findings letter are excellent opportunities to hand deliver a reply instead of simply mailing it. The verbal discussion with the reply may make the content much more meaningful and will allow both sides to learn more about the intent and seriousness with which the subject is being approached.

Prepare for inspections. When the FDA investigator walks into your manufacturing facility or corporate offices, there should be a procedure established that everyone is familiar with as to who is called, who escorts the investigator through the facility, who is available to make copies of records requested, etc. A corollary to this suggestion is to be prepared to deal with adverse inspectional findings or other communications from the agency that indicates the FDA has found violations, a serious health hazard, or other information that requires high-level company knowledge and decision-making.

Take seriously 483's and letters. Many regulatory actions are processed with no apparent indication that a firm seriously considered the violations noted by the agency.

Keep up with current events and procedures of the FDA. This will minimize the changes or surprise interpretations that could have an effect on a firm's operations and will allow for advance planning for new FDA requirements. The agency publishes much of its new program information in bulletins and other broad distribution documents, but much more can be learned from obtaining copies of FDA's *Compliance Policy Guides and Compliance Programs*.

Let the FDA know of your firm's opinions on issues, whether they are in the development state at the agency or are policies or programs established and in operation. The agency does recognize that the firms it regulates are the true experts in device manufacturing and distribution, and their views are important. The agency also recognizes that the regulation of manufacturers in not the only bottom line solving public health problems is equally or more important, and there are generally many ways to solve those problems.

EXERCISES

1. Meat industry is, despite efforts since 1906, still a major concern for some individuals. Find and discuss any recently reported hamburger spoilage problem.
2. Feedback mechanisms are of value in electronic and mechanical control systems. Why are they not useful for material flows such as foodstuffs.
3. Perform a Web search for patent medicines, document two interesting examples.
4. Why did the FDA not have intrastate control in 1906?
5. Find, using the FDA Web site, any warning letter of interest to you and report on it.
6. Use the manufacturers and users device experience (MAUDE) database to do a search for device = bed, outcome = death, for any recent year. Report on your results.
7. Select a medical device used by one of your acquaintances, do a MAUDE search to determine if there have been any negative outcomes in the past 5 years.
8. Perform a Web search for quack medical devices. Report on one.
9. Much recent television advertising exaggerates claims for nonmedical drugs or devices (such as Viagra and muscle stimulators). Find and report on an example, discuss how truth is being bent. What would it take to get the FDA involved?

BIBLIOGRAPHY

Banta, H.D., The regulation of medical devices, *Preventive Medicine*, 19(6), November 1990.
Basile, E., Overview of current FDA requirements for medical devices, in *The Medical Device Industry: Science, Technology, and Regulation in a Competitive Environment*. New York: Marcel Dekker, 1990.

Basile, E.M. and Alexis, J.P., Compiling a successful PMA application, in *The Medical Device Industry: Science, Technology, and Regulation in a Competitive Environment*. New York: Marcel Dekker, 1990.

Bureau of Medical Devices, Office of Small Manufacturers Assistance, *Regulatory Requirements for Marketing a Device*. Washington, DC: U.S. Department of Health and Human Services, 1982.

Center for Devices and Radiological Health, *Device Good Manufacturing Practices Manual*. Washington, DC: U.S. Department of Health and Human Services, 1987.

Food and Drug Administration, *Federal Food, Drug and Cosmetic Act, as Amended January, 1979*. Washington, DC: U.S. Government Printing Office, 1979.

Food and Drug Administration, *Guide to the Inspection of Computerized Systems in Drug Processing*. Washington, DC: Food and Drug Administration, 1983.

Food and Drug Administration, *FDA Policy for the Regulation of Computer Products (Draft)*. Washington, DC: Federal Register, 1987.

Food and Drug Administration, *Software Development Activities*. Washington, DC: Food and Drug Administration, 1987.

Food and Drug Administration, *Medical Devices GMP Guidance for FDA Inspectors*. Washington, DC: Food and Drug Administration, 1987.

Food and Drug Administration, *Reviewer Guidance for Computer-Controlled Medical Devices (Draft)*. Washington, DC: Food and Drug Administration, 1988.

Food and Drug Administration, *Investigational Device Exemptions Manual*. Rockville MD: Center for Devices and Radiological Health, 1992.

Food and Drug Administration, *Premarket Notification 510(k): Regulatory Requirements for Medical Devices*. Rockville MD: Center for Devices and Radiological Health, 1992.

Food and Drug Administration, *Premarket Approval (PMA) Manual*. Rockville MD: Center for Devices and Radiological Health, 1993.

Food and Drug Administration, *Design Control Guidance for Medical Device Manufacturers*. Draft. Rockville, MD: Center for Devices and Radiological Health, 1996.

Food and Drug Administration, *Do It By Design: An Introduction to Human Factors in Medical Devices*. Draft. Rockville, MD: Center for Devices and Radiological Health, 1996.

Fries, R.C., *Medical Device Quality Assurance and Regulatory Compliance*. New York: Marcel Dekker, 1998.

Fries, R.C., *Handbook of Medical Device Design*. New York: Marcel Dekker, 2001.

Fries, R.C., *Reliable Design of Medical Devices*, 2nd ed. New York: Marcel Dekker, 2005.

Fries, R.C. et al., Software regulation, in *The Medical Device Industry: Science, Technology, and Regulation in a Competitive Environment*. New York: Marcel Dekker, 1990.

Ginzburg, H.M., Protection of research subjects in clinical research, in *Legal Aspects of Medicine*. New York: Springer-Verlag, 1989.

Gundaker, W.E., FDA's regulatory program for medical devices and diagnostics, in *The Medical Device Industry: Science, Technology, and Regulation in a Competitive Environment*. New York: Marcel Dekker, 1990.

Holstein, H.M., How to submit a successful 510(k), in *The Medical Device Industry: Science, Technology, and Regulation in a Competitive Environment*. New York: Marcel Dekker, 1990.

Jorgens, J. III and Burch, C.W., FDA regulation of computerized medical devices, *Byte*, 7, 1982.

Jorgens, J. III, Computer hardware and software as medical devices, *Medical Device and Diagnostic Industry*. May 1983.

Jorgens, J. III and Schneider, R., Regulation of medical software by the FDA, *Software in Health Care*. April–May 1985.

Kahan, J.S., Regulation of computer hardware and software as medical devices, *Canadian Computer Law Reporter*, 6(3), January 1987.

King, P.H. and Richard, C.F., *Design of Biomedical Devices and Systems*. New York: Marcel Dekker, 2002.

Munsey, R.R. and Howard, M.H., FDA/GMP/MDR inspections: Obligations and rights, in *The Medical Device Industry: Science, Technology, and Regulation in a Competitive Environment*. New York: Marcel Dekker, 1990.

Office of Technology Assessment, *Federal Policies and the Medical Devices Industry*. Washington, DC: U.S. Government Printing Office, 1984.

Sheretz, R.J. and Stephen, A.S., Medical devices. Significant risk vs nonsignificant risk, *Journal of the American Medical Association*, 272(12), September 28, 1994.

Trull, F.L. and Barbara, A.R., The animal testing issue, in *The Medical Device Industry: Science, Technology, and Regulation in a Competitive Environment*. New York: Marcel Dekker, 1990.

U.S. Congress, House Committee on Interstate and Foreign Commerce. *Medical Devices. Hearings before the Subcommittee on Public Health and the Environment*. October 23–24, 1973. Serial Numbers 93–61 Washington DC: U.S. Government Printing Office, 1973.

Wholey, M.H. and Jordan, D.H., An introduction to the Food and Drug Administration and how it evaluates new devices: Establishing safety and efficacy, *CardioVascular and Interventional Radiology*, 18(2), March/April 1995.

17 Regulations and Standards

The Lord's Prayer is 66 words, the Gettysburg Address is 286 words, and there are 1,322 words in the Declaration of Independence. Yet government regulations on the sale of cabbage total 26,911 words.

David McIntosh

The degree to which formal standards and regulations are applied to product development varies from company to company. In many cases, standards are dictated by customers or regulatory mandate. In other situations, standards are self-imposed. If formal standards do exist, assurance activity must be established to assure that they are being followed. An assessment of compliance to standards may be conducted as part of a formal technical review or by audit.

The European Community's (EC) program on the completion of the internal market has, as the primary objective for medical devices, to assure Community-wide free circulation of products. The only means to establish such free circulation, in view of quite divergent national systems, regulations governing medical devices, and existing trade barriers, was to adopt legislation for the Community, by which the health and safety of patients, users, and third persons would be ensured through a harmonized set of device-related protection requirements. Devices meeting the requirements and those sold to members of the Community are identified by means of a CE mark.

The Active Implantable Medical Devices Directive (AIMDD) adopted by the Community legislator in 1990 and the Medical Devices Directive (MDD) in 1993 cover more than 80% of medical devices for use with human beings. After a period of transition, that is, a period during which the laws implementing a Directive coexist with preexisting national laws, these directives exhaustively govern the conditions for placing medical devices on the market. Through the agreements on the European Economic Area (EEA), the relevant requirements and procedures are the same for all EC member states and European Free Trade Association (EFTA) countries that belong to the EEA, an economic area comprising more than 380 million people.

17.1 DEFINITION OF A MEDICAL DEVICE

The various MDD define a medical device as

any instrument, appliance, apparatus, material or other article, whether used alone or in combination, including the software necessary for its proper application, intended by the manufacturer to be used for human beings for the purpose of:

- diagnosis, prevention, monitoring, treatment or alleviation of disease
- diagnosis, monitoring, alleviation of or compensation for an injury or handicap
- investigation, replacement or modification of the anatomy or of a physiological process
- control of conception,

and which does not achieve its principal intended action in or on the human body by pharmacological, immunological or metabolic means, but which may be assisted in its function by such means.

One important feature of the definition is that it emphasizes the "intended use" of the device and its "principal intended action." This use of the term "intended" gives manufacturers of certain products some opportunity to include or exclude their product from the scope of the particular Directive.

Another important feature of the definition is the inclusion of the term "software." The software definition will probably be given further interpretation, but is currently interpreted to mean that (1) software intended to control the function of a device is a medical device, (2) software for patient records or other administrative purposes is not a device, (3) software which is built into a device, for example, software in an electrocardiograph monitor used to drive a display, is clearly an integral part of the medical device, and (4) a software update sold by the manufacturer, or a variation sold by a software house, is a medical device in its own right.

17.2 MDD

17.2.1 MDD Process

The process of meeting the requirements of the MDD is a multistep approach, involving the following activities:

- Analyze the device to determine which directive is applicable.
- Identify the applicable essentials requirements list.
- Identify any corresponding harmonized standards.
- Confirm that the device meets the essential requirements/harmonized standards and document the evidence.
- Classify the device.
- Decide on the appropriate conformity assessment procedure.
- Identify and choose a Notified Body.
- Obtain conformity certifications for the device.
- Establish a declaration of conformity.
- Apply for the CE mark.

This process does not necessarily occur in a serial manner, but iterations may occur throughout the cycle. Each activity in the process will be examined in detail.

17.2.2 Choosing the Appropriate Directive

Because of the diversity of current national medical device regulations, the Commission decided that totally new Community legislation covering all medical devices was needed. Software or a medical device containing software may be subject to the requirements of the AIMDD or the MDD. Three directives are envisaged to cover the entire field of medical devices.

17.2.2.1 AIMDD

This directive applies to a medical device which depends on a source of electrical energy or any source of power other than that directly generated by the human body or gravity, which is intended to be totally or partially introduced, surgically or medically, into the human body or by medical intervention into a natural orifice, and which is intended to remain after the procedure. This directive was adopted in June 1990, implemented in January 1993, and the transition period ended in January 1995.

17.2.2.2 MDD

This directive applies to all medical devices and accessories, unless they are covered by the AIMDD or the In Vitro Diagnostic Medical Devices Directive (IVDMDD). It was adopted in June 1993, was implemented in January 1995 and the transition period ended in June 1998.

17.2.2.3 IVDMDD

This directive applies to any medical device that is a reagent, reagent product, calibrator, control kit, instrument, equipment, or system intended to be used in vitro for the examination of samples derived from the human body for the purpose of providing information concerning a physiological state of health or disease or congenital abnormality, or to determine the safety and compatibility with potential recipients.

17.2.3 IDENTIFYING THE APPLICABLE ESSENTIAL REQUIREMENTS

The major legal responsibility the Directives place on the manufacturer of a medical device requires the device meet the essential requirements set out in Annex I of the Directive which applies to them, taking into account the intended purpose of the device. The essential requirements are written in the form of (1) general requirements which always apply and (2) particular requirements, only some of which apply to any particular device.

The general requirements for the essential requirements list take the following form:

- Device must be safe. Any risk must be acceptable in relation to the benefits offered by the device.
- Device must be designed in such a manner that risk is eliminated or minimized.
- Device must perform in accordance with the manufacturer's specification.
- Safety and performance must be maintained throughout the indicated lifetime of the device.
- Safety and performance of the device must not be affected by normal conditions of transport and storage.
- Any side effects must be acceptable in relation to the benefits offered.

The particular requirements for the essential requirements list address the following topics:

- Chemical, physical, and biological properties
- Infection and microbial contamination
- Construction and environmental properties
- Devices with a measuring function
- Protection against radiation
- Requirements for devices connected to or equipped with an energy source
- Protection against electrical risks
- Protection against mechanical and thermal risks
- Protection against the risks posed to the patient by energy supplies or substances
- Information supplied by the manufacturer

The easiest method of assuring the essential requirements are met to establish a checklist of the essential requirements from Appendix I of the appropriate Directive, which then forms the basis of the technical dossier. Figure 17.1 is an example of an essential requirements checklist.

The essential requirements checklist includes (1) a statement of the essential requirements, (2) an indication of the applicability of the essential requirements to a particular device, (3) a list of the standards used to address the essential requirements, (4) the activity that addresses the essential requirements, (5) the clause(s) in the standard detailing the applicable test for the particular essential requirement, (6) an indication of whether the device passed/or failed the test and (7) a statement of the location of the test documentation or certificates.

Essential Requirement	A or N/a	Standards	Activity	Test Clause	Pass/Fail	Document Location
1. The device must be designed and manufactured in such a way that when used under the conditions and for the purposes intended, they will not compromise the clinical condition or the safety of patients, users, and where applicable, other persons. The risks associated with devices must be reduced to an acceptable level compatible with a high level of protection for health and safety.	A	Internal	Risk analysis Safety review			Design history file Design history file
2. The solutions adopted by the manufacturer for the design and construction of the devices must comply with safety principles and also take into account the generally acknowledged state of the art.	A	Internal	Specification reviews Design reviews Safety review			Design history file Design history file Design history file

FIGURE 17.1 Example of essential requirements checklist.

17.2.4 Identification of Corresponding Harmonized Standards

A harmonized standard is a standard produced under a mandate from the European Commission by one of the European standardizations such as CEN (the European Committee for Standardization) and CENELEC (the European Committee for Electrotechnical Standardization), and which has its reference published in the *Official Journal of the European Communities*.

The essential requirements are worded such that they identify a risk and state that the device should be designed and manufactured so that the risk is avoided or minimized. The technical detail for assuring these requirements is to be found in harmonized standards. Manufacturers must therefore identify the harmonized standards corresponding to the essential requirements that apply to their device.

With regard to choosing such standards, the manufacturer must be aware of the hierarchy of standards that have been developed:

- Horizontal standards: Generic standards covering fundamental requirements common to all, or a very wide range of medical devices.
- Semihorizontal standards: Group standards that deal with requirements applicable to a group of devices.
- Vertical standards: Product-specific standards that give requirements to one device or a very small group of devices.

Manufacturers must give particular attention to the horizontal standards, because of their general nature they apply to almost all devices. As these standards come into use for almost all products, they will become extremely powerful.

Semihorizontal standards may be particularly important as they have virtually the same weight as horizontal standards for groups of devices, such as orthopedic implants, in vitro devices (IVDs), or x-ray equipment.

Vertical standards might well be too narrow to cope with new technological developments when a question of a specific feature of a device arises.

Table 17.1 lists some common harmonized standards for medical devices and medical device electromagnetic compatibility (EMC) standards.

TABLE 17.1
Common Harmonized Standards for Medical Devices

Standard	Areas Covered
EN 60 601 series	Medical electrical equipment
EN 29000 series	Quality systems
EN 46000 series	Quality systems
EN 55011 (CISPR 11)	EMC/emission
EN 60801 series	EMC/immunity
EN 540	Clinical investigation of medical devices
EN 980	Symbols on medical equipment
IEC 601-1-2	Medical device emission and immunity
IEC 801-2	Electrostatic discharge
IEC 801-2	Immunity to radiated radio frequency electromagnetic fields
IEC 801-4	Fast transients/burst
IEC 801-5	Voltage surge immunity

17.2.5 ASSURANCE THAT THE DEVICE MEETS THE ESSENTIAL REQUIREMENTS AND HARMONIZED STANDARDS AND DOCUMENTATION OF THE EVIDENCE

Once the essential requirements list has been developed and the harmonized standards chosen, the activity necessary to address the essential requirements list must be conducted. Taking the activity on the essential requirements checklist from Figure 17.1, the following activity may be conducted to assure the requirements are met.

17.2.5.1 Essential Requirement 1

This requirement is concerned with the device not compromising the clinical condition or the safety of patient, users, and where applicable, other persons. The methods used to meet this requirement are the conduction of a hazard analysis and a safety review.

17.2.5.1.1 Hazard Analysis
A hazard analysis is the process, continuous throughout the product development cycle that examines the possible hazards that could occur due to equipment failure and helps the designer to eliminate the hazard through various control methods. The hazard analysis is conducted on hardware, software, and the total system during the initial specification phase and is updated throughout the development cycle. The hazard analysis is documented on a form similar to that shown in Figure 17.2.

The hazard analysis addresses the following issues:

Potential hazard	Identifies possible harm to patient, operator, or system.
Generic cause	Identifies general conditions that can lead to the associated potential hazard.
Specific cause	Identifies specific instances that can give rise to the associated generic cause.
Probability	Classifies the likelihood of the associated potential hazard according to Table 17.2.
Severity	Categorizes the associated potential hazard according to Table 17.3.
Control mode	Means of reducing the probability and severity of the associated potential hazard.
Control method	Actual implementation to achieve the associated control mode.
Comments	Additional information, references, etc.

Potential Hazard	Generic Cause	Specific Cause	Probability	Severity	Control Mode	Control Method	Comments

FIGURE 17.2 Example of a hazard analysis sheet.

When the hazard analysis is initially completed, the probability and severity refer to the potential hazard before it is being controlled. As the device is designed to minimize or eliminate the hazard, and control methods are imposed, the probability and severity will be updated.

An organization separate from R&D, such as quality assurance, reviews the device to assure it is safe and effective for its intended use. The device, when operated according to specification, must not cause a hazard to the user or the patient. In the conduction of this review, the following may be addressed.

17.2.5.1.2 Safety Review
- Pertinent documentation such as drawings, test reports, and manuals
- Sample of the device
- Checklist specific to the device, which may include
  - Voltages
  - Operating frequencies
  - Leakage currents
  - Dielectric withstand
  - Grounding impedance
  - Power cord and plug
  - Electrical insulation
  - Physical stability
  - Color coding
  - Circuit breakers and fuses
  - Alarms, warnings, and indicators
  - Mechanical design integrity

The checklist is signed by the reviewing personnel following the analysis.

TABLE 17.2

Hazard Analysis Probability Classification

Classification Indicator	Classification Rating	Classification Meaning
1	Frequent	Likely to occur often
2	Occasional	Will occur several times in the life of the system
3	Reasonably remote	Likely to occur sometime in the life of the system
4	Remote	Unlikely to occur, but possible
5	Extremely remote	Probability of occurrence indistinguishable from zero
6	Physically impossible	

TABLE 17.3
Hazard Analysis Severity Classification

Severity Indicator	Severity Rating	Severity Meaning
I	Catastrophic	May cause death or system loss
II	Critical	May cause severe injury, severe occupational illness or severe system damage
III	Marginal	May cause minor injury, minor occupational illness, or minor system damage
IV	Negligible	Will not result in injury, illness, or system damage

17.2.5.2   Essential Requirement 2

This requirement is concerned with the device complying with safety principles and the generally acknowledged state of the art. The methods used to meet this requirement are peer review and safety review.

17.2.5.2.1   Peer Review

Peer review of the product specification, design specification, software requirements specification, and the actual design are conducted using qualified individuals not directly involved in the development of the device. The review is attended by individuals from design, reliability, quality assurance, regulatory affairs, marketing, manufacturing, and service. Each review is documented with issues discussed and action items. After the review, the project team assigns individuals to address each action item and a schedule for completion.

17.2.5.2.2   Safety Review

This was already discussed under Section 17.2.5.1.

17.2.5.3   Essential Requirement 3

This requirement is concerned with the device achieving the performance intended by the manufacturer. The methods used to meet this requirement are the various specification reviews and the validation of the device to meet these specifications.

17.2.5.3.1   Specification Reviews

This was discussed under Section 17.2.5.2.

17.2.5.3.2   Validation Testing

This activity involves assuring that the design and the product meet the appropriate specifications that were developed at the beginning of the development process. Testing is conducted to address each requirement in the specification and the test plan and test results were documented. It is helpful to develop a requirements matrix to assist in this activity.

17.2.5.4   Essential Requirement 4

This requirement is concerned with the device being adversely affected by stresses which can occur during normal conditions of use. The methods used to meet this requirement are environmental testing, environmental stress screening (ESS), and use/misuse evaluation.

17.2.5.4.1   Environmental Testing

Testing is conducted according to environmental specifications listed for the product. Table 17.4 lists the environmental testing to be conducted and the corresponding standards and methods employed. Test results are documented.

TABLE 17.4

List of Environmental Testing

Environmental Test	Specification Range	Applicable Standard
Operating temperature	5°C–35°C	IEC 68-2-14
Storage temperature	−40°C to +65°C	IEC 68-2-1-Ab
		IEC 68-2-2-Bb
Operating humidity	15%–95% RH noncondensing	IEC 68-2-30
Operating pressure	500–797 mm Hg	IEC 68-2-13
Storage pressure	87–797 mm Hg	IEC 68-2-13
Radiated electrical emissions	System: 4 dB margin	CISPR 11
	Subsystem: 15 dB	
Radiated magnetic emissions	System: 4 dB margin	VDE 871
	Subsystem: 6 dB	
Line conducted emissions	System: 2 dB margin	CISPR 11
	Subsystem: 2 dB	VDE 871
Electrostatic discharge	Contact: 7 kV	EN 60601-2
	Air: 10 kV	EN 1000-4-2
Radiated electric field immunity	5 V/m at 1 kHz	EN 60601-2
		EN 1000-4-3
Electrical fast transient immunity	Power mains: 2.4 kV	EN 60601-2
	Cables >3 m: 1.2 kV	EN 1000-4-4
Stability		UL 2601
Transportation		NSTA preshipment
Transportation		NSTA overseas

17.2.5.4.2 Environmental Stress Screening

The device is subjected to temperature and vibration stresses beyond which the device may ordinarily see to precipitate failures. The failure may then be designed out of the device before it is produced. ESS is conducted according to a specific protocol which is developed for the particular device. Care must be taken in preparing the protocol to avoid causing failures which would not ordinarily be anticipated. Results of the ESS analysis are documented.

17.2.5.4.3 Use/Misuse Evaluation

Whether through failure to properly read the operation manual or through improper training, medical devices are going to be misused and even abused. There are many stories of product misuse, such as the handheld monitor that was dropped into a toilet bowl, the physician that hammered a 9 V battery in backwards and then reported the device was not working, or the user that spilled a can of soda on and into a device.

Practically, it is impossible to make a device completely immune to misuse, but it is highly desirable to design around the misuse situations than can be anticipated. These include

- Excess application of cleaning solutions
- Physical abuse
- Spills
- Excess weight applied to certain parts of the device
- Excess torque applied to controls or screws
- Improper voltages, frequencies, or pressures
- Interchangeable electrical or pneumatic connections

Each potential misuse situation should be evaluated for its possible result on the device and a decision should be made whether the result can be designed out. Activities similar to these are carried out to complete the remainder of the essential requirements checklist for the device.

17.2.6 CLASSIFICATION OF THE DEVICE

It is necessary for the manufacturer of a medical device to have some degree of proof that a device complies with the essential requirements before the CE marking can be applied. This is defined as a "conformity assessment procedure." For devices applicable to the AIMDD, there are two alternatives for the conformity assessment procedure. For devices applicable to the IVDMDD, there is a single conformity assessment procedure. For devices applicable to the MDD, there is no conformity assessment procedure that is suitable for all products, as the Directive covers all medical devices. Medical devices are therefore divided into four classes, which have specific conformity assessment procedures for each of the four classes.

It is crucial for manufacturers to determine the class into which each of their devices falls. This demands careful study of the classification rules given in Annex IX of the Directive. As long as the intended purpose, the implementing rules, and the definitions are clearly understood, the classification process is straightforward and the rules, which are laid out in a logical order, can be worked out in succession from rule 1. If the device is used for more than one intended purpose, then it must be classified according to the one which gives the highest classification.

The rules for determining the appropriate classification of a medical device include

Rule	Type of Device	Class
1–4	Noninvasive devices are in Class I except those used for	IIa
	Storing body fluids connected to an active medical device in Class IIa or higher	
	Modification of body fluids	IIa/IIb
	Some wound dressings	IIa/IIb
5	Devices invasive with respect to body orifices	
	Transient use	I
	Short-term use	IIa
	Long-term use	IIb
6–8	Surgically invasive devices	
	Reusable surgical instruments	I
	Transient or short-term use	IIa
	Long-term use	IIb
	Contact with CCS or CNS	III
	Devices that are absorbable or have a biological effect	IIb/III
	Devices that deliver medicines	IIb/III
	Devices applying ionizing radiation	IIb
13	Devices incorporating medicinal products	III
14	Contraceptive devices	IIb/III
15	Chemicals used for cleaning or disinfecting	
	Medical devices	IIa
	Contact lenses	IIb
16	Devices specifically intended for recording x-ray images	IIa
17	Devices made from animal tissues	III
18	Blood bags	IIb

In the cases of active devices, the rules are based mainly on the purpose of the device, that is, diagnosis or therapy, and the corresponding possibility of absorption of energy by the patient.

Rule	Type of Device	Class
9	Therapeutic devices administering or exchanging energy	IIa
	If operating in a potentially hazardous way	IIb
10	Diagnostic devices	
	Supplying energy other than illumination	IIa
	Imaging radiopharmaceuticals in vivo	IIa
	Diagnosing/monitoring vital functions	IIa
	Monitoring vital functions in critical care conditions	IIb
	Emitting ionizing radiation	IIb
11	Active devices administering/removing medicines/body substances	IIa
	If operating in a potentially hazardous way	IIb
12	All other active devices	I

In order to use the classification system correctly, manufacturers must have a good understanding of the implementing rules and definitions. The key implementing rules include

- Application of the classification rules is governed by the intended purpose of the device.
- If the device is intended to be used in combination with another device, the classification rules are applied separately to each device.
- Accessories are classified in their own right, separately from the device with which they are used.
- Software, which drives a device or influences the use of a device, falls automatically in the same class as the device itself.

17.2.7 DECISION ON THE APPROPRIATE CONFORMITY ASSESSMENT PROCEDURE

17.2.7.1 MDD

There are six conformity assessment annexes (II–VII) to the MDD. Their use for the different classes of devices is specified in Article 11 of the Directive.

17.2.7.1.1 Annex II
This annex describes the system of full quality assurance covering both the design and manufacture of devices.

17.2.7.1.2 Annex III
This annex describes the type examination procedure according to which the manufacturer submits full technical documentation of the product, together with a representative sample of the device to a Notified Body.

17.2.7.1.3 Annex IV
This annex describes the examination by the Notified Body of every individual product, or one or more samples from every production batch, and the testing which may be necessary to show the conformity of the products with the approved/documented design.

17.2.7.1.4 Annex V
This annex describes a production quality system which is to be verified by a Notified Body as assuring that devices are made in accordance with an approved type, or in accordance with technical documentation describing the device.

17.2.7.1.5 Annex VI

This annex describes a quality system covering final inspection and testing of products to ensure that devices are made in accordance with an approved type, or in accordance with technical documentation.

17.2.7.1.6 Annex VII

This annex describes the technical documentation that the manufacturer must compile to support a declaration of conformity for a medical device, where there is no participation of a Notified Body in the process. The class to which the medical device is assigned has an influence on the type of conformity assessment procedure chosen.

17.2.7.1.6.1 Class I

Compliance with the essential requirements must be shown in technical documentation compiled according to Annex VII of the Directive.

17.2.7.1.6.2 Class IIa

The design of the device and its compliance with the essential requirements must be established in technical documentation described in Annex VII. However, for this class, agreement of production units with the technical documentation must be assured by a Notified Body according to one of the following alternatives:

Sample testing	Annex IV
An audited production quality system	Annex V
An audited product quality system	Annex VI

17.2.7.1.6.3 Class IIb

The design and manufacturing procedures must be approved by a Notified Body as satisfying Annex II, or the design must be shown to conform to the essential requirements by a type examination (Annex III) carried out by a Notified Body.

17.2.7.1.6.4 Class III

The procedures for this class are similar to Class IIb, but significant differences are that when the quality system route is used, a design dossier for each type of device must be examined by the Notified Body. Clinical data relating to safety and performance must be included in the design dossier or the documentation presented for the type examination.

17.2.7.2 AIMDD

For devices following the AIMDD, Annexes II through V cover the various conformity assessment procedures available. There are two alternative procedures.

17.2.7.2.1 Alternative 1

A manufacturer must have in place a full quality assurance system for design and production and must submit a design dossier on each type of device to the Notified Body for review.

17.2.7.2.2 Alternative 2

A manufacturer submits an example of each type of his device to a Notified Body, satisfactory production must be assured by either the quality system at the manufacturing site must comply with EN 29002 + EN 46002 and must be audited by a Notified Body, or samples of the product must be tested by a Notified Body.

17.2.7.3 IVDMDD

For devices adhering to the IVDMDD, the conformity assessment procedure is a manufacturer's declaration. In vitro devices for self-testing, must additionally have a design examination by a Notified Body, or be designed and manufactured in accordance with a quality system.

In choosing a conformity assessment procedure it is important to remember that (1) it is essential to determine the classification of a device before deciding on a conformity assessment procedure, (2) it may be more efficient to operate one conformity assessment procedure throughout a manufacturing plant, even though this procedure may be more rigorous than strictly necessary for some products, and (3) tests and assessments carried out under current national regulations can contribute toward the assessment of conformity with the requirements of the Directives.

17.2.8 TYPE TESTING

A manufacturer of Class IIb or Class III medical devices can choose to demonstrate that his device meets the essential requirements by submitting to a Notified Body for a type examination as described in Annex III of the Directive. The manufacturer is required to submit technical documentation on his device together with an example of the device. The Notified Body will then carry out such tests as it considers necessary to satisfy itself, before issuing the EC type examination certificate.

Type testing of many kinds of medical devices, particularly electromedical equipment, is required under some current national regulations. Manufacturers who are familiar with this process and who have established relations with test houses which are, or will be, appointed as Notified Bodies, are likely to find this a more attractive procedure than the design control procedures of EN 29001/EN 46001. Existing products which have already been type tested under current national procedures are likely to meet most of the essential requirements and may require little or no further testing. Testing by one of the nationally recognized test houses may also gain entitlement to national or proprietary marks which can be important in terms of market acceptance.

A major issue in type examination is the handling of design and manufacturing changes. Annex III states that the manufacturer must inform the Notified Body of any significant change made to an approved product, and that the Notified Body must give further approval if the change could affect conformity with the essential requirements. The meaning of significant change must be negotiated with the Notified Body but clearly for certain products or for manufacturers with a large number of products, the notification and checking of changes could impose a serious burden.

When a change could have an effect on the compliance with the essential requirements, the manufacturer should make his own assessment, including tests, to determine that the device still complies and submit updated drawings and documentation, together with the test results. The Notified Body must be informed of all changes made as a result of an adverse incident.

When the assessment is that the changes are not liable to have an effect, they should be submitted to the Notified Body "for information only." The manufacturer must, in such cases, keep records of the change and of the rationale for the conclusion that the change could not have an effect.

17.2.9 IDENTIFICATION AND CHOICE OF A NOTIFIED BODY

Identifying and choosing a Notified Body is one of the most critical issues facing a manufacturer. A long-term and close relationship should be developed, and time and care spent in making a careful choice of a Notified Body should be viewed as an investment in the future of the company.

Notified bodies must satisfy the criteria given in Annex XI of the MDD, namely

* Independence from the design, manufacture, or supply of the devices in question
* Integrity

- Competence
- Staff who are trained, experienced, and able to report
- Impartiality of the staff
- Possession of liability insurance
- Professional secrecy

In addition, the bodies must satisfy the criteria fixed by the relevant harmonized standards. The relevant harmonized standards include those of the EN 45000 series dealing with the accreditation and operation of certification bodies. The tasks to be carried out by Notified Bodies include

- Audit manufacturers; quality systems for compliance with Annexes II, V, and VI.
- Examine any modifications to an approved quality system.
- Carry out periodic surveillance of approved quality systems.
- Examine design dossiers and issue EC design examination certificates.
- Examine modifications to an approved design.
- Carry out type examinations and issue EC type examination certificates.
- Examine modifications to an approved type.
- Carry out EC verification.
- Take measures to prevent rejected batches from reaching the market.
- Agree with the manufacturer time limits for the conformity assessment procedures.
- Take into account the results of tests or verifications already carried out.
- Communicate to other Notified Bodies (on request) all relevant information about approvals of quality systems issued, refused, and withdrawn.
- Communicate to other Notified Bodies (on request) all relevant information about EC type approval certificates issued, refused, and withdrawn.

Notified Bodies must be located within the EC in order that effective control may be applied by the competent authorities that appointed them, but certain operations may be carried out on behalf of Notified Bodies by subcontractors who may be based outside the EC. Competent authorities will generally notify bodies on their own territory, but they may notify bodies based in another member state provided that they have already been notified by their parent Competent Authority. There are several factors to be taken into account by a manufacturer in choosing a Notified Body, including

- Experience with medical devices
- Range of medical devices for which the Notified Body has skills
- Possession of specific skills, for example, EMC or software
- Links with subcontractors and subcontractor skills
- Conformity assessment procedures for which the body is notified
- Plans for handling issues, such as clinical evaluation
- Attitude to existing certifications
- Queue times/processing times
- Costs
- Location and working languages

Experience with medical devices is limited to a small number of test houses and their experience is largely confined to electromedical equipment. Manufacturers should probe carefully the competence of the certification body to assess their device. Actual experience with a product of a similar nature would be reassuring. The certification body should be pressed to demonstrate sufficient understanding of the requirements, particularly where special processes are involved (e.g., sterilization) or where there is previous experience.

Certain devices demand specific skills that may not be found in every Notified Body. Clearly, the Notified Body must have, or be able to obtain, the skills required for the manufacturer's devices.

Many Notified Bodies will supplement their in-house skills by the use of specialist subcontractors. This is perfectly acceptable as long as all the rules of subcontracting are followed. Manufacturers should verify themselves for the reputation of the subcontractor and the degree of supervision applied by the Notified Body.

The main choice open to manufacturers is full quality system certification or type examination combined with one of the less rigorous quality system certifications. Some Notified Bodies have a tradition of either product testing or systems evaluation and it therefore makes sense to select a Notified Body with experience in the route chosen.

A clinical evaluation is required for some medical devices, especially Class III devices and implants. Although this will be a key aspect of demonstrating conformity, it will be important for manufacturers to know how the Notified Body intends to perform this function.

In preparing the MDD, the need to avoid reinventing the wheel has been recognized. In order to maximize this need, companies whose products have already been certified by test houses that are likely to become Notified Bodies may wish to make use of the organizations with whom they have previously worked. It will be important to verify with the Notified Body the extent to which the testing previously performed is sufficient to meeting the essential requirements.

At the time of this writing, most Notified Bodies seem to be able to offer fairly short lead times. The time for actually carrying out the examination or audit should be questioned. It must be remembered that manufacturers will have to pay Notified Bodies for carrying out the conformity assessment procedures. There will certainly be competition and this may offer some control over costs. Although it will always be a factor, the choice of a Notified Body should not be governed by cost alone bearing in mind the importance of the exercise.

For obvious reasons of expense, culture, convenience, and language there will be a tendency for European manufacturers to use a Notified Body situated in their own country. Nevertheless, this should not be the principal reason for selection and account should be taken of the other criteria discussed here. For manufacturers outside the EC, the geographical location is less important. Of greater significance to them, particularly U.S. companies, is the existence of overseas subsidiaries or subcontractors of some of the Notified Bodies. Manufacturers should understand that the Notified Body must be a legal entity established within the member state which has notified it. This does not prevent the Notified Body subcontracting quite significant tasks to a subsidiary.

Article 11.12 states that the correspondence relating to the conformity assessment procedures must be in the language of the member state in which the procedures are carried out or in another Community language acceptable to the Notified Body. Language may thus be another factor affecting the choice of Notified Body, although most of the major certification bodies will accept English and other languages.

The most significant factor of all is likely to be existing good relations with a particular body. Notified Bodies will be drawn from existing test and certification bodies and many manufacturers already use such bodies, either as part of a national approval procedure, or as part of their own policy for ensuring the satisfactory quality of their products and processes.

Another consideration which could become significant is that of variations in the national laws implementing the Directives. Notified Bodies will have to apply the law of the country in which they are situated and some differences in operation could be introduced by this means.

17.2.10 ESTABLISHING A DECLARATION OF CONFORMITY

Of all documents prepared for the MDD, the most important may be the declaration of conformity. Every device, other than a custom-made or clinical investigation device, must be covered by a declaration of conformity.

Declaration of Conformity

We   Company name
          Company address

declare that the product(s) listed below

Product(s) to be declared

hereby conform(s) to the European Council Directive 93/42/EEC, Medical Device Directive, Annex II, Article 3. This declaration is based on the Certification of the Full Quality Assurance System by name of Notified Body, Notified Body No. XXXX.

Name (print or type)   _____
Title   _____
Signature   _____
Date   _____

FIGURE 17.3   Sample declaration of conformance.

The general requirement is that the manufacturer shall draw up a written declaration that the products concerned meet the provisions of the Directive that apply to them. The declaration must cover a given number of the products manufactured. A strictly literal interpretation of this wording would suggest that the preparation of a declaration of conformity is not a one-and-for-all event with an indefinite coverage, but rather a formal statement that products which have been manufactured and verified in accordance with the particular conformity assessment procedure chosen by the manufacturer do meet the requirements of the Directive. Such an interpretation would impose severe burdens on manufacturers, and the Commission is understood to be moving to a position where a declaration of conformity can be prepared in respect of future production of a model of device for which the conformity assessment procedures have been carried out. The CE marking of individual devices after manufacture can then be regarded as a short-form expression of the declaration of conformity in respect of that individual device. This position is likely to form part of future Commission guidance.

Even so, the declaration remains a very formal statement from the manufacturer and accordingly, must be drawn up with care. The declaration must include serial numbers or batch numbers of the products it covers and manufacturers should give careful thought to the appropriate coverage of a declaration. In the extreme, it may be that a separate declaration should be prepared individually for each product or batch.

A practical approach is probably to draw up one basic declaration that is stated to apply to the products whose serial (batch) numbers are listed in an Appendix. The Appendix can then be added to at sensible intervals. A suggested format is shown in Figure 17.3.

17.2.11 APPLICATION OF THE CE MARK

The CE marking (Figure 17.4) is the symbol used to indicate that a particular product complies with the relevant essential requirements of the appropriate Directive, and as such, that the product has achieved a satisfactory level of safety and thus may circulate freely throughout the Community.

It is important to note that it is the manufacturer or his authorized representative who applies the CE marking to the product, and not the Notified Body. The responsibility for ensuring that each and every product conforms to the requirements of the Directive is that of the manufacturer and the affixing of the CE marking constitutes the manufacturer's statement that an individual device conforms.

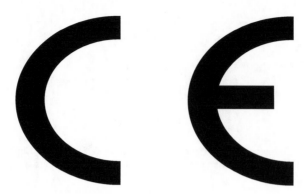

FIGURE 17.4   Example of CE mark.

The CE marking should appear on the device itself, if practicable on the instructions for use and on the shipping packaging. It should be accompanied by the identification number of product(s) listed below: the Notified Body that has been involved in the verification of the production of the device. It is prohibited to add other marks which could confuse or obscure the meaning of the CE marking.

The XXXX noted in Figure 17.3 is the identification number of the Notified Body.

17.2.12   CONCLUSION

Compliance with the new EC Directives will imply major changes for medical device manufacturers. Such changes relate to the requirements to be met in view of the design and manufacture of medical devices as well as to the procedures to be followed by manufacturers before and after placing medical devices on the European market. Manufacturers who wish to market medical devices in western Europe are therefore faced with a quite far-reaching and rather complex decision-making process.

17.3   UNITED STATES DOMESTIC STANDARDS

Standards simplify communication, promote consistency and uniformity, and eliminate the need to invent yet another solution to the same problem. They also provide vital continuity so that we are not forever reinventing the wheel. They are ways of preserving proven practices above and beyond the inevitable staff changes within organizations. Standards, whether official or merely agreed upon, are especially important when talking to customers and suppliers, but it is easy to underestimate their importance when dealing with different departments and disciplines within our own organization.

17.3.1   DOMESTIC STANDARDS ORGANIZATIONS

17.3.1.1   Association for the Advancement of Medical Instrumentation

The Association for the Advancement of Medical Instrumentation (AAMI) is an alliance of health care professionals, united by the common goal of increasing the understanding and beneficial use of medical devices and instrumentation. In meeting this goal, AAMI distributes information in the form of various publications, including voluntary standards. AAMI is a highly respected and widely recognized national and international consensus standards organization. AAMI is accredited by the American National Standards Institute (ANSI) and is one of the principal voluntary standards organizations in the world.

17.3.1.2 ANSI

The ANSI not only creates standards, but also is responsible for U.S. representation at the International Electrotechnical Commission and is the U.S. representative for the International Organization for Standards.

17.3.1.3 American Society for Quality Control

The American Society for Quality Control (ASQC) is a worldwide network of more than 83,000 individual members and over 600 sustaining members in the quality field. Coverage ranges from the fundamentals of quality technology to total quality management.

17.3.1.4 American Society for Testing and Materials

The American Society for Testing and Materials (ASTM) is a scientific and technical organization formed for the development of standards on characteristics and performance of materials, products, systems, and services. ASTM is the world's largest source of voluntary consensus standards.

17.3.1.5 Institute of Electrical and Electronic Engineers

The Institute of Electrical and Electronic Engineers (IEEE) was founded in 1884 and is one of the oldest societies in the United States. It is an organization that develops standards on a variety of topics relating to electrical and electronic equipment. In recent years, primary focus for the standards organization has been the areas of software development and software quality assurance. Some of their software standards have been accredited by the ANSI and have been primarily used for the development and validation of military software. Recently, these standards have been referenced by the Food and Drug Administration (FDA) in the development of guidelines on medical software.

17.3.1.6 Institute of Environmental Sciences

The Institute of Environmental Sciences (IES) is a technical society that covers space simulation, contamination control practices, solar and nuclear energy, military environmental testing, reliability testing, and ESS of components and systems.

17.3.1.7 IPC

The Institute for Interconnecting and Packaging Electronic Circuits (IPC) is known for work on printed circuit boards, specifications, and standards, including general requirements for soldered connections, component packaging, interconnecting and mounting, surface-mount land patterns, and studies such as the Impact of Moisture on Plastic IC Packaging Cracking (IPC-SM-786).

17.3.1.8 National Electrical Manufacturers Association

The National Electrical Manufacturers Association (NEMA) publishes standards including power circuits, plugs, receptacles, and sockets.

17.3.1.9 National Fire Protection Association

The National Fire Protection Association (NFPA) is organized to assure the appointment of technically competent committees, with balanced representation, to establish criteria to minimize the hazards of fire, explosion, and electricity in health care facilities. These criteria include

- Performance
- Maintenance
- Testing

- Safe practices
- Material
- Equipment
- Appliances

The NFPA does not approve, inspect, or certify any installation, procedure, equipment, or material. NFPA has no authority to police or enforce compliance to their standards. However, installations may base acceptance of a device on compliance with their standards.

17.3.1.10   Occupational Safety and Health Administration

The Occupational Safety and Health Administration (OSHA) was established in 1970 and is responsible for regulating workplace health and safety.

17.3.1.11   Underwriters Laboratory

The Underwriters Laboratory (UL) is an independent, not for profit testing laboratory organized for the purpose of investigating materials, devices, products, equipment construction, methods, and systems with respect to hazards affecting life and property. It tests devices in six different areas:

1. Burglary protection and signaling
2. Casualty and chemical hazards
3. Electrical
4. Fire protection
5. Heating, air conditioning and refrigeration
6. Marine

UL inspection services personnel visit companies unannounced to verify that products that bear the UL mark comply with applicable UL safety requirements. The registered UL mark on a device is a means by which a manufacturer, distributor, or importer can show that samples of the product have been verified for compliance with safety standards. Many hospitals require that the medical devices they purchase comply with applicable UL standards.

17.3.2   Software Standards and Regulations

There are a myriad of software standards to assist the developer in designing and documenting his program. IEEE standards cover documentation through all phases of design. Military standards describe how software is to be designed and developed for military use. There are also standards on software quality and reliability to assist developers in preparing a quality program. The international community has produced standards, primarily dealing with software safety. In each case, the standard is a voluntary document that has been developed to provide guidelines for designing, developing, testing, and documenting a software program.

In the United States, the FDA is responsible for assuring the device utilizing software or the software as a device is safe and effective for its intended use. The FDA has produced several drafts of reviewer guidelines, auditor guidelines, software policy, and good manufacturing practices (GMP) regulations addressing both device and process software. In addition, guidelines for FDA reviewers have been prepared as well as training programs for inspectors and reviewers. The new version of the GMP regulation addresses software as part of the design phase.

The United States is ahead of other countries in establishing guidelines for medical software development. There is, however, movement within several international organizations to develop regulations and guidelines for software and software-controlled devices. For example, ISO 9000-3 specifically addresses software development in addition to what is contained in ISO 9001. Canadian Standards Association (CSA) addresses software issues in four standards covering new and previously developed software in critical and noncritical applications. International Electrotechnical Commission (IEC) has a software document currently in development.

17.4 REST OF THE WORLD STANDARDS

No country regulates medical devices as consistently and thoroughly as the United States. However, there is a trend toward regulation in other industrialized counties, especially in Europe. France requires registration and evaluation of medical devices for public hospitals. Germany passed a law about 1987 that requires the registration of all medical devices linked to approval by defined testing organizations. England's Department of Health and Social Security is active in evaluating selected devices. And Italy also has a law, passed in 1986, that requires registration of all medical devices marketed in that country.

In Europe, an important international organization working in the area of devices is the European Commission. The EC has directives dealing with medical equipment. For example, under EC directives, all governments are required to develop standards for x-ray machines and x-ray therapy. Under another EC directive, issued by the IEC, requires member states to set standards for electrical safety. An EC working group on biomedical engineering focuses on the safety of medical equipment. At present, the group is examining such technologies as perinatal monitoring, chromosome analysis, technology for sensory impairment, aids to the disabled, replacement of body function, quantitative electrocardiography, imaging, especially NMR, blood flow measurement by ultrasound, medical telemetry, and accelerated fracture healing.

The World Health Organization, especially the European Office in Copenhagen, has become increasingly involved in medical devices, especially promoting the idea of international exchange of information. International cooperation and communication could make much more information on medical equipment available and save evaluation resources of all countries.

17.4.1 INTERNATIONAL NOTION OF STANDARDS

The British Standards Institute defines a standard as

> A technical specification or other document available to the public, drawn up with the cooperation and consensus or general approval of all interests affected by it, based on the consolidated results of science, technology and experience, aimed at the promotion of optimum community benefits and approved by a body on the national, regional or international level.

While this definition gives some way to saying what a standard is, it says nothing about the subject matter or purpose, apart from stating that the objectives of the standard must in some way be tied to community benefits. Standards, however, have a definite subject matter. They include

- Standardizing of particular processes
- Providing consistent and complete definition of a commodity or process
- Recording good practice regarding the development process associated with the production of commodities
- Encoding good practice for the specification, design, manufacture, testing, maintenance and operation of commodities

One of the primary requirements of a standard is that it should be produced in such a way that conformance to the standard can be unambiguously determined. A standard is devalued if conformance cannot be easily determined or if the standard is so loosely worded that it becomes a matter of debate and conjecture as to whether the standard has been complied with. Standards also exist in various types:

- De facto and de jure standards: These are usually associated with the prevailing commercial interests in the market place. These de facto standards are often eventually subject to the standardization process.
- Reference models: These provide a framework within which standards can be formulated.
- Product versus process standards: Some standards relate to specific products while others relate to the process used to produce products.
- Codes of practice, guidelines, and specifications: These terms relate to the manner in which a standard may be enforced. Codes of practice and guidelines reflect ways of working that are deemed to be good or desirable, but for which conformance is difficult to determine. Specifications are far more precise and conformance can be determined by analysis or test.
- Prospective and retrospective standards: It is clearly undesirable to develop a standard before the subject matter is well understood scientifically, technically, and through practice. However, it may be desirable to develop a standard alongside the evolving technology.

17.4.2 International Regulatory Scene

The production and adoption of software standards is very much the responsibility of international and national standards organizations and, in the case of the European context, bodies set up to represent a number of national organizations. Progressively, it is becoming the case that standards are developed by the international bodies and then adopted by the national bodies. Some of the international bodies include:

17.4.2.1 British Standards Institute

The British Standards Institute (BSI) is the United Kingdom's national standards making organization. In performing its duties, it collaborates with industry, government agencies, other standard bodies, professional organizations, etc.

17.4.2.2 Comité Européen de Normalisation

The Comité Européen de Normalisation (CEN) (European Committee for Standardization) is composed of members drawn from the European Union (EU) and the EFTA. The role of CEN is to produce standards for use within Europe and effectively covers the area addressed by ISO.

17.4.2.3 Comité Européen de Normalisation Electronique

The Comité Européen de Normalisation Electronique (CENELAC) (European Committee for Electrotechnical Standardization) is made up of representatives from the National Electrotechnical Committees, the majority of whom are represented on the IEC. Its responsibilities are for electrical and electronic standards within Europe and it has close links with the activities of the IEC.

17.4.2.4 CISPR

The International Special Committee on Radio Interference is a committee under the auspices of the IEC and run through a Plenary Assembly consisting of delegates from all the member bodies, including the United States. The committee is headquartered in Geneva, Switzerland and is composed of seven subcommittees, including

- Radio interference measurement and statistical methods
- Interference from industrial, scientific, and medical radio frequency apparatus

- Interference from overhead power lines, high-voltage equipment, and electric traction systems
- Interference related to motor vehicles and internal combustion engines
- Interference characteristics of radio receivers
- Interference from motor, household appliances, lighting apparatus, etc.
- Interference from information technology equipment

17.4.2.5 Canadian Standards Association

The CSA is a membership association that brings people and ideas together to develop services that meet the needs of businesses, industry, governments, and consumers. Among the many services available are standards development, testing and application of the CSA mark to certified products, testing to international standards, worldwide inspection, and related services.

17.4.2.6 Deutsches Institut für Normung

The Deutsches Institut für Normung (DIN) (German Standardization Institute) is the committee that sets German standards.

17.4.2.7 Department of Health

The Department of Health (DOH) has the same responsibility in England that the FDA has in the United States. DOH sets forth standards for medical devices and has established a GMP for medical equipment, similar to that of the FDA. DOH is headquartered in London. Currently DOH has reciprocity with the FDA, meaning the FDA will accept DOH inspection data as their own and DOH will accept FDA inspection data. This is particularly applicable for companies with facilities in both England and the United States.

17.4.2.8 IEC

The IEC was established in 1906 with the responsibility for developing international standards within the electrical and electronics field. By agreement with the International Standards Organization, the IEC has sole responsibility for these standards.

17.4.2.9 Institution of Electrical Engineers

The Institution of Electrical Engineers (IEE) is the main United Kingdom professional body responsible for electrical and electronic engineering. It is responsible for the production of a wide range of standards in the electrical engineering field and is progressively widening its interests to include software engineering.

17.4.2.10 International Standards Organization

The International Standards Organization (ISO) was established in 1947 and its members are drawn from the national standards bodies of its members. ISO is responsible for standardization in general, but with the exception of electrical and electronic standards which are the responsibility of the IEC.

17.4.2.11 Japanese Standards Association

The Japanese Standards Association (JSA) was established as a public institution for the promotion of industrial standardization on December 6, 1945, under government authorization. JSA has no true performance standards, but tends to follow IEC 601-1. JSA does have a complicated approval process that can be very lengthy (up to 9 months). This process can delay distribution of products in Japan. JSA activities include

- Standards and document publishing
- Seminars and consulting services
- Research on standardization
- National sales agent for foreign national standard bodies

17.4.2.12   Other Japanese Standards Organizations

The unified national system of industrial standardization began to function by the setup of the Japanese Engineering Standards Committee (JESC) in 1921. This group undertook the establishment of national standards. In 1949, the Industrial Standardization Law was promulgated and the Japanese Industrial Standards Committee (JISC) was established under the law as an advisory organization of competent ministers in charge of the elaboration of Japanese Industrial Standards (JIS) and the designation of the JIS mark to products.

17.4.3   TickIT Program

The TickIT project came from two studies commissioned by the Department of Trade and Industry (DTI) which showed that the cost of poor quality in software in the United Kingdom was very considerable and that quality system certification was desired by the market. The studies undertook extensive research into the respective subjects and included a broad consultative process with users, suppliers, in-house developers, and purchasers with a primary task being to identify options for harmonization. The reports made a number of significant recommendations, including

- All quality management system standards in common use were generically very similar.
- Best harmonization route was through ISO 9001.
- Action was required to improve market confidence in third party certifications of quality management systems.
- There was an urgent need to establish an accredited certification body or bodies for the software sector.

These principal recommendations were accepted by the DTI and further work was commissioned with the British Computer Society (BCS) to set up an acceptable means to gain accredited certification of quality management systems (QMS) by auditors with necessary expertise. Draft guidance material for an acceptable certification scheme was developed. The onward development from this draft material has become known as the TickIT project.

TickIT is principally a certification scheme, but this is not its primary purpose. The main objectives are to stimulate developers to think about what quality really is and how it may be achieved. Unless certification is purely a by-product of these more fundamental aims, much of the effort will be wasted. To stimulate thinking, TickIT includes some quality themes that give direction to the setting up of a QMS and the context of certification.

Generally, TickIT certification applies for information technology systems supply where software development forms a significant or critical part. The main focus of TickIT is software development because this is the component that gives an information system its power and flexibility. It is also the source of many of the problems.

17.4.4   Software Quality System Registration Program

The Software Quality System Registration (SQSR) Committee was established in 1992. The Committee's charter was to determine whether a program should be created in the United States for ISO 9001 registration of software design, production, and supply. A comparable program, TickIT, had been operational in the United Kingdom for over a year and was gaining European acceptance. To ensure mutual recognition and to leverage the experience of the worldwide software industry, the

SQSR program preserves ISO 9001 as the sole source for requirements and ISO 9000-3 as a source of official guidance for software registrants.

The SQSR program is designed for ISO 9001 registration of suppliers who design and develop software as a significant or crucial element in the products they offer. The SQSR program addresses the unique requirements of software engineering and provides a credible technical basis to allow the Registrar Accreditation Board (RAB) to extend its current programs for accrediting ISO 9000 registrars, certifying auditors, and accrediting specific courses and course providers.

The program is intended to ensure that ISO registration is an effective, enduring indicator of a software supplier's capability. The effectiveness of the SQSR program is based on three factors: mutual recognition, guidance, and an administrative infrastructure tailored to the U.S. marketplace.

17.4.5 ISO GUIDANCE DOCUMENTS FOR ISO 9001 AND 9002

ISO Technical Committee 210 has recently developed two guidelines relating ISO 9001 and 9002 to medical devices. The 1994 version of ISO 9001 and 9002 are intended to be general standards defining quality system requirements. ISO 13485 provides particular requirements for suppliers of medical devices that are more specific than the general requirements of ISO 9001. ISO 13488 provides particular requirements for suppliers of medical devices that are more specific than the general requirements of ISO 9002.

In conjunction with ISO 9001 and 9002, these International Standards define requirements for quality systems relating to the design, development, production, installation, and servicing of medical devices. They embrace all the principles of the GMP used in the production of medical devices. They can only be used in conjunction with ISO 9001 and 9002 and are not stand-alone standards.

They specify the quality system requirements for the production and, where relevant, installation of medical devices. They are applicable when there is a need to assess a medical device supplier's quality system or when a regulatory requirement specifies that this standard shall be used for such assessment. As part of an assessment by a third party for the purpose of regulatory requirements, the supplier may be required to provide access to confidential data to demonstrate compliance with one of these standards. The supplier may be required to exhibit these data, but is not obliged by the standard to provide copies for retention.

Particular requirements in a number of clauses of these standards are covered in detail in other International Standards. Suppliers should review the requirements and consider using the relevant International Standards in these areas.

To assist in the understanding of the requirements of ISO 9001, 9002, ISO 13485, and 13488, an international guidance standard is being prepared. The document provides general guidance on the implementation of quality systems for medical devices based on ISO 13485. Such quality systems include those of the EU MDD and the GMP requirements currently in preparation in Canada, Japan, and the United States. It may be used for systems based on ISO 13488 by the omission of subclause 4.4.

The guidance given in this document is applicable to the design, development, production, installation, and servicing of medical devices of all kinds. The document describes concepts and methods to be considered by medical device manufacturers who are establishing and maintaining quality systems. This document describes examples of ways in which the quality system requirements can be met, and it is recognized that there may be alternative ways that are better suited to a particular device/manufacturer. It is not intended to be directly used for assessment of quality systems.

17.4.6 PROPOSED REGULATORY REQUIREMENTS FOR CANADA

In February 1991, a review of the Medical Devices Regulatory Program was initiated when the minister of National Health and Welfare established the Medical Devices Review Committee to formulate recommendations concerning the regulation of medical devices and associated activities.

This committee was established in recognition of the increased volume and complexity of new medical devices used in health care and the need for timely availability of safe and effective devices in the next decade.

In May 1993, a development plan for an Improved Medical Devices Regulatory Program was published. The plan is based on two principles. First, the level of scrutiny afforded on a device is dependent upon the risk the device presents. Secondly, the safety and effectiveness of medical devices can be best assessed through a balance of quality systems, premarket scrutiny, and postmarket surveillance. To enshrine these two principals into the Medical Device Regulatory Program, it was evident that a reengineering of the program was necessary. This reengineering activity is currently ongoing.

The document describes a plan to establish a regulatory mark for Canada, similar to the CE mark currently being used in the European Union. A manufacturer would need to be audited by a Canadian third party for the successful implementation of a quality system as well as meeting the requirements for the regulatory mark.

17.4.7 ISO 14000 Series

ISO formed Technical Committee 207 in 1993 to develop standards in the field of environmental management tools and systems. The work of ISO TC 207 encompasses seven areas:

- Management systems
- Audits
- Labeling
- Environmental performance evaluation
- Life cycle assessment
- Terms and definitions
- Environmental aspects in product standards

The ISO 14000 standards are neither product standards, nor do they specify performance or pollutant/effluent levels. They specifically exclude test methods for pollutants and do not set limit values regarding pollutants or effluents.

The ISO 14000 standards are intended to promote the broad interests of the public and users, be cost-effective, nonprescriptive and flexible, and be more easily accepted and implemented. The goal is to improve environmental protection and quality of life.

ISO 14000 provides for the basic tenets of an environmental management system (EMS). An EMS is the management system which addresses the environmental impact of a company's processes and product on the environment. The EMS provides a formalized structure for ensuring that environmental concerns are addressed and met, and works to both control a company's significant environmental effects and achieve regulatory compliance. The certification process for ISO 14000 has six steps:

- Quality documentation review
- Initial visit, preassessment or checklist
- On-site audit
- Follow-up audits to document corrective action
- Periodic audits to document compliance
- Renewal audit every 3–5 years

Currently, there is a limited correlation between ISO 14000 and ISO 9000, but the requirements of the two series may become more harmonized in the future. Under certain conditions, the ISO 14000 audit and the ISO 9000 audit can be combined into one. It has been estimated that the cost of complying with ISO 14000 would be comparable to that for certification to ISO 9000. The registration process itself could take up to 18 months to complete.

It is expected that ISO 14000 will be in print and official by midsummer, 1996. Many European countries have already accepted the draft version as their EMS standard and have begun issuing accreditation. The Japanese Ministry of International Trade and Industry is asking companies to prepare new environmental management plans that conform to ISO 14000 by the end of 1996.

In the United States, ANSI has established a national program to accredit ISO 14000 registrars, auditor certifiers, and training providers. The ISO 14000 registrars are likely to come from the registrars currently performing certifications in ISO 9000.

The creation of a universal single set of EMS standards will help companies and organizations to better manage their environmental affairs, and show a commitment to environmental protection. It should also help them avoid multiple registrations, inspections, permits, and certifications of products exchanged among countries. In addition, it should concentrate worldwide attention on environmental management. The World Bank and other financial institutions may qualify their loans to less developed countries and being to use the 14000 standards as an indicator of commitment to environmental protection.

In the United States, implementation of ISO 14000 could become a condition of business loans to companies that are not even involved in international trade. Insurance companies may lower premiums for those who have implemented the standard. It may become a condition of some supplier transactions, especially in Europe and with the U.S. government. Evidence of compliance could become a factor in regulatory relief programs, the exercise of prosecutorial and sentencing discretion, consent decrees and other legal instruments, and multilateral trade agreements. U.S. government agencies considering the ISO 14000 standards include the

- Environmental Protection Agency
- Department of Defense
- Department of Energy
- Food and Drug Administration
- National Institute of Standards and Technology
- Office of the U.S. Trade Representative
- Office of Science, Technology, and Policy

17.4.8 MEDICAL INFORMATICS

The real world is perceived as a complex system characterized by the existence of various parallel autonomous processes evolving in a number of separate locations, loosely coupled, cooperating by the interchange of mutually understandable messages. Due to the fact that medical specialties, functional areas, and institutions create, use, and rely on interchanged information; they should share a common basic understanding to cooperate in accordance with a logical process constrained under an administrative organization, a medical heuristics and approach to care.

A health care framework is a logical mapping between the real world, in particular the health care environment, and its health care information systems architecture. This framework, representing the main health care subsystems, their connections, rules, etc., is the basis for an evolutionary development of heterogeneous computer-supported health care information and communications systems. A key feature of the framework is its reliance on the use of abstractions. In this way, the framework, at its most abstract level, reflects the fundamental and essential features of health care processes and information, and can be seen as applicable to all health care entities. It defines the general information structure, enables the exchangeability of the information.

The European health care framework will maintain and build upon the diversity of national health care systems in the European countries. A harmonized description/structure of planning documentation will be provided to ensure comparisons between European countries.

The main rationale for a standardized health care information framework is to

- Act as a contract between the users and procures on the one hand, and the developers and providers of information systems on the other.
- Ensure that all applications and databases are developed to support the health care organization as a whole as opposed to just a single organization or department.
- Obtain economies of scale, originating from enhanced portability, as health care information systems are expensive to develop and maintain, and tend to be installed on an international basis.
- Define a common basic understanding that allows all health care information systems to interchange data.

To this end, CEN/CENELAC has tasked a committee with creating the health care framework model.

EXERCISES

1. Compare the EU and the FDA definition of a medical device. What is similar? What differs?
2. Perform a Web search with ISO 9000 as the search term. You likely will turn up several companies that offer ISO 9000 and related services. What are the companies really offering (guidance/advice/consulting)? Justify your answer.
3. You have developed a portable device that monitors the EEG of patients prone to grand mal seizures. If one is predicted, your device automatically injects a drug to stop the impending seizure. How would this device be classified in the United States? In the European Union?
4. Same as question 3, but the device only warns the patient.
5. Visit the Web site: http://www.ghtf.org/. Briefly report on the purposes of the four study groups listed. Why do you think such a group is needed?
6. You manufacture a device currently accepted by the FDA. Why would you wish to get CE certification?

BIBLIOGRAPHY

AIMDD, *The Active Implantable Medical Devices Directive*. 90/385/EEC, June 20, 1990.

Association for the Advancement of Medical Instrumentation, *ISO/DIS 13485, Medical Devices—Particular Requirements for the Application of ISO 9001*. Arlington, VA: Association for the Advancement of Medical Instrumentation, 1995.

Association for the Advancement of Medical Instrumentation, *ISO/DIS 13488, Medical Devices—Particular Requirements for the Application of ISO 9002*. Arlington, VA: Association for the Advancement of Medical Instrumentation, 1995.

Association for the Advancement of Medical Instrumentation, *First Draft of Quality Systems—Medical Devices—Guidance on the Application of ISO 13485 and ISO 13488*. Arlington, VA: Association for the Advancement of Medical Instrumentation, 1995.

Banta, H.D., The regulation of medical devices, *Preventive Medicine*, 19(6), November 1990.

Cascio, J., Ed., *The ISO 114000 Handbook*. Milwaukee, WI: The American Society for Quality Control, 1996.

CEN/CENELAC, *Directory of the European Standardization Requirements for Health Care Informatics and Programme for the Development of Standards*. Brussels, Belgium: European Committee for Standardization, 1992.

CEN/CENELAC, *EN 46001*. Brussels, Belgium: European Committee for Standardization, 1995.

CEN/CENELAC, *EN 46002*. Brussels, Belgium: European Committee for Standardization, 1995.

Department of Trade and Industry, *Guide to Software Quality Management System Construction and Certification Using EN 29001*. London: DISC TickIT Office, 1987.

Donaldson, J., U.S. companies gear up for ISO 14001 certification, *InTech*, 43(5), May 1996.

Draft Proposal for a Council Directive on In Vitro Diagnostic Medical Devices, Working Document. III/D/4181/93, April 1993.

ECRI, *Health Care Standards: 1995.* Plymouth Meeting, PA: ECRI, 1995.

Fries, R.C. and Graber, M., Designing medical devices for conformance with harmonized standards, *Biomedical Instrumentation and Technology*, 29(4), July/August 1995.

Fries, R.C., *Medical Device Quality Assurance and Regulatory Compliance.* New York: Marcel Dekker, 1998.

Fries, R.C., *Handbook of Medical Device Design.* New York: Marcel Dekker, 2001.

Fries, R.C., *Reliable Design of Medical Devices.* Boca Raton, FL: CRC Press/Taylor & Francis Group, 2006.

Higson, G.R., *The Medical Devices Directive—A Manufacturer's Handbook.* Brussels: Medical Technology Consultants Europe, 1993.

Higson, G.R., *Medical Device Safety: The Regulation of Medical Devices for Public Health and Safety.* Bristol, UK: Institute of Physics Publishing, 2002.

International Standard Organization, *Quality Management and Quality Assurance Standards—Part 3: Guidelines for the Application of ISO 9001 to the Development, Supply and Maintenance of Software.* Geneva, Switzerland: International Organization for Standardization, 1991.

Jorgens, J. III and Burch, C.W., FDA regulation of computerized medical devices, *Byte*, 7, 1982, p. 204.

Jorgens, J. III, Computer hardware and software as medical devices, *Medical Device and Diagnostic Industry*, May 1983.

Jorgens, J. III and Schneider, R., Regulation of medical software by the FDA, *Software in Health Care*, April–May 1985.

Kahan, J.S., Regulation of computer hardware and software as medical devices, *Canadian Computer Law Reporter*, 6(3), January 1987.

King, P.H. and Fries, R.C., *Design of Biomedical Devices and Systems.* New York: Marcel Dekker, 2002.

The Medical Devices Directive. 93/42/EEC, June 14, 1993.

Meeldijk, V., *Electronic Components: Selection and Application Guidelines.* New York: John Wiley & Sons, 1996.

Shaw, R., Safety critical software and current standards initiatives, *Computer Methods and Programs in Biomedicine*, 44(1), July 1994.

SWBC, Organization for the European conformity of products, *CE-Mark: The New European Legislation for Products.* Milwaukee, WI: ASQC Quality Press, 1996.

Tabor, T. and Feldman, I., *ISO 14000: A Guide to the New Environmental Management Standards.* Milwaukee, WI: The American Society for Quality Control, 1996.

Working Group No. 4, *Draft of Proposed Regulatory Requirements for Medical Devices Sold in Canada.* Ottawa, Ontario: Environmental Heal Directorate, 1995.

18 Licensing, Patents, Copyrights, and Trade Secrets

Ultimately property rights and personal rights are the same thing.

Calvin Coolidge

The march of invention has clothed mankind with powers of which a century ago the boldest imagination could not have dreamt.

Henry George

Intellectual property (IP) is a generic term used to describe the products of the human intellect that have economic value. IP is a property because a body of laws has been created over the last 200 years that gives owners of such works legal rights similar in some respects to those given to owners of real estate or tangible personal property. IP may be owned, bought, leased (licensed), and sold the same as other type of property.

There are four separate bodies of law that may be used to protect IP: patent law, copyright law, trademark law, and trade secret law. Each of these bodies of law may be used to protect different aspects of IP, although there is a great deal of overlap among them.

18.1 PATENTS

A patent is an official document, issued by the U.S. government or another government, which describes an invention and confers on the inventors a monopoly over the disposition of the invention. The monopoly allows the patent owner to go to court to stop others from making, selling, or using the invention without the patent owner's permission.

Generally, an invention is any device or process that is based on an original idea conceived by one or more inventors and is useful in getting something done or solving a problem. An invention may also be a nonfunctional unique design or a plant. But when the word "invention" is used out in the technical world, it almost always means a composition, device or process. In order for an invention to be patentable, it must meet three criteria: novelty, nonobviousness, and usefulness. Many inventions, while extremely clever, do not qualify for patents, primarily because they are not considered to be sufficiently innovative in light of previous developments. The fact that an invention is not patentable does not necessarily mean that it has no value for its owner.

There are three types of patents that can be created: utility, design, and plant patents. Table 18.1 compares the three types of patents and the monopoly each type grants to the author.

18.1.1 WHAT QUALIFIES AS A PATENT?

An invention must meet several basic legal tests to qualify as a patent. These include

- Patentable subject matter
- Usefulness
- Novelty

TABLE 18.1

Patent Monopolies

Type of Patent	Legal Test	Length of Monopoly (Years)
Utility	Useful/nonobvious/ improvement/novel design of process/machine/matter	20
Design	New, nonobvious design or appearance	14
Plant	New or discovered and asexually reproduced plant or variety, nontuberous	20

- Nonobviousness
- Improvement over an existing invention
- Design
- Plant

We will concentrate primarily on the utility patent here. The plant patent would be of concern for a design text in agricultural engineering or bioengineering; the design patent is more of concern for those working in industrial design. An example of a plant patent would be for a new variety of rose. A design example would be a uniquely shaped bumper on a new car.

18.1.1.1 Patentable Subject Matter

The most fundamental qualification for a patent is that the invention consists of patentable subject matter. The patent laws define patentable subject matter as inventions that are one of the following:

- Process or method
- Machine or apparatus
- Article of manufacture
- Composition of matter
- New and useful improvement of an invention in any of these classes

Computer software is included in the above category, and has been since the early 1990s.

18.1.1.2 Usefulness

Almost always, an invention must be useful in some way to qualify for a patent. Fortunately, this is almost never a problem, since virtually everything can be used for something.

18.1.1.3 Novelty

As a general rule, no invention will receive a patent unless it is different in some important way from previous inventions and developments in the field, whether patented or not. To use legal jargon, the invention must be novel over the prior art. The basic test for this criterion is has anyone (including the inventor) reported this invention before (in written or oral form) in a manner that will allow someone else to duplicate the invention (enabling disclosure). Thus it is recommended that one is to be careful about conference proceedings, poster presentations, publications, etc. in advance of patent filing. Once disclosed, you have 1 year to file a patent (in the United States). In the rest of the world, countries depend on absolute novelty; that is, the invention has not been disclosed before the patent application.

18.1.1.4 Nonobviousness

In addition to being novel, an invention must have a quality that is referred to as nonobviousness. This means that the invention would have been surprising or unexpected to someone who is familiar with the field of the invention. Moreover in deciding whether an invention is nonobvious, the U.S. Patent and Trademark Office (PTO) may consider all previous developments (prior art) that existed when the invention was conceived. Obviousness is a quality that is difficult to define, but supposedly patent examiners know it when they see it.

As a general rule, an invention is considered nonobvious when it does one of the following:

- Solves a problem that people in the field have been trying to solve for some time.
- Does something significantly quicker than was previously possible.
- Performs a function that could not be performed before.

18.1.1.5 Improvement of an Existing Invention

Earlier we noted that to qualify a patent, an invention must fit into at least one of the statutory classes of matter entitled to a patent: For example, process, machine, manufacture, composition of matter, or an improvement of any of these. As a practical matter, this statutory classification is not very important since even an improvement on an invention in any one of the statutory classes will also qualify as an actual invention in that class. In other words, an invention will be considered as patentable subject matter as long as it fits within at least one of the other four statutory classes, whether or not it is viewed as an improvement or an original invention. However, the improvement must still meet the three tests for patentability: novelty, nonobviousness, and usefulness.

18.1.1.6 Design Patent

Design patents are granted to new, original, and ornamental designs that are a part of articles of manufacture. Articles of manufacture are in turn defined as anything made by the hands of humans. In the past, design patents have been granted to items such as truck fenders, chairs, fabric, athletic shoes, toys, tools, and artificial hip joints. The key to understanding this type of patent is the fact that a patentable design is required to be primarily ornamental and an integral part of an item made by humans.

A design patent provides a 14 year monopoly to industrial designs that have no functional use. That is, contrary to the usefulness rule discussed above, designs covered by design patents must be purely ornamental. The further anomaly of design patents is that while the design itself must be primarily ornamental, as opposed to primarily functional, it must at the same time be embodied in something artificial. Design patents are easy to apply for, as they do not require much written description. They require drawings for the design, a short description of each figure or drawing, and one claim that says little more than the inventor claims the ornamental design depicted on the attached drawings. In addition the design patent is less expensive to apply for than a utility patent, lasts for 14 rather than 20 years, and requires no maintenance fees.

18.1.2 PATENT PROCESS

The best method for achieving patentable material is to adequately document your design process as you go through your design process. Before you begin your project, purchase a composition (aka design notebook) book with bound pages for keeping your design notes. Start each entry with the date, and include all details of problem identification and solutions. Use drawings or sketches of your idea. Never remove any pages. If you do not like an entry or have made a mistake, simply make an X through the entry or write error. Sign all entries and have a witness sign and date them as frequently as possible. Your witness should be someone you trust who understands your idea and will maintain confidentiality. If you were to do a project on the improvement of an equipment or device, the design process and the related patent information documentation process would consist of the following steps:

- Note all problems caused by equipment, supplies, or nonexisting devices when performing a task.
- Focus on the problem every time you perform the task or use the item.
- Concentrate on solutions.
- Keep a detailed, dated diary of problems and solutions; include drawings and sketches.
- Record the benefits and usefulness of your idea.
- Evaluate the marketability of your idea. If it does not have a wide application, it may be more advantageous to abandon the idea and focus on another.
- Do not discuss your idea with anyone except one person you trust who will maintain confidentially.
- Prepare an application with a patent attorney.
- Have a search done; first a computer search, then a hand search. It is strongly suggested that the U.S. PTO files, either at the patent office or at various repositories be consulted.

Be sure to understand your rights as an inventor, as your rights depend on your employment agreement if you are developing patentable material within a company, or your status as a student developing material for hire or for an educational experience (see Section 18.1.4).

Once you have your patent material at hand, you and your advisors (patent lawyer, campus technology transfer office, etc.) must decide between filing a provisional patent or a full patent. A provisional patent allows one a low-cost way of establishing an early date of filing, which allows one to use patent pending on any literature relating to the design, and may give one the elbow room to determine if there exists a big enough market to make a full patent worthy of pursuit. A written description of your device and drawing (if necessary) are typically all that needed to be filed at this point. The provisional patent application has a lifetime of 1 year, within which it must be converted to a complete patent application or the provisional application is considered abandoned.

Should you pursue a full patent, the patent document that may come of your effort as filed with the PTO will contain

- Title for the invention and the names and addresses of the inventors
- Ownership (assignee) of the patent
- Details of the patent search made by the PTO
- Abstract that concisely describes the key aspects of the invention
- Drawings or flowcharts of the invention
- Very precise definitions of the invention covered by the patent (called the patent claims)
- Summary of the invention

Taken together, the various parts of the patent document provide a complete disclosure of every important aspect of the covered invention. When a U.S. patent is issued, all the information in the patent is readily accessible to the public in the PTO and in patent libraries across the United States and through online patent database services.

As an example of the above, patent number 3,359,806, titled "Multicrystal tomographic scanner for mapping thin cross section of radioactivity in an organ of the human body" has the following as an abstract.

ABSTRACT: A multicrystal tomographic scanner is utilized for mapping thin slices or cross sections of radioactivity in an organ of the human body which has been injected or injected with a suitable radioisotope. A plurality of radiation detectors are arranged in a cylindrical monoplanar array with each detector focused in such a manner that the fields of view of all of the detectors intersect at a common point, and the detectors are driven mechanically such that this common point of the detectors is caused to move in a rectilinear raster of about 8 inches square. The distribution of radioactivity measured by the detectors due to the amount of radioisotope within the area being scanned is stored in a computer memory and reproduced on an oscilloscope display, as the section is being examined.

Section 23.1 discusses issues relating to this patent further.

U.S. patents are obtained by submitting to the PTO, a patent application and an application fee. Once the application is received, the PTO assigns it to an examiner who is supposed to be knowledgeable in the technology underlying the invention. The patent examiner is responsible for deciding whether the invention qualifies for a patent, and assuming it does, what the scope of the patent should be. Usually, back-and-forth communications, called patent prosecution, occurs between the applicant and the examiner regarding these issues. Clearly, the most serious and hard-to-fix issue is whether the invention qualifies for a patent.

Eventually, if all of the examiner's objections are overcome by the applicant, the invention is approved for a patent. A patent issue fee is paid and the applicant receives an official copy of the patent deed. Three additional fees must be paid over the life of the patent to keep it in effect. Note that, though one or more individuals may be listed as inventors, the assignee is the patent holder of record. For the above-mentioned patent, four inventors were listed, but the U.S. Atomic Energy Commission is the assignee, as the invention was done in part with funds from that agency.

18.1.3 PATENT CLAIMS

Patent claims are the part of the patent application that precisely limits the scope of the invention—where it begins and where it ends. Perhaps it will help understand what the patent claims do if you analogize them to real estate deeds. A deed typically includes a description of the parcel's parameters precise enough to map the exact boundaries of the plot of land in question, which in turn, can be used as the basis of a legal action to toss out any trespassers.

In patents, the idea is to similarly draw with the patent claims a clear line around the property of the inventor so that any infringer can be identified and dealt with. Patent claims have an additional purpose. Because of the precise way in which they are worded, claims also are used to decide whether, in light of previous developments, the invention is patentable in the first place.

Unfortunately, to accomplish these purposes, all patent claims are set forth in an odd, stylized format. But the format has a big benefit. It makes it possible to examine any patent application or patent granted by the PTO and get a pretty good idea about what the invention covered by the patent consists of. While the stylized patent claim language and format have the advantage of lending a degree of precision to a field that badly needs it, there is an obvious and substantial downside to the use of the arcane patentspeak. Mastering it amounts to climbing a fairly steep learning curve.

For the above-mentioned patent, there exists 10 claims, which, when coupled with the technical drawings and proof of concept, serve to describe the invention in very fine detail, and serves to delimit the range of additional patents that might arise from a study of this particular patent. One claim is, for example, "5. The scanner set forth in claim 4, wherein the number of radiation detectors in said monopolar array is at least eight," would seem to limit one's ability to claim a similar patent with 9 or 10 detectors.

It is when you set out to understand a patent claim that the rest of the patent becomes crucially important. The patent's narrative description of the invention, set out in the patent specification, with all or many of the invention's possible uses, and the accompanying drawings or flowcharts, usually provide enough information in combination to understand any particular claim. And of course, the more patent claims you examine, the more adept you will become in deciphering them.

18.1.4 PROTECTING YOUR RIGHTS AS AN INVENTOR

In the United States, if two inventors apply for a patent at the same time, the patent will be awarded to the inventor who came up with the invention first; this is the so-called first-to-invent rule. This may or may not be the inventor who was first to file a patent application. In the rest of the world, the inventor who files the invention first is eligible for the patent; this is the so-called first-to-file rule. For this reason, it is vital that one carefully document the inventive activities. If two or more pending patent applications by different inventors claim the same invention, the PTO will ask the inventors to establish the date each of them first conceived the invention and the ways in which they then showed diligence in reducing the invention to practice.

Inventors can reduce the invention to practice in two ways: (1) by making a working model (a prototype) which works as the idea of the invention dictates it should or (2) by constructively reducing it to practice—that is, by describing the invention in sufficient detail for someone else to build it—in a document that is then filed as a patent application with the PTO.

The inventor who conceived the invention first will be awarded the patent if he or she also showed diligence in either building the invention or filing a patent application. If the inventor who was second to conceive the invention was the first one to reduce it to practice—for instance by filing a patent application—that inventor may end up with the patent.

It is often the quality of the inventor's documentation (dated, written in a notebook, showing the conception of the invention and the steps that were taken to reduce the invention to practice) that determines which invention ends up with the patent.

You especially should be aware that you can unintentionally forfeit your right to obtain patent protection. This can happen if you disclose your invention to others, such as a company interested in the invention, and then do not file an application within 1 year from that disclosure date. Any public disclosure, such as in a speech, poster session, or publication (Web or paper) begins a 1 year countdown after which patents are precluded in the United States. The same 1 year period applies if you offer your invention, or a product made by your invention, for sale. You must file your patent application in the United States within 1 year from any offer of sale.

Even more confusing is the fact that most other countries do not allow this 1 year grace period. Any public disclosure before you file your first application will prevent you from obtaining patent protection in nearly every country other than the United States.

18.1.5 PATENT INFRINGEMENT

Patent infringement occurs when someone makes, uses, or sells a patented invention without the patent owner's permission. Defining infringement is one thing, but knowing when it occurs in the real world is something else. Even with common technologies, it can be difficult for experienced patent attorneys to tell whether patents have been infringed.

There are multiple steps in deciding whether infringement of a patent has occurred:

- Identify the patent's independent apparatus and method claims.
- Break these apparati and method claims into their elements.
- Compare these elements with the alleged infringing device or process and decide whether the claim has all of the elements that constitute the alleged infringing device or process. If so, the patent has probably been infringed. If not, proceed to the next step.
- If the elements of the alleged infringing device or process are somewhat different than the elements of the patent claim, ask if they are the same in structure, function, and result. If yes, you probably have infringement. Note that for infringement to occur, only one claim in the patent needs to be infringed.

A patent's independent claims are those upon which usually one or more claims immediately following depend. A patent's broadest claims are those with the fewest words and that therefore provide the broadest patent coverage. The patent's broadest claims are its independent claims. To infringe a patent, only one of its claims (independent or dependent) need be infringed; it is not necessary to infringe multiple or all claims. In this sense, each patent claim conveys a stand-alone patent right.

In apparatus (machine) claims, the elements are usually conceptualized as the a, b, c, etc. parts of the apparatus that are listed, interrelated, and described in detail following the word "comprising" at the end of the preamble of the claim. Elements in method (process) claims are the steps of the method and subparts of those steps.

If each and every element of the patent's broadest claims is in the infringing device, the patent is probably infringed. The reason you start by analyzing the broadest claim is that by definition, that claim has the fewest elements and it is therefore easier to find infringements.

Even if infringement cannot be found on the basis of the literal language in the claims, the courts may still find infringement if the alleged infringing device's elements are equivalent to the patent claims in structure, function, and result. Known as the doctrine of equivalents, this rule is difficult to apply in practice.

Defending a patent or initiating a patent infringement lawsuit can be extremely costly; consultation with a good patent lawyer is advised before any legal actions.

18.1.6 WORD OR FOUR OF WARNING

In terms of the invention disclosure, irrespective of who goes through the process of obtaining a patent, it is advisable to start by informing the responsible office in your school of your work through an invention disclosure form. Your advisor can point you to the correct office where the disclosure form should be submitted. Such a disclosure form typically contains the following sections:

1. Invention title
2. Inventors name(s), contribution(s), contact information
3. Applicable contract or grant numbers
4. Are any IP agreements in place on this work?
5. Were any other parties involved?
6. Earliest date of idea conception
7. Invention description
8. Prior disclosures, nature, dates
9. Invention function (commercial)
10. List of possible licensees
11. Potential impediments to commercialization of invention
12. Signatures of inventor(s) and witnesses

Filling out such a form has several advantages. In addition, it helps you think about your invention in a critical manner, it also helps answer ownership questions and provides the basis to determine whether or not to pursue patent protection.

This brings up the question of costs associated with patent protection. Typically, it costs in the neighborhood of $10,000–$20,000 to get an allowed U.S. patent; thus decisions to pursue patents are nontrivial, regardless of who foots the bill. Non-U.S. patent protection can cost significantly more than the above, depending primarily on the geographic scope of coverage. In addition, all patents have to be maintained in good standing by paying periodic fees which vary with the jurisdiction. Given the associated costs, the decision to proceed with patent protection is primarily based on economic considerations. In an academic setting, the only reason to proceed with a patent is to be able to prevent others from making, using or selling products and services based on the patent claims in the absence of a license from the academic institution. Thus an academic institution will base its decisions on the probability of obtaining a license (in a reasonable time frame) on a patentable invention. In a commercial setting, in addition to providing monopoly, patent protection can also provide a competitive edge.

Two considerations in patents are often comingled and confused; these are inventorship and ownership. In general, it is fair to say that inventorship is a matter of law while ownership is a matter of contract. In terms of inventorship, only those individuals who make an inventive contribution to one or more claims of a patent are eligible to be inventors. There are several ramifications of this concept.

1. As claims from some of the original patent application submitters may not eventually issue, thereby preventing some of them from being inventors, strictly speaking, true inventorship can be determined only after patent claims are allowed by the U.S. PTO.
2. Individual needs to be an inventor on a single claim to be a named inventor on the patent.
3. If an individual has proven inventorship and is purposefully denied inventorship status by the other inventors, this is a serious matter and can be grounds for invalidation of an otherwise issued and valid patent.
4. Individual must have inventory contribution to at least one claim to be a named inventor in a patent. Simply being the person to state the problem that the patent solves does not make one an inventor; neither does being in a superior or supervisory role in a project or institution.

On the other hand, ownership of a patent is a matter of contract. In other words, the ownership of a patent (assignee) does not necessarily lie with the inventor(s). The ownership can be vested in an employer, research sponsor or other entity depending on the specific circumstances. In an academic setting, most faculties are required to grant ownership of IP to their parent institution. For students, it is advisable to check the institution's policies; ownership can very from the inventor to the institution depending on the specifics of the circumstances. These rules and policies are usually available on the university's Web site.

18.2 COPYRIGHTS

A copyright is a legal device that provides the creator of a work of authorship the right to control how the work is used. If someone wrongly uses material covered by a copyright, the copyright owner can sue and obtain compensation for any losses suffered, as well as an injunction requiring the copyright infringer to stop the infringing activity.

A copyright is a type of tangible property. It belongs to its owner and the courts can be asked to intervene if anyone uses it without permission. Like other forms of property, a copyright may be sold by its owner, or otherwise exploited by the owner for economic benefit.

The Copyright Act of 1976 grants creators many intangible, exclusive rights over their work, including reproduction rights, the right to make copies of a protected work; distribution rights, the right to sell or otherwise distribute copies to the public; the right to create adaptations, the right to prepare new works based on the protected work; performance and display rights, the right to perform a protected work or display a work in public. Copyright law is evolving, the most recent revisions occurred in 1998.

Copyright protects all varieties of original works of authorship, including

- Literary works
- Motion pictures, videos, and other audiovisual works
- Photographs, sculpture, and graphic works
- Sound recordings

- Pantomimes and choreographic works
- Architectural works

18.2.1 What Can Be Copyrighted?

Not every work of authorship receives copyright protection. A program or other work is protected if it satisfies all three of the following requirements:

1. Fixation
2. Originality
3. Minimal creativity

The work must be fixed in a tangible medium of expression. Any stable medium from which the work can be read back or heard, either directly or with the aid of a machine or device, is acceptable.

Copyright protection begins the instant you fix your work. There is no waiting period and it is not necessary to register the copyright. Copyright protects both completed and unfinished works, as well as works that are widely distributed to the public or never distributed at all.

A work is protected by copyright only if, and to the extent, it is original. But this does not mean that copyright protection is limited to works that are novel, that is, new to the world. For copyright purposes, a work is original if at least a part of the work owes its origin to the author. A work's quality, ingenuity, aesthetic merit, or uniqueness is not considered.

A minimal amount of creativity over and above the independent creation requirement is necessary for copyright protection. Works completely lacking creativity are denied copyright protection even if they have been independently created. However, the amount of creativity required is very slight. It is important to recognize that, unlike patent protection, copyright protection extends only to the expression of the idea and not the idea itself.

In the past, some courts held that copyright protected works that may have lacked originality and or creativity if a substantial amount of work was involved in their creation. Recent court cases have outlawed this sweat of the brow theory. It is now clear that the amount of work put in to create a work of authorship has absolutely no bearing on the degree of copyright protection it will receive. Copyright only protects fixed, original, minimally creative expressions, not hard work.

Perhaps the greatest difficulty with copyrights is determining just what aspects of any given work are protected. All works of authorship contain elements that are protected by copyright and elements that are not protected. Unfortunately, there is no system available to precisely identify which aspects of a given work are protected. The only time we ever obtain a definitive answer as to how much any particular work is protected is when it becomes the subject of a copyright infringement lawsuit. However, there are two tenets which may help in determining what is protected and what is not. The first tenet states that a copyright only protects expressions, and not ideas, systems, or processes. The second tenet states that the scope of copyright protection is proportional to the range of expression available. Let us look at both in detail.

Copyright only protects the tangible expression of an idea, system, or process and not the idea, system or process itself. Copyright law does not protect ideas, procedures, processes, systems, mathematical principles, formulas, titles, algorithms, methods of operation, concepts, facts, and discoveries. Remember, copyright is designed to aid the advancement of knowledge. If the copyright law gave a person a legal monopoly over ideas, the progress of knowledge would be impeded rather than helped.

The scope of copyright protection is proportional to the range of expression available. The copyright law only protects original works of authorship. Part of the essence of original authorship is the making of choices. Any work of authorship is the end result of a whole series of choices made by its creator. For example, the author of a novel expressing the idea of love must choose the novel's plot, characters, locale, and the actual words used to express the story. The author of such a novel has a nearly limitless array of choices available. However, the choices available to the creators of many works of authorship are severely limited. In these cases, the idea or ideas underlying the work and the way they are expressed by the author are deemed to merge. The result is that the author's expression is either treated as if it was in the public domain or protected only against virtually verbatim or slavish copying.

18.2.2 Copyright Process

18.2.2.1 Copyright Notice

Before 1989, all published works had to contain a copyright notice (the © symbol followed by the publication date and copyright owner's name) to be protected by copyright. This is no longer necessary. Use of copyright notices is now optional in the United States. Even so, it is always a good idea to include a copyright notice on all work distributed to the public so that potential infringers will be informed of the underlying claim to copyright ownership. In addition, copyright protection is not available in some 20 foreign countries unless a work contains a copyright notice.

There are strict technical requirements as to what a copyright notice must contain. A valid copyright must contain three elements:

1. Copyright symbol: Use the familiar © symbol, that is, lowercase letter "c" completely surrounded by a circle. The word "copyright" or the abbreviation "Copr." are also acceptable in the United States, but not in many foreign countries. So if your work might be distributed outside the United States, always use the © symbol.
2. Year in which the work was published: You only need to include the year the work was first published.
3. Name of the copyright owner: The owner is (1) the author or authors of the work, (2) the legal owner of a work made for hire, or (3) the person or entity to whom all the author's exclusive copyright rights have been transferred.

Although the three elements of a copyright notice need not appear in a particular order, it is common to list the copyright symbol, followed by the date and owners.

According to copyright office regulations, the copyright notice must be placed so as not be concealed from an ordinary user's view upon reasonable examination. A proper copyright notice should be included on all manuals and promotional materials. Notices on written works are usually placed on the title page or the page immediately following the title page.

18.2.2.2 Copyright Registration

Copyright registration is a legal formality by which a copyright owner makes a public record in the U.S. Copyright Office in Washington, DC of some basic information about a protected work, such as the title of the work, who wrote it and when, and who owns the copyright. It is not necessary to register to create or establish a copyright. Since original works are born copyrighted, registration of copyright is optional.

Copyright registration is a relatively easy process. You must fill out the appropriate preprinted application form, pay an application fee, and mail the application and fee to the copyright office in Washington, DC along with two copies of the work being registered.

18.2.3 Copyright Duration

One of the advantages of copyright protection is that it is long lasting. The copyright in a protectable work created after 1977 by an individual creator lasts for the life of the creator plus an additional 50 years. If there is more than one creator, the life plus 50 term is measured from the date the last creator dies. Many classical novels are now out of copyright, and may be found in the public domain. The copyright in works created by employees for their employers last for 75 years from the date of publication, or 100 years from the date of creation, whichever occurs first.

18.2.4 PROTECTING YOUR COPYRIGHT RIGHTS

The exclusive rights granted by the Copyright Act initially belong to a work's author. There are four ways to become an author:

1. Individual may independently author a work.
2. Employer may pay an employee to create the work, in which case, the employer is the author under the work made for hire rule.
3. Person or business entity may specially commission an independent contractor to create the work under a written work made for hire contract, in which case, the commissioning party becomes the author.
4. Two or more individuals or entities may collaborate to become joint authors.

The initial copyright owner of a work is free to transfer some or all copyright rights to other people or businesses, who will then be entitled to exercise the rights transferred.

18.2.5 INFRINGEMENT

Copyright infringement occurs when a person other than the copyright owner exploits one or more of the copyright owner's exclusive rights without the owner's permission. A copyright owner who wins an infringement suit may stop any further infringement, obtain damages from the infringer and recover other monetary losses. This means, in effect, that a copyright owner can make a copyright infringer restore the author to the same economic position they would have been in had the infringement never occurred. In this respect, works likely to have significant economic value should be registered since the damage awards from infringement of registered copyrights are significantly higher than unregistered copyrights.

Copyright infringement is usually proven by showing that the alleged infringer had access to the copyright owner's work and that the protected expression in the two works is substantially similar. In recent years, the courts have held that the person who claims his work was infringed upon must subject his work to a rigorous filtering process to find out which elements of the work are and are not protected by copyright. In other words, the plaintiff must filter out from his work ideas, elements dictated by efficiency or external factors, or taken from the public domain. After this filtration process is completed, there may or may not be any protectable expression left.

18.3 TRADEMARKS

A trademark is a work, name, symbol, or a combination used by a manufacturer to identify its goods and distinguish them from others. Trademark rights continue indefinitely as long as the mark is not abandoned and is properly used.

A federal trademark registration is maintained by filing a declaration of use during the sixth year after its registration and by renewal every 20 years, as long as the mark is still in use. The federal law provides that nonuse of a mark for two consecutive years is ordinarily considered abandonment, and the first subsequent user of the mark can claim exclusive trademark rights. Trademarks therefore must be protected or they will be lost. They must be distinguished in print form from other words and must appear in a distinctive manner.

Trademarks should be followed by a notice of their status. If it has been registered in the U.S. PTO, the registration notice® or Reg. U.S. Pat Off, should be used. Neither should be used; however, if the trademark has not been registered, but the superscripted letter TM should follow the mark, or an asterisk can be used to refer to a footnote starting "a trademark of xxx." The label compliance manager should remember that trademarks are proper adjectives and must be accompanied by the generic name for the product they identify. Trademarks are neither to be used as possessives nor in the plural form.

A trademark is any visual mark that accompanies a particular tangible product, or line of goods, and serves to identify and distinguish it from products sold by others and it indicates its source. A trademark may consist of letters, words, names, phrases, slogans, numbers, colors, symbols, designs, or shapes. As a general rule, to be protected from unauthorized use by others, a trademark must be distinctive in some way.

Trademark is also a generic term used to describe the entire broad body of state and federal law that covers how businesses distinguish their products and services from the competition. Each state has its own set of laws establishing when and how trademarks can be protected. There is also a federal trademark law, called the Lanham Act, which applies in all 50 states. Generally, state trademark laws are relied upon for marks used only within one particular state, while the Lanham Act is used to protect marks for products that are sold in more than one state or across territorial or national borders.

18.3.1 SELECTING A TRADEMARK

Not all trademarks are treated equally by the law. The best trademarks are distinctive, that is, they stand out in a customer's mind because they are inherently memorable. The more distinctive the trademark is, the stronger it will be and the more legal protection it will receive. Less distinctive marks are weak and may be entitled to little or no legal protection.

Generally, selecting a mark begins with brainstorming for general ideas. After several possible marks have been selected, the next step is often to use formal or informal market research techniques to see how the potential marks will be accepted by customers. Next, a trademark search is conducted. This means that an attempt is made to discover whether the same or similar marks are already in use.

18.3.1.1 What Is a Distinctive Trademark?

A trademark should be created that is distinctive rather than descriptive. A trademark is distinctive if it is capable of distinguishing the product to which it is attached from competing products. Certain types of marks are deemed to be inherently distinctive and are automatically entitled to maximum protection. Others are viewed as not inherently distinctive and can be protected only if they acquire secondary meaning through use.

Arbitrary, fanciful, or coined marks are deemed to be inherently distinctive and are therefore very strong marks. These are words or symbols that have absolutely no meaning in the particular trade or industry before their adoption by a particular manufacturer for use with its goods or services. After use and promotion, these marks are instantly identified with a particular company and product, and the exclusive right to use the mark is easily asserted against potential infringers.

Fanciful or arbitrary marks consist of common words used in an unexpected or arbitrary way so that their normal meaning has nothing to do with the nature of the product or service they identify. Some examples would be Apple Computer and Peachtree Software. Coined words are words made up solely to serve as trademarks, such as ZEOS or Intel.

Suggestive marks are also inherently distinctive. A suggestive mark indirectly describes the product it identifies but stays away from literal descriptiveness. That is, the consumer must engage in a mental process to associate the mark with the product it identifies. For example, WordPerfect and VisiCalc are suggestive marks.

Descriptive marks are not considered to be inherently distinctive. They are generally viewed by the courts as weak and thus not deserving of much, if any, judicial protection unless they acquire a secondary meaning, that is, become associated with a product in the public's mind through long and continuous use. There are three types of descriptive marks: (1) marks that directly describe the nature or characteristics of the product they identify (e.g., Quick Mail), (2) marks that describe the geographic location from which the product emanates (e.g., Oregon Software), and (3) marks consisting primarily of a person's last name (e.g., Norton Utilities). A mark that is in continuous and exclusive use by its owner for a 5 year period is presumed to have acquired secondary meaning and qualifies for registration as a distinctive mark.

A generic mark is a word(s) or symbol that is commonly used to describe an entire category or class of products or services, rather than to distinguish one product or service from another. Generic marks are in the public domain and cannot be registered or enforced under the trademark laws. Some examples of generic marks include computer, mouse, and RAM. A term formerly protected as a trademark may lose such protection if it becomes generic. This often occurs when a mark is assimilated into common use to such an extent that it becomes the general term describing an entire product category (e.g., Escalator and Xerox).

18.3.2 TRADEMARK PROCESS

A trademark is registered by filing an application with the PTO in Washington, DC. Registration is not mandatory. Under both federal and state law, a company may obtain trademark rights in the states in which the mark is actually used. However, federal registration provides many important benefits including

- Trademark's owner is presumed to have the exclusive right to use the mark nationwide.
- Everyone in the country is presumed to know that the mark is already taken.
- Trademark owner obtains the right to put an ® symbol after the mark.
- Anyone who begins using a confusingly similar mark after the mark has been registered will be deemed a willful infringer.
- Trademark owner obtains the right to make the mark incontestable by keeping it in continuous use for 5 years.

To qualify for federal trademark registration, a mark must meet several requirements. The mark must

- Actually be used in commerce
- Be sufficiently distinctive to reasonably operate as a product identifier
- Not be confusingly similar to an existing, federally registered trademark

A mark you think will be good for your product could already be in use by someone else. If your mark is confusingly similar to one already in use, its owner may be able to sue you for trademark infringement and get you to change it and even pay damages. Obviously, you do not want to spend time and money on marketing and advertising a new mark only to discover that it infringes on another preexisting mark and must be changed. To avoid this, state and federal trademark searches should be conducted to attempt to discover if there are any existing similar marks. You can conduct a trademark search yourself, either manually or with the aid of computer databases. You may also pay a professional search firm to do so (advisable). It is worth noting that the same mark can be used for completely different classes of products and services.

18.3.3 INTENT-TO-USE REGISTRATION

If you seriously intend to use a trademark on a product in the near future, you can reserve the right to use the mark by filing an intent-to-use registration. If the mark is approved, you have 6 months to actually use the mark on a product sold to the public. If necessary, this period may be increased by 6 month intervals up to 24 months if you have a good explanation for the delay. No one else may use the mark during this interim period. You should promptly file intent-to-use registration as soon as you have definitely selected a trademark for a forthcoming product.

18.3.4 PROTECTING YOUR TRADEMARK RIGHTS

The owner of a valid trademark has the exclusive right to use the mark on its products. Depending on the strength of the mark and whether and where it has been registered, the trademark owner may be able to bring a court action to prevent others from using the same or similar marks on competing or related products (e.g., Johnson & Johnson has as its trademark, the red cross, on all of its products).

Trademark infringement occurs when an alleged infringer uses a mark that is likely to cause consumers to confuse the infringer's products with the trademark owner's products. A mark need not be identical to one already in use to infringe upon the owner's rights. If the proposed mark is similar enough to the earlier mark to risk confusing the average consumer, its use will constitute infringement.

Determining whether an average consumer might be confused is the key to deciding whether infringement exists. The determination depends primarily on whether the products or services involved are related, and, if so, whether the marks are sufficiently similar to create a likelihood of consumer confusion.

If a trademark owner is able to convince a court that infringement has occurred, he or she may be able to get the court to order the infringer to stop using the infringing mark and to pay monetary damages. Depending on whether the mark was registered, such damages may consist of the amount of the trademark owner's losses caused by the infringement or the infringer's profits. In cases of willful infringement, the courts may double or triple the damages award.

A trademark owner must be assertive in enforcing its exclusive rights. Each time a mark is infringed upon; it loses strength and distinctiveness and may eventually die by becoming generic.

18.4 TRADE SECRETS

Trade secrecy is basically a do-it-yourself form of IP protection. It is based on the simple idea that by keeping valuable information secret, one can prevent competitors from learning about and using it. Trade secrecy is by far the oldest form of IP, dating back at least to ancient Rome. It is as useful now as it was then.

A trade secret is any formula, pattern, physical device, idea, process, compilation of information or other information that (1) is not generally known by a company's competitors, (2) provides a business with a competitive advantage, and (3) is treated in a way that can reasonably be expected to prevent the public or competitors from learning about it, absent improper acquisition or theft.

Trade secrets may be used to

- Protect ideas that offer a business a competitive advantage.
- Keep competitors from knowing that a program is under development and from learning its functional attributes.
- Protect source code, software development tools, design definitions and specifications, manuals, and other documentation.
- Protect valuable business information such as marketing plans, cost and price information, and customer lists.

Unlike copyrights and patents, whose existence is provided and governed by federal law that applies in all 50 states, trade secrecy is not codified in any federal statute. Instead, it is made up of individual state laws. Nevertheless, the protection afforded to trade secrets is much the same in every state. This is partly because some 26 states have based their trade secrecy laws on the Uniform Trade Secrecy Act (1995), a model trade secrecy law designed by legal scholars.

18.4.1 WHAT QUALIFIES FOR TRADE SECRECY?

Information that is public known or generally known cannot be a trade secret. Things that everybody knows cannot provide anyone with a competitive advantage. However, information comprising a trade secret need not be novel or unique. All that is required is that the information not be generally known by people who could profit from its disclosure and use.

18.4.2 TRADE SECRECY AUTHORSHIP

Only the person that owns a trade secret has the right to seek relief in court if someone else improperly acquires or discloses the trade secret. Only the trade secret owner may grant others a license to use the secret.

As a general rule, any trade secrets developed by an employee in the course of employment belong to the employer. However, trade secrets developed by an employee on their own time and with their own equipment can sometimes belong to the employee. To avoid possible disputes, it is a very good idea for employers to have all the employees who may develop new technology sign an employee agreement that assigns in advance all trade secrets developed by the employee during their employment to the company.

18.4.3 HOW TRADE SECRETS ARE LOST?

A trade secret is lost if either the product in which it is embodied is made widely available to the public through sales or displays on an unrestricted basis, or the secret can be discovered by reverse engineering or inspection.

18.4.4 DURATION OF TRADE SECRETS

Trade secrets have no definite term. A trade secret continues to exist as long as the requirements for trade secret protection remain in effect. In other words, as long as secrecy is maintained, the secret does not become generally known in the industry and the secret continues to provide a competitive advantage, it will be protected.

18.4.5 PROTECTING YOUR TRADE SECRET RIGHTS

A trade secret owner has the legal right to prevent the following two groups of people from using and benefiting from its trade secrets or disclosing them to other without the owner's permission. People who

- Are bound by a duty of confidentiality not to disclose or use the information
- Steal or otherwise acquire the trade secret through improper means

A trade secret owner's rights are limited to the two restricted groups of people discussed above. In this respect, a trade secret owner's rights are much more limited than those of a copyright owner or patent holder.

A trade secret owner may enforce their rights by bringing a trade secret infringement action in court. Such suits may be used to

- Prevent another person or business from using the trade secret without proper authorization.
- Collect damages for the economic injury suffered as a result of the trade secret's improper acquisition and use.

All persons responsible for the improper acquisition and all those who benefited from the acquisition are typically named as defendants in trade secret infringement actions. To prevail in a trade secret infringement suit, the plaintiff must show that the information alleged to be secret is actually a trade secret. In addition, the plaintiff must show that the information was either improperly acquired by the defendant or improperly disclosed, or likely to be so, by the defendant.

There are two important limits on trade secret protection. It does not prevent others from discovering a trade secret through reverse engineering, nor does it apply to persons who independently create or discover the same information.

18.4.6 Trade Secrecy Program

The first step in any trade secret protection program is to identify exactly what information and material is a company trade secret. It makes no difference in what form a trade secret is embodied. Trade secrets may be stored on hard disks or floppies, written, or memorized by the employees.

Once a trade secret has been established, the protection program should include the following steps:

- Maintain physical security
- Enforce computer security
- Mark confidential documents "Confidential"
- Use nondisclosure agreements

Nondisclosure agreements are generally simple documents, which indicate that you will be(come) privy to some knowledge about a product or procedure. Minimal documents basically indicate that

> The undersigned reader acknowledges that the information provided by XXX inc. in these product declarations is confidential; therefore, readers agrees not to disclose it without the express written permission of XXX inc. or YYY, director of XXX.
>
> It is acknowledged by reader that information to be furnished in these product requirements is confidential in nature, other than that which is in the public domain through others means and that any disclosure or use of same by reader may cause harm to XXX, inc.

This section would typically be signed and dated by you, the one who is seeking the information.

A nondisclosure agreement is a legal document. Be sure to understand your company/university policies on this matter.

18.4.7 Use of Trade Secrecy with Copyrights and Patents

Trade secrecy is a vitally important protection for any medical device, but because of its limitations listed above, it should be used in conjunction with copyright and, in some cases, patent protection.

18.4.7.1 Trade Secrets and Patents

The federal patent laws provide the owner of a patentable invention with far greater protection than that available under trade secrecy laws. Trade secret protection is not lost when a patent is applied for. The patent office keeps patent applications secret unless or until a patent is granted. However, once a patent is granted and an issue feed is paid, the patent generally becomes public record. Then all the information disclosed in the patent application is no longer a trade secret. This is so even if the patent is later challenged in court and invalidated.

If, for example, a software program is patented, the software patent applies only to certain isolated elements of the program. The remainder need not be disclosed in the patent and can remain a trade secret.

18.4.7.2 Trade Secrets and Copyrights

Trade secrecy and copyright are not incompatible. To the contrary, they are typically used in tandem to provide the maximum legal protection available.

EXERCISES

1. What are the basic differences between a patent, a copyright, and a trademark? (To include: brief definition, rights included/excluded, lifespan of each.)
2. Give a specific example of material in which the creator would seek a patent, a copyright, and a trademark.
3. One method of patent searching is to sift through the many patents in the U.S. PTO database. Go to http://www.uspto.gov. Do a patent search of a medical device of your choice. Write a summary of your search information. Of the six types of subject matter included under a patent, under which category can your material be classified? (Summary to include: exact definition of the item patented, who patented the device, application number, and date filed.) Hint: there are many ways to do a search on the U.S. PTO home page. One way is to click on patents, issued years and patent numbers, search, patent database, Boolean, then enter Medical into the query. However, you are not bound by these steps to search the database.
4. What are some alternatives to this method (sifting through the U.S. PTO home page) of patent searching?
5. Do a U.S. patent search using your last name as a search term. Write up a patent found (no result, use Smith). What does it do?
6. Do a copyright search similar to question 4.
7. One of the authors of this text holds patent number 3,591,806. How many other patents refer to this patent as prior art?
8. Draft an IP agreement with your advisor. Draft an IP disclosure for your work.
9. Outline your patent application for your design project.
10. Find and briefly report on the topics of service marks and mask works.

BIBLIOGRAPHY

American Intellectual Property Law Association, *How to Protect and Benefit from Your Ideas*. Arlington, VA: American Intellectual Property Law Association, 1988.

Banner & Allegretti, *An Overview of Changes to U.S. Patent and Trademark Office Rules and Procedures Effective June 8, 1995*. Chicago, IL: Banner & Allegretti, 1995.

Fishman, S., *Software Development—A Legal Guide*. Berkeley, CA: Nolo Press, 1994.

Fries, R.C., *Reliable Design of Medical Devices*. New York: Marcel Dekker, 1997.

Noonan, W.D., Patenting medical technology, *Journal of Legal Medicine*, 11(3), September 1990.

Rebar, L.A., The nurse as inventor—Obtain a patent and benefit from your ideas, *AORN Journal*, 53(2), February 1991.

Sherman, M., Developing a labeling compliance program, in *The Medical Device Industry*. New York: Marcel Dekker, 1990.

U.S. Department of Commerce, *General Information Concerning Patents*. Washington, DC: U.S. Government Printing Office, 1993.

Web site information from the Library of Congress, http://www.loc.gov/.

Web site information for patents and trademarks, http://www.uspto.gov/.

19 Manufacturing and Quality Control

Amateurs work until they get it right. Professionals work until they can't get it wrong.

Anonymous

The Food and Drug Administration (FDA) promulgated the good manufacturing practices (GMPs) for medical devices regulations in 1978, drawing authority from the Medical Devices Amendments to the Federal Food, Drug, and Cosmetic (FFD&C) Act of 1976. The GMP regulations represented a total quality assurance program intended to control the manufacture and distribution of devices. It allows the FDA to periodically inspect medical device manufacturers for compliance to the regulations.

Manufacturers must operate in an environment in which the manufacturing process is controlled. Manufacturing excellence can only be achieved by designing products and processes to address potential problems before they occur. Manufacturers must also operate in an environment that meets GMP regulations. This requires proof of control over manufacturing processes.

19.1 HISTORY OF GMPs

Two years after the Medical Device Amendments of 1976 were enacted; FDA issued its final draft of the medical device GMP regulation, a series of requirements that prescribed the facilities, methods, and controls to be used in the manufacturing, packaging, and storage of medical devices. Except for an update of organizational references and revisions to the critical device list included in the 1978 final draft's preamble, these regulations have remained virtually unchanged since they were published in the *Federal Register* on July 21, 1978. That does not mean that their interpretation has not changed.

Several key events since that date have influenced the way FDA has interpreted and applied these regulations. The first occurred in 1987 with FDA's publication of the *Guidelines on General Principles of Process Validation*, which not only provided guidance but adviced industry that device manufacturers must validate other processes when necessary to assure that these processes would consistently produce acceptable results.

In 1989, FDA published a notice of availability for design control recommendations titled Preproduction Quality Assurance Planning: Recommendations for Medical Device Manufacturers. These recommendations fulfilled a promise made by the CDRH director to a congressional hearing committee to do something to prevent device failures that were occurring due to design defects, resulting in some injuries and deaths. It was also a warning to industry that FDA was moving to add design controls to the GMP regulation.

The next year, FDA moved closer to adding design controls, publishing the Suggested Changes to the Medical Device Good Manufacturing Practices Regulation Information Document, which described the changes the agency was proposing to make to the GMP regulation. Comments asserted that FDA did not have the authority to add design controls to the GMPs, a point that became moot later that year when the Safe Medical Devices Act (SMDA) of 1990 became law. SMDA amended section 520(f) of the FFD&C Act to add preproduction design validation controls to the device GMP regulation.

SMDA also added the Food, Drug, and Cosmetic Act a new section 803, which encouraged FDA to work with foreign countries toward mutual recognition agreements for the GMP and other regulations. Later, FDA began to actively pursue the harmonization of GMP requirements on a global basis.

Over the following 2 years, FDA took steps to assure that manufacturers with device applications under review at the agency were also in compliance with GMPs. The first step was taken in 1991, when CDRH established its reference list program for manufacturers with pending premarket approval (PMA) applications, ensuring that no PMA would be approved while the device maker had significant GMP violations on record. In 1992, the program was extended to all 510(k)s. Under this umbrella program, 510(k)s would not be processed if there was evidence on hand that the site where the 510(k) device would be manufactured was not in compliance with GMPs.

On November 23, 1993, FDA acted on comments it had received 3 years earlier regarding its "Suggested Changes" document, publishing a proposed revision of the 1978 GMPs in the *Federal Register*. The proposal incorporated almost the entire 1987 version of ISO 9001, the quality systems (QSs) standard compiled by the International Organization for Standardization. While supporting adoption of ISO 9001, most of the comments received from industry objected to the addition of proposals such as applying the GMP regulation to component manufacturers.

In July 1995, FDA published a working draft of the proposed final revised GMP regulation. As stated in that draft, the two reasons for the revision were to bring about the addition of design and servicing controls, and to ensure that the requirements were made compatible with those of ISO 9001 and EN 46001 (ISO 13485), the quality standard that manufacturers must meet if they select the European Union directives' total QS approach to marketing.

Among the proposals in this version that drew the most fire from industry were the application of GMPs to component manufacturers and used of the term "end of life," which was intended to differentiate between servicing and reconditioning. FDA agreed to delete most but not all of the objectionable requirements during an August 1995 FDA-industry meeting and the GMP Advisory Committee meeting in September 1995. The end of life concept was deleted from the GMPs, but was retained in the medical device reporting regulation.

As the 1995 working draft now stands, it is very similar to the proposed ISO 13485 standard. To further harmonize the two documents, FDA's July 1995 working draft includes additions that incorporate the requirements of the 1994 version of ISO 9001 that were not in the 1987 version. FDA expects to publish a final GMP regulation by the summer of 1996. Production requirements of the revised GMPs will probably be effective from 60 to 90 days after publication of the final regulation, with the design and service requirements probably effective a year after publication.

In addition, FDA has indicated that GMP inspections might be made by third parties. If this happens, these inspections would probably begin on a small scale with third parties doing follow-up to nonviolative inspections. But eventually, third parties could play an important role in mitigating delays from FDA's reference list, which, while not now referred to by that name, is still in effect and not likely to be dropped by the agency. Although review of a 510(k) is not affected by the manufacturer being on the list, a 510(k) will not be approved until the manufacturing site is found to be in GMP compliance. The availability of the third-party auditors to inspect those sites might speed the review process under those circumstances.

Also in the future, is a training course for GMP specialists being prepared by the Association for the Advancement of Medical Instrumentation. If this course were incorporated into the FDA investigator certification training, it could help assure that the GMP regulation is interpreted and applied uniformly by FDA, consultants, and the device industry.

19.2   GMP REGULATION

The latest draft of the GMP regulation was published for comment in the *Federal Register* on November 23, 1993. They were established to replace quality assurance program requirements with

QS requirements that include design, purchasing, and servicing controls, clarify record-keeping requirements for device failure and complaint investigations, clarify requirements for qualifying, verifying, and validating processes and specification changes, and clarify requirements for evaluating quality data and correcting quality problems. In addition, the FDA has also revised the current good manufacturing practice (CGMP) requirements for medical devices to assure they are compatible with specifications for QSs contained in international quality standard ISO 9001. The following changes were made from previous regulations.

19.2.1 DESIGN CONTROLS

Over the past several years, the FDA has identified lack of design controls as one of the major causes of device recalls. The intrinsic quality of devices, including their safety and effectiveness, is established during the design phase. The FDA believes that unless appropriate design controls are observed during preproduction stages of development, a finished device may not be safe or effective for its intended use. Based on experience with administering the CGMP regulations, which currently do not include preproduction design controls, the FDA is concerned that the current regulations provide less than an appropriate level of assurance that devices will be safe and effective. Therefore, the FDA is proposing to add general requirements for design controls to the device CGMP regulations for all Class III and II devices, and several Class I devices.

19.2.2 PURCHASING CONTROLS

The quality of purchased product and services is crucial to maintaining the intrinsic safety and effectiveness of a device. Many device failures due to problems with components that result in recall are due to unacceptable components provided by suppliers. The FDA has found during CGMP inspections that the use of unacceptable components is often due to the failure of the manufacturer of finished devices to adequately establish and define requirements for the device's purchased components, including quality requirements. Therefore, the FDA believes that the purchasing of components, finished devices, packaging, labeling, and manufacturing materials must be conducted with the same level of planning, control, and verification as internal activities. The FDA believes the appropriate level of control should be achieved through a proper mixture of supplier and in-house controls.

19.2.3 SERVICING CONTROLS

The FDA has found, as a result of reviewing service records that the data resulting from the maintenance and repair of medical devices provide valuable insight into the adequacy of the performance of devices. Thus, the FDA believes that service data must be included among the data manufacturers use to evaluate and monitor the adequacy of the device design, the QS, and the manufacturing process. Accordingly, the FDA is proposing to add general requirements for the maintenance of servicing records and for the review of these records by the manufacturer. Manufacturers must assure that the performance data obtained as a part of servicing product are fed back into the manufacturer's QS for evaluation as part of the overall device experience data.

19.2.4 CHANGES IN CRITICAL DEVICE REQUIREMENTS

The FDA is proposing to eliminate the critical component and critical operation terminology contained in the present CGMP regulation. The increased emphasis on purchasing controls and on establishing the acceptability of component suppliers assures that the intent of the present critical component requirement is carried forward into the revised CGMP. The addition of a requirement to validate and document special processes further ensures that the requirements of the present critical operation requirements are retained. FDA is proposing to retain the distinction between critical and noncritical devices for one regulatory purpose. Traceability will continue to be required only for critical devices.

19.2.5 HARMONIZATION

The FDA is proposing to reorganize the structure of the device CGMP regulations and modify some of their language to harmonize them with international quality standards. FDA is proposing to relocate and combine certain requirements to better harmonize the requirements with specifications for QSs in the ISO 9001 quality standard and to use as much common language as possible to enhance conformance with ISO 9001 terminology. By requiring all manufacturers to design and manufacture devices under the controls of a total QS, the FDA believes the proposed changes in the CGMP regulations will improve the quality of medical devices manufactured in the United States for domestic distribution or exportation as well as devices imported from other countries. The proposed changes should ensure that only safe and effective devices are distributed in conformance with the act. Harmonization means a general enhancement of CGMP requirements among the world's leading producers of medical devices.

19.3 DESIGN FOR MANUFACTURABILITY

Design for manufacturability (DFM) assures that a design can be consistently manufactured while satisfying the requirements for quality, reliability, performance, availability, and price. One of the fundamental principles of DFM is reducing the number of parts in a product. Existing parts should be simple and add value to the product. All parts should be specified, designed, and manufactured to allow 100% usable parts to be produced. It takes a concerted effort by design, manufacturing, and vendors to achieve this goal.

DFM is desirable due to its low cost. The reduction in cost is due to

- Simpler design with fewer parts
- Simple production processes
- Higher quality and reliability
- Easier to service

19.3.1 DFM PROCESS

The theme of DFM is to eliminate nonfunctional parts, such as screws or fasteners, while also reducing the number of functional parts. The remaining parts should each perform as many functions as possible. The following questions help in determining if a part is necessary:

- Must the part move relative to its mating part?
- Must the part be of a different material than its mating part or isolated from all other parts?
- Must the part be separate for disassembly or service purposes?

All fasteners are automatically considered candidates for elimination.

A process that can be expected to have a defect rate of no more than a few parts per million consists of

- Identification of critical characteristics.
- Determine product elements contributing to critical characteristics.
- For each identified product element, determine the step or process choice that affects or controls required performance.

- Determine a nominal value and maximum allowable tolerance for each product component and process step.
- Determine the capability for parts and process elements that control required performance.
- Assure that the capability index (Cp) is greater than or equal to 2, where Cp is the specification width/process capability.

19.4   DESIGN FOR ASSEMBLY

Design for assembly (DFA) is a structured methodology for analyzing product concepts or existing products for simplification of the design and its assembly process. Reduction in parts and assembly operations, and individual part geometry changes to ease assembly are the primary goals. The analysis process exposes many other life cycle cost and customer satisfaction issues which can then be addressed. Design and assembly process quality are significantly improved by this process.

Most textbook approaches to DFA discuss elimination of parts. While this is a very important aspect of DFA, there are also many other factors that affect product assembly. A few rules include

1. Overall design concept
   (a) Design should be simple with a minimum number of parts.
   (b) Assure the unit is lightweight.
   (c) System should have a unified design approach, rather than look like an accumulation of parts.
   (d) Components should be arranged and mounted for the most economical assembly and wiring.
   (e) Components that have a limited shelf life should be avoided.
   (f) Use of special tools should be minimized.
   (g) Use of wiring and harnesses to connect components should be avoided.
2. Component mounting
   (a) Preferred assembly direction is top down.
   (b) Repositioning of the unit to different orientations during assembly should be avoided.
   (c) All functional internal components should mount to one main chassis component.
   (d) Mating parts should be self-aligning.
   (e) Simple, foolproof operations should be used.
3. Test points
   (a) Pneumatic test point shall be accessible without removal of any other module.
   (b) Electrical test points shall include, but not be limited to
      (i) Reference voltages
      (ii) Adjustments
      (iii) Key control signals
      (iv) Power supply voltages
   (c) All electronic test points shall be short-circuit protected and easily accessible.
4. Stress levels and tolerances
   (a) Lowest possible stress levels should be used.
   (b) Maximum possible operating limits and mechanical tolerances should be maximized.
   (c) Operations of known capability should be used.
5. PCBs
   (a) Adequate clearance should be provided around circuit board mounting locations to allow for tools.
   (b) Components should be soldered, not socketed.
   (c) PCBs must be mechanically secured and supported.
   (d) There must be unobstructed access to test and calibration points.
   (e) Exposed voltages should be less than 40 V.

6. Miscellaneous
   (a) All air intakes should be filtered and an indication that the filter needs to be changed should be given to the user.
   (b) Device shall be packed in a recyclable container so as to minimize the system installation time.

19.4.1 Design for Assembly Process

Develop a multifunctional team before the new product architecture is defined. This team should foster a creative climate which will encourage ownership of the new product's design and delivery process.

Establish product goals through a benchmarking process or by creating a model, drawing, or a conception of the product.

Perform a design for assembly analysis of the product. This identifies possible candidates for elimination or redesign, as well as highlighting high-cost assembly operations.

Segment the product architecture into manageable modules or levels of assembly.

Apply design for assembly principles to these assembly modules to generate a list of possible cost opportunities.

Apply creative tools, such as brainstorming, to enhance the emerging design and identify further design improvements.

As a team, evaluate and select the best ideas, thus narrowing and focusing the team's goals.

Make commodity and material selections. Start early supplier involvement to assure economical production.

With the aid of cost models or competitive benchmarking, establish a target cost for every part in the new design.

Start the detailed design of the emerging product. Model, test, and evaluate the new design for form, fit, and function.

Reapply the process at the next logical point.

Share the results.

19.5 MANUFACTURING PROCESS

The process of producing new product may be said to be a multiphased process consisting of

- Preproduction activity
- Pilot run build
- Production run
- Delivery to the customer

19.5.1 Preproduction Activity

Before the first manufacturing build, manufacturing is responsible for completing a myriad of activity.

Manufacturing and engineering should work together to identify proposed technologies and to assure that the chosen technology is manufacturable.

The selection of suppliers should begin by consulting the current approved suppliers listing to determine if any of the existing suppliers can provide the technology and parts. A new supplier evaluation would be necessary if a supplier is being considered as a potential source for a component, subassembly, or device.

A pilot run plan must be developed in such a way that it specifies the quantity of units to be built during the pilot run, the yield expectations and contingency plans, the distribution of those units, the feedback mechanism for problems, the intended production location, staffing requirements, training plan, postproduction evaluation, and any other key issues specific to the project.

The manufacturing strategy needs to be developed. The strategy must be documented and communicated to appropriate personnel to ensure it is complete, meets the business objectives, and ultimately is reflected in the design for the product. Developing a strategy for producing the product involves work on five major fronts:

1. Production plan
2. Quality plan
3. Test plan
4. Materials plan
5. Supplier plan

The production plan details how manufacturing will produce the product. The first step is defining the requirements of the production process. Some of these requirements will be found in the business proposal and product specification. A bill of materials structure is developed for the product which best meets the defined requirements. On the basis of bill of materials, a process flow diagram can be developed along with specific details of inventory levels and locations, test points, skills, resources, tooling required, and processing times.

The quality plan details the control through all phases of manufacture, procurement, packaging, storage, and shipment, which collectively assures that the product meets the required specifications. The plan should cover not only initial production, but also how the plan will be matured over time, using data collected internally and from the field.

The test plan specifies the "how" of the quality plan. This document must have enough technical detail to assure that the features are incorporated in the product design specification. Care must be taken to ensure that the manufacturer's test strategies are consistent with those of all suppliers.

The materials plan consists of defining the operating plan by which the final product, parts, accessories, and service support parts will be managed logistically to meet the launch plans. This involves product structure, lead times, inventory management techniques, inventory phasing/impact estimates, and identification of any special materials considerations that must be addressed. Any production variants which will be in production as well as potentially obsolete product would be detailed.

The supplier plan consists of a matrix of potential suppliers versus evaluation criteria. The potential suppliers have been identified using preliminary functional component specifications. The evaluation criteria should include business stability, QSs, cost, engineering capabilities, and test philosophy.

The design for manufacture and assembly (DFMA) review should be held when a representative model is available. This review should be documented, with action item assigned.

19.5.2 Pilot Run Build

The objective of this phase is to complete the pilot run and validate the manufacturing process against the objectives set forth in the manufacturing strategy and the product specification.

The pilot run build is the first build of devices using the manufacturing documentation. It is during this phase that training of the assembly force takes place. All training should be documented so no employee is given a task without the appropriate training before the task.

The pilot run build will validate the manufacturing process against the strategy and the manufacturing documentation. The validation will determine if manufacturing has met its objectives, including

- Standard cost
- Product quality

- Documentation
- Tooling
- Training
- Process control

The validation will also determine if the production testing is sufficient to ensure that the product meets the specified requirements.

The pilot run build also validates the supplier plan and supplier contracts. The validation will determine if the manufacturing plan is sufficient to control the internal processes of the supplier. The method and ground rules for communication between the two companies must be well defined to ensure that both parties keep each other informed of developments which impact the other. It should also confirm that all points have been addressed in the supplier contract and that all the controls and procedures required by the agreement are in place and operated in correct manner.

Internal failure analysis and corrective action takes place, involving investigating to the root cause for all failures during the pilot run. The information should be communicated to the project team in detail and in a timely manner. The project team determines the appropriate corrective action plans.

A pilot run review meeting is held to review all aspects of the build, including the manufacturing documentation. All remaining issues must be resolved and documentation corrected. Sufficient time should be allowed in the project schedule for corrective action to be completed before the production run.

19.5.3 PRODUCTION RUN

The objective of this phase is to produce high quality product on time, while continuing to fine tune the process using controls which have been put in place. During this phase, the first production order of units and service parts are manufactured. The training effort continues, as new employees are transferred in or minor refinements are made to the process. Line failures at any point in the process should be thoroughly analyzed and the root cause determined. Product cost should be verified at this time.

19.5.4 CUSTOMER DELIVERY

The objective of this phase is to deliver the first production units to the customer, refine the manufacturing process based on lessons learned during the first build, and finally to monitor field unit performance to correct any problems.

Following production and shipment of product, continued surveillance of the production process should take place to measure its performance against the manufacturing strategy. The production process should be evaluated for effectiveness as well as unit field performance. Feedback from the field on unit problems should be sent to the project team, where it may be disseminated to the proper area.

EXERCISES

1. Year 2000 (Y2k) problems were a concern for medical device manufacturers, especially those that dealt with imbedded microprocessors. Investigate this statement using a Web search. How might the GMP regulation have avoided this problem?
2. Visit the Web site http://www.fda.gov/cdrh/dsma/gmp_man.html and briefly look at the manual listed here. How does this differ from the GMP regulation?
3. Perform a Web search for DFMA. Report on the best site you can find.
4. Find and report on at least one good example of DFA or DFMA.

5. Related term involves design for the environment. Find information on this type of process and report on its value.
6. Related activity involves design for life cycle. Report on this concept.
7. Investigate a typical blood pressure unit that may be purchased at your corner drug store. What improvements can you suggest with respect to DFA?
8. As in problem 7 in Exercises, but investigate an in-the-ear temperature unit.

BIBLIOGRAPHY

Boothroyd Dewhurst, *Design for Manufacture and Assembly/Service/Environment and Concurrent Engineering (Workshop Manual)*. Wakefield, RI: Boothroyd Dewhurst, 1966.

Food and Drug Administration, 21 CFR Part 820 medical devices; current good manufacturing practice (CGMP regulations; proposed revisions), November 23, 1993.

Fries, R.C., *Reliability Assurance for Medical Devices, Equipment and Software*. Buffalo Grove, IL: Interpharm Press, 1991.

Fries, R.C., *Reliable Design of Medical Devices*. New York: Marcel Dekker, 1997.

Fries, R.C., *Handbook of Medical Device Design*. New York: Marcel Dekker, 2001.

Hooten, W.F., A brief history of FDA good manufacturing practices, *Medical Device & Diagnostic Industry*, 18(5), May 1996.

King, P.H. and Fries, R.C., *Design of Biomedical Devices and Systems*. New York: Marcel Dekker, 2002.

20 Miscellaneous Issues

When you get right down to it, one of the most important tasks of a leader is to eliminate his people's excuse for failure.

Robert Townsend

20.1 LEARNING FROM FAILURE

It is important to recognize that the engineering profession has learned from failures and the study of their causes. An exposure to classical failures such as the Tacoma Narrows Bridge, the Shuttle Challenger, the Hyatt Regency Walkway Collapse, Three Mile Island, the Bhopal Chemical Plant disaster, the World Trade Center disaster, and more recently the collapse of the I35W Bridge in Minneapolis, amongst others, should be a part of every engineer's education.* Most disasters are the result of a combination of unexpected circumstances, poor design, and ethical failures, but not all (Hendley, 1998).

A brief mention of a number of biomedical engineering related failures should help point out some of the considerations that students in bioengineering should consider in the process of design activities. Other examples are in various sections of this text to assist in a sensitization to the need for safe design procedures.

A computer programming glitch on a Therac-25 radiation therapy machine allowed a technician to deliver over 125 times the required therapeutic dose of radiation to a patient. The error message "Malfunction 54" did not convey the correct message that the technician should not repeat the dose. Needless to say, the patient died (Casey, 1993).

Technicians in an ambulance taking a heart attack victim to a hospital lost use of their heart machine every time they attempted to use their radio transmitter. (The fault was caused by unshielded radio frequency [RF] interference.) The patient died (Geddes, 1998).

Toxic shock syndrome plagued some users of super absorbency tampons in the late 1970s. It also caused some deaths. There had been no Food and Drug Administration (FDA) or other guidelines as to the composition, degree of absorbency, or recommendations on length of use (time) for these products until this occurred.

Thalidomide was sold in Europe in the late 1950s, causing over 8000 births of malformed children. The drug had not been tested adequately before market release.

Laetrile, a substance that can be extracted (or synthesized) from apricot seeds has been touted as a cancer cure since the 1960s. Banned in the United States by the FDA, it can still be obtained in Mexico.

In 1938, 107 deaths of (primarily) children were caused due to ingestion of elixir of sulfanilamide, a toxic combination of diethylene glycol and sulfa. This disaster is one of the prime initiators of the early FDA drug (especially patent drug) enforcement activities.

Quack medical devices have plagued the U.S. population for years. Most advertisers claim for devices or drugs that have claims for medical benefits come under the scrutiny of the FDA, which has the power to fine and recall for false claims.

---

* See, for example, http://www.matscieng.sunysb.edu/disaster/.

20.2   DESIGN FOR FAILURE

It is important to consider, when designing systems and devices that sometimes you must consider and plan for failure. One must often be proactive, rather than reactive, when considering failure. Designing for failure can be for the purposes of safety and convenience.

Safety considerations are paramount in many design problems and an understanding of several examples is important. A few examples are as follows:

- Fuses—current flow through a fine wire or a low-temperature melt point wire causes it to vaporize or melt, protecting the circuit beyond the fuse point.
- Shear pins—many devices have a section that will break rather than ruin the entire system. Many lawnmowers have a shear pin, which breaks before the main crankshaft can.
- Sprinkler systems—the increase in temperature due to a fire causes melting of a metal plug and the opening of a sprinkler or gas quenching system puts out the fire.
- Coating on a medicine "lasts long enough" in the stomach to deliver a drug to the intestines, where it is needed or causes no harm compared to direct stomach delivery (enteric coatings, a variation on the M&M melts in your mouth, not your hands philosophy).
- Individually bubble packed drugs stay isolated from the atmosphere (generally used with hydroscopic drugs) until the bubble is burst.
- Humidification/heating system is allowed to operate until a bimetallic element snaps a vent shut at a given temperature (too hot or too cold).
- In the event of a power failure, a lead shield drops in front of a cobalt therapy delivery unit.
- Current limiter is placed between a patient and a medical device; the patient is protected from excessive currents.
- Bottle tops can be fashioned to require a minimal amount of squeezing and/or manipulation before they open, thus protecting the weak or young (typically for dispensing of medicines).
- Plastic or real peanuts used for packing deform during impacts, protect the packaged item.
- Graphite rods are designed to drop into reactors to quell runaway reactions.
- Feathers protect a bird but pull out in order to enable a bird to escape a predator.
- Eggshells protect an embryo but can be shattered from within by a chick ready to hatch.
- Pine seed can sit dormant for years, opening after a fire when there then exists a chance for sunlight and growth.

For systems such as computer security systems, the goal of a safe design would be to do the following items: deter intrusion, detect intrusions, delay intrusions, warn of intrusions, and perhaps redirect intrusions to a "honey pot" system which can collect information on the intruder. Military systems can be designed so that in case of the failure of an outer system, an inner ring picks up on the challenge.

20.3   DESIGN FOR CONVENIENCE

Many items are designed to fail in a particular manner only for the convenience of the user, a quick listing of a few of these items include

- Postage stamps—sheets of postage stamps typically contain individual stamps separated by perforations. A slightly skilled user can easily separate out an individual stamp by causing failure along the perforations. Obviously this same concept has been applied to toilet paper, paper towels, and checkbooks.

- Waffles in family and other packs typically are packed two or four to a sheet, the connections between the waffles being much thinner than any other part in order to allow ease of separations.
- Scoring of a surface to enhance breakage is a common way to ensure easy opening of bags of coffee and pop-top cans of various designs.

20.4   UNIVERSAL DESIGN

The term universal design refers to a mind-set that is inclusive in nature when one is considering the design of new environments and products.* Briefly, it gives special consideration to design elements that may be altered to include persons with abilities not falling within the "norm," without calling special attention to the fact that such a design change has been made. These abilities might include problems including hearing, vision, balance, strength, attention, memory, etc. A reminder system for pill dispensing for the elderly thus might have a louder than normal alarm and visual indicators to indicate that a drug has not been taken. Wider than normal doors, for example, is a case of universal design (to allow ease of wheelchair entry.)

It must be noted that the Americans with Disabilities Act (ADA) preceded this endeavor. The ADA prohibits discrimination and ensures equal opportunity for persons with disabilities in employment, state and local government services, public accommodations, commercial facilities, and transportation. It also mandates the establishment of telecommunication display devices (TDD)/ telephone relay services. There are a number of published ADA standards for accessible design that have been codified in our legal system (see, for example, 28*CFR* Part 36, ADA Standards for Accessible Design.)

20.5   DESIGN FOR ASSEMBLY

The term design for assembly refers to a technique whereby the design and manufacturing considerations involved in an assembled device are optimized for ease of assembly of the device. Special attention is given to tolerances of parts that fit together, requirements for each part (in terms of motions to place the part in the assembly), and standardization of the assembly materials. Design for assembly methods are credited with major cost savings in several industries.

20.6   PREVENTION THROUGH DESIGN

A meeting was held in 2007, sponsored by the National Institute of Occupational Safety and Health (NIOSH), on the topic of "Prevention through Design." The premise of the meeting/workshop was that many accidents are due to faulty design at the outset, and that design techniques need to be developed which consider prevention of potential future accidents. The audience for this workshop consisted of personnel from a multitude of industries, from mining to health care. One of the major questions addressed by some of the speakers was the necessity to prove to management that money spent at the outset for improved safety paid for itself many times over in long term costs of doing business. This group's efforts will likely bring new legislation to bear with respect to design education.

20.7   DESIGN FOR THE ENVIRONMENT

Another "design for" term potentially of interest is design for the environment. This term applies to the "clean" manufacturing and recycling of devices, including packaging. Devices are to be built using a minimum of energy, with minimum emissions and scrap and by-products. As much as is possible, packaging materials are also meant to be reused or recycled.

---

* Winters, J.M. *Medical Instrumentation, Accessibility and Usability Considerations*, Boca Raton, FL: CRC Press, 2007.

20.8 POKA-YOKE

Poka-Yoke is the name given to a methodology to foolproof a process. It was developed as a part of the Toyota production system. Generally speaking, one of the three types of techniques is used: a contact method (has contact been made properly?), a fixed-value method (have the proper movements been made?) or motion-step method (have the prescribed steps been done in the correct order?) The use of the method involves studying the process that needs to be improved, then deciding how to mistake proof the process. This mistake proofing step can involve prevention of an error being made (machine will not work without the part that needs to be added), warning the worker involved that an error has been made, or a more serious flagging of the situation (audible alarm, for example).

20.9 PRODUCT LIFE ISSUES

The goal of the product development process is to put a safe, effective, and reliable medical device in the hands of a physician or other medical personnel where it may be used to improve health care. The device has been designed and manufactured to be safe, effective, and reliable. The manufacturer warranties the device for a certain period of time, usually 1 year. Is this the end of the manufacturer's concern about the device? It should not be. There is too much valuable information to be obtained.

Analysis of field data is the means of determining how a product is performing in actual use. It is a means of determining the reliability growth over time. It is a measure of how well the product was specified, designed, and manufactured. It is a source of information on the effectiveness of the shipping configuration. It is also a source for information for product enhancements or new designs. Field information may be obtained in any of several ways, including

- Analysis of Field Service Reports (FSR)
- Failure analysis of failed units
- Warranty analysis

20.9.1 ANALYSIS OF FSR

The type of data necessary for a meaningful analysis of product reliability is gathered from FSR. The reports contain vital information such as

- Type of product
- Serial number
- Date of service activity
- Symptom of the problem
- Diagnosis
- List of parts replaced
- Labor hours required
- Service representative

The type of product allows classification by individual model. The serial number allows a history of each individual unit to be established and traceability to the manufacturing date. The date of service activity helps to indicate the length of time until the problem occurred.

The symptom is the problem, as recognized by the user. The diagnosis is the description of the cause of the problem from analysis by the service representative. The two may be mutually exclusive, as the cause of the problem may be remote from the user's original complaint. The list of parts replaced is an adjunct to the diagnosis and can serve to trend parts usage and possible vendor problems. The diagnosis is then coded, where it may later be sorted.

The required labor hours help in evaluating the complexity of a problem, as represented by the time involved in repair. It, along with the name of the service representative, acts as a check on the efficiency of the individual representative, as average labor hours for the same failure code may be compared on a representative to representative basis. The labor hours per problem may be calculated to assist in determining warranty cost as well as determining the efficiency of service methods.

The only additional data, which is not included in the FSR, is the date of manufacture of each unit and the length of time since manufacture that the problem occurred. The manufacturing date is kept on file in the device history record. The length of time since manufacture is calculated by subtracting the manufacturing date from the date of service.

20.9.1.1 Database

FSR are sorted by-product upon receipt. The report is scanned for completeness. Service representatives may be contacted where clarification of an entry or lack of information would lead to an incomplete database record. The diagnoses are coded, according to a list of failures, as developed by reliability assurance, design engineering, and manufacturing engineering (Figure 20.1). Manufacturing date and the length of time since manufacture are obtained. The data is then ready to be entered into the computer.

The data is entered into a computer database, where it may be manipulated to determine the necessary parameters. Each FSR is input to a single database record, unless the service report contains multiple failure codes. Figure 20.2 shows a sample database record.

The data is first sorted by service date, so trending can be accomplished by a predetermined time period, such as a fiscal quarter. Data within that time frame is then sorted by problem code, indicating the frequency of problems during the particular reporting period. A Pareto analysis of the problems can then be developed. Data is finally sorted by serial number, which gives an indication of which device experienced multiple service call and or experienced continuing problems.

Percentages of total problems are helpful in determining primary failures. Spread sheets are developed listing the problems versus manufacturing dates and the problems versus time since manufacturing. The spreadsheet data can then be plotted and analyzed.

Failure Code Base Machine	Failure
101	Missing parts
102	Shipping damage
103	Circuit breaker wiring damage
104	Regulator defect
105	Shelf latch broken
Monitor	
201	Display problems
202	Control cable defect
203	Power board problem
204	Control board problem
205	Unstable reference voltage

FIGURE 20.1 List of failure codes. (From Fries, R.C., *Reliable Design of Medical Device*, Marcel Dekker, New York, 1997.)

Field	Field Content
1	Service date
2	Device serial number
3	Manufacturing date
4	Time in use (hours)
5	Failure code
6	Failed parts 1
7	Failed parts 2
8	Failed parts 3
9	Failed parts 4
10	Failed parts 5
11	Time to repair (hours)
12	Service representative ID

FIGURE 20.2 Sample database record. (From Fries, R.C., *Reliable Design of Medical Device*, Marcel Dekker, New York, 1997.)

20.9.1.2 Data Analysis

The most important reason for collecting the field data is to extract the most significant problem information and put it in such a form that the cause of product problems may be highlighted, trended and focused upon. The cause of the problem must be determined and the most appropriate solution implemented. A band-aid solution is unacceptable. Company response to problems involving any problem worthy of reporting in the FDA manufacturers and users device experience (MAUDE) database is critical. Companies must show that they are responsive to user complaints and have a process in place for complaint correction. The use of a consistent response to user complaints, such as "user error" is not a legally defensible position.

Pareto analysis is used to determine what the major problems are. The individual problems are plotted along the x-axis and the frequency on the y-axis. The result is a histogram of problems, where the severity of the problem is indicated, leading to the establishment of priorities in addressing solutions. Similar plots versus day of week may indicate personnel problems, versus supplier may indicate supply problems.

Several graphical plots are helpful in analyzing problems. One is the plot of particular problems versus length of time since manufacturing (sometimes termed a "run plot"). This plot is used to determine the area of the life cycle in which the problem occurs. Peaks of problem activity indicate infant mortality, useful life or wearout, depending on the length of time since manufacture. An example plot may be seen in Figure 20.3, the data is taken from a complaint investigation done in 2007. One may surmise that the units are in the "infant mortality" stage of their life, as the complaint rate has not yet risen after 6 years. This is an interesting case, as the "guarantee" on the unit is only 1 year.

A second plot of interest is that of a particular problem versus the date of manufacture. This plot 0i0s a good indication of the efficiency of the manufacturing process. It shows times where problems occur, for example, the rush to ship product at the end of a fiscal quarter, lot of problems on components, or vendor problems. The extent of the problem is an indication of the correct or incorrect solution. An example of this type of data may be seen in Figure 20.4. One may infer from the plot that complaints start immediately after manufacture and distribution (infant mortality) and continue throughout the life of the devices. Vertical gaps are indicative of product recalls (2005 and 2006.) A general lightening of complaints left to right may be indicative of improved maintenance or withdrawal of old product from the market.

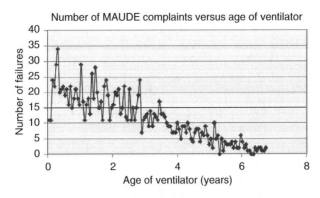

FIGURE 20.3 MAUDE complaints versus age of ventilator.

Another useful plot is that of the total number of problems versus the date of manufacture. The learning curve for the product is visible at the peaks of the curve. It can also be shown how the problems for subsequent builds decrease as manufacturing personnel become more familiar and efficient with the process.

Trending of problems, set against the time of reporting is an indicator of the extent of a problem and how effective the correction is. Decreasing numbers indicate the solution is effective. Reappearing high counts indicate the initial solution did not address the cause of the problem.

The database is also useful for analyzing warranty costs. The data can be used to calculate warranty expenses, problems per manufactured unit and warranty costs as a percentage of sales. A similar table can be established for installation of devices.

20.9.2 FAILURE ANALYSIS OF FIELD UNITS

Most failure analysis performed in the field is done at the board level. Service representatives usually solve problems by board swapping, since they are not equipped to troubleshoot at the component level. Boards should be returned to be analyzed to the component level. This not only yields data for trending purposes, but also highlights the real cause of the problem. It also gives data on problem parts or problem vendors.

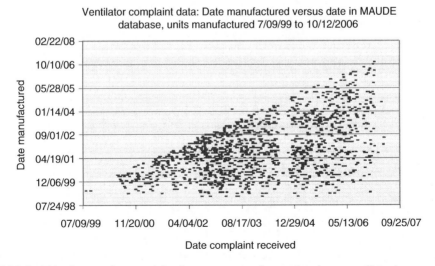

FIGURE 20.4 This plot contains complaint data versus manufacture date (same ventilators).

   The most important process in performing field failure analysis is focusing on the cause of the problem, based on the symptom. It does no good to develop a fix for a symptom, if the cause is not known. To do so only creates additional problems. Analysis techniques, such as fault tree analysis or failure mode and effects analysis may help to focus on the cause.

   Once the component level analysis is completed, Pareto charts may be made, highlighting problem areas and prioritizing problem solutions. The major problems can be placed in a spread sheet and monitored over time. Graphical plots can also be constructed to monitor various parameters over time.

20.9.3   WARRANTY ANALYSIS

Warranty analysis is an indication of the reliability of a device in its early life, usually the first year. Warranty analysis (Figure 20.5) is a valuable source of information on parameters such as warranty cost as a percentage of sales, warranty cost per unit, installation cost per unit and percentage of shipped units experiencing problems. By plotting this data, a trend can be established over time.

20.10   PRODUCT TESTING ISSUES

Analysis of field data is also a significant means of reviewing the testing completed during the product development cycle to determine if it was sufficient for the intended use of the device. If field reports indicate a litany of problems, the types and severity of the testing performed need to be reviewed. HALT (highly accelerated life testing) testing may have to be performed, as this type of testing may indicate problems early in the testing that would take some time to occur in the field. The severity of the test parameters needs to be reviewed to determine if more severe parameters could have indicated a problem was present. If the failure was caused by customer misuse of the product, the type and severity of the misuse testing needs to be reviewed.

Product Code	Parameters	Cost 1/95	Cost 2/95	Cost Year to Date
xxxxx	Normal warranty	$	$	$
xxxxx	Recall warranty	$	$	$
xxxxx	Total warranty	$	$	$
xxxxx	Setup cost	$	$	$
xxxxx	Total cost	$	$	$
xxxxx	Sales	$	$	$
	Warranty/sales			
	Setup/sales			
	Total/sales			
	Number of units shipped			
	Number of units setup			
	Number warranty units			
	Number of recall units			
xxxxx	Warranty/unit	$	$	$
xxxxx	Recall/unit	$	$	$
xxxxx	Setup/unit	$	$	$
xxxxx	Total/unit	$	$	$

FIGURE 20.5   Warranty analysis. (From Fries, R.C., *Reliable Design of Medical Device*, Marcel Dekker, New York, 1997.)

When reviewing the tests that were performed, it is important to analyze test severity, as you want the test parameters to be severe enough to indicate a weakness in the design or component, yet you do not want the parameters so severe that they cause problems that would not occur under ordinary use of the device.

EXERCISES

1. Find and report on heart valve failure history. What valves and valve types are still in the development phase?
2. Find information on the health effects of the Chernobyl accident. Report on the current state of this event.
3. Your Volvo hits a guardrail at high speed. How many systems are involved in the incident as a design to fail device? Detail these (at least three).
4. There was a significant social outcry associated with the lack of patient informed consent in a long-term study of Syphilis in the South of the United States in the 1900s. Find and discuss information on this event.
5. Illegal medical experimentation was detected during World War II. Find information on this and report on the outcomes.
6. There are a few excellent Web sites dealing with ethical issues in the United States. Find one and document what is at the site.
7. Find and discuss at least one good university Web site relating to medical ethics.
8. Find and report on the Bhopal incident. What elementary safety rule was violated in this case?
9. Find and discuss at least one new design for failure example.
10. Find and discuss at least one design for convenience example.

BIBLIOGRAPHY

Association for the Advancement of Medical Instrumentation (AAMI), *Guideline for Establishing and Administering Medical Instrumentation*. Arlington, VA: AAMI, 1984.

Casey, S., *Set Phasers on Stun*, Santa Barbara, CA: Aegean Publishing, 1993.

Fries, R.C., *Reliable Design of Medical Devices*. New York: Marcel Dekker, 1997.

Fries, R.C. et al., A reliability assurance database for analysis of medical product performance, *Proceedings of the Symposium on Engineering of Computer-Based Medical Systems*. New York: The Institute of Electrical and Electronic Engineers, 1988.

Geddes, L., Medical Device Accidents, Boca Raton, FL: CRC Press, 1998, P. 27.

Hendley, V., The importance of failure, *ASEE Prism*, October 1998: 18–23.

MIL-HDBK-472, *Maintainability Prediction*. Washington, DC: Department of Defense, 1966.

21 Product Issues

An error doesn't become a mistake until you refuse to correct it.

Orlando A. Battista

21.1 PRODUCT SAFETY AND LEGAL ISSUES

When designing for safety, there are two aspects to consider. The first is risk assessment that addresses the questions: What failure could cause harm to the patient or user? What misuse of the device could cause harm? These failures must be analyzed using methods such as fault tree analysis or failure mode analysis and must be designed out of the device.

The second aspect of safety is liability assessment. This addresses the questions: Have all possible failure modes been explored and designed out? Have all possible misuse situations been addressed? Court cases have special punitive judgments for companies that have knowledge about an unsafe condition and do nothing about it.

21.1.1 DEFINITION OF SAFETY

Safety may be defined as freedom from accidents or losses. Some people have argued that there is no such thing as absolute safety, and therefore safety should be defined in terms of acceptable losses. Using this argument, an alternative definition of safety would be a judgment of the acceptability of risk, with risk, in turn, as a measure of the probability and severity of harm to human health.

A product is safe if its attendant risks are judged to be acceptable. This definition of safety implies that hazards cannot be eliminated, when they often can. While in most instances, all hazards cannot be eliminated, specific hazards can be totally eliminated from a product or system.

System safety is a subdiscipline of systems engineering that applies scientific, management, and engineering principles to ensure adequate safety throughout the system life cycle, without constraints of operational effectiveness, time, and cost. Although safety has been defined as freedom from those conditions that can cause death, injury, occupational illness, or damage to or loss of equipment or property, it is generally recognized that this is unrealistic. By this definition, any system that presents an element of risk is unsafe. But almost any system that produces personal, social, or industrial benefits contains an indispensable element of risk.

The problem is complicated by the fact that attempts to eliminate risk often result in risk displacement rather than risk elimination. Benefits and risks often have trade-offs, such as trading off the benefits of improved medical diagnosis capabilities against the risks of exposure to diagnostic x-rays. Unfortunately, the question "How safe is safe enough?" has no simple answer.

Safety is also relative in that nothing is completely safe under all conditions. There is always some case in which a relatively safe material or piece of equipment becomes hazardous. The act of drinking water, if done to excess, can cause kidney failure. Thus, safety is a function of the situation in which it is measured. One definition might be that safety is a measure of the degree of freedom from risk in any environment. To understand safety better, it is helpful to consider the nature of accidents in general.

An accident is traditionally defined by safety engineers as an unwanted and unexpected release of energy. However, release of energy is not involved in some hazards associated with new technologies and potentially lethal chemicals. Therefore, the term mishap is often used to denote an unplanned event or series of events that result in death, injury, occupational illness, damage to or loss of equipment or property, or environmental harm. The term mishap includes both accidents and harmful exposures.

Mishaps are almost always caused by multiple factors and the relative contribution of each factor is usually not clear. A mishap can be thought of as a set of events combining in random fashion, or alternatively, as a dynamic mechanism that begins with the activation of a hazard and flows through the system as a series of sequential and concurrent events in a logical sequence until the system is out of control and a loss is produced. The high frequency of complex, multifactorial mishaps may arise from the fact that the simpler potential mishaps have been anticipated and handled. However, the very complexity of the events leading up to a mishap implies that there may be many opportunities to interrupt the sequences.

Mishaps often involve problems in subsystem interfaces. It appears to be easier to deal with failures of components than failures in the interfaces between components.

How do engineers deal with safety problems? The earliest approach to safety, called operational or industrial safety, involves examining the system during its operational life and correcting what are deemed to be unacceptable hazards. In this approach, accidents are examined, causes are determined, and corrective and preventive actions are initiated. In some complex systems, however, a single accident can involve such a great loss as to be unacceptable. The goal of system safety is to design an acceptable safety level into the system before actual production or operation.

System safety engineering attempts to optimize safety by applying scientific and engineering principles to identify and control hazards through analysis, design, and management procedures.

21.1.2 Safety and Reliability

There is some confusion in the industry about the difference between safety and reliability. Both are good things to which systems should aspire. They remain, however, distinct concepts. They may at times even be conflicting concerns. The literature has muddied the picture by using these terms imprecisely, particularly the term safety.

A safe system is one that does not incur too much risk to persons or equipment. A risk is an event or condition that can occur, but is undesirable. Risk is measured both in terms of severity and probability. Safety only concerns itself with failures that introduce hazards. The probability of failure of a device to meet its requirements defines its reliability. Safety takes a broader view—it is possible to write requirements so that they do not consider all safety concerns. The concept of safety is not defined in terms of meeting requirements, but on a level of risk.

A safe system is one in which damage to persons or property does not happen often or, when it does, the damage is minor. If the damage potential is small, then it can happen more frequently and still be considered safe. If the damage potential is great, then the chance for a mishap must be correspondingly small for the system to be safe. Note that the availability of the system does not appear in the definition of safety. A system can fail all the time, but provided that it fails in a safe way, that is, in a way that does not lead to mishaps, the system is still safe. Conversely, a system can be up and running all the time and consistently put people at risk. Such a system is reliable, but not safe.

Consider the example of a pacemaker. For the vast majority of pacemaker patients, the pacemaker provides only assistance. When the sinoatrial (SA) node, the normal pacemaker of the heart, fails to function properly, some other area of cardiac tissue takes over its role. However, the SA node provides the best rate for physiological control, typically 60–80 beats/min. When other portions of the heart assume the pacing function, the rates are typically much less and can be as low as 30 beats/min. Patients with this condition have a reduced cardiac output and have difficulty in performing tasks that require increased cardiac flow, such as climbing stairs.

A pacemaker solves this problem by artificially pacing the heart at some minimum programmed rate; a pacemaker that paces at 110 beats/min continuously no matter what is very reliable. However, if the patient is in cardiac failure, a high pacing rate is medically inappropriate. Thus, this is a reliable but unsafe device.

An unreliable pacemaker would be one that did not always pace at the programmed rate. However, not pacing is not a safety concern, except for a small minority of patients. In this case, we have an unreliable but safe device.

Hardware components and subsystems usually have known failure histories and there are published values for such reliability measures as mean time to failure and mean time between failures (MTBF). When a system is entirely composed of components whose reliability statistics are known, the reliability of the entire system can be estimated by combining the reliability of the components according to the mathematical laws of probability. Such calculations are the source of assertions that certain safety-critical systems, such as aircraft controls, have a very low probability of failure.

For software, there are no sound foundations for quantitative statistical failure estimates, such as MTBF. Software faults are design errors, not random equipment failures. Control software is customized for each product. A new control program is therefore a unique artifact with no performance history of any kind. That is why object-oriented design and programming have suggested the reuse of software objects that have been tested and used in the field and thus have a history of success or failure.

Failure data should be collected during development and early field experience. It is usual to discover many faults early in a product's lifetime. The fault discovery rate gradually declines as more subtle problems are unearthed. There exist statistical software reliability models that attempt to predict the number of undiscovered faults remaining, based on the past history of failures, but these models are necessarily less trustworthy than statistics gathered from mass-produced items. In practice, it is not possible to predict when a program will next fail, and it is not realistic to assign failure probabilities or measures to programs just entering service. The main practical lessons of the software reliability models are that a program that has recently exhibited many failures is likely to continue to fail. It is also important to note that a new version of a program may need to be considered as a completely new program from the point of view of failure history.

The concerns for safety and reliability can be at odds with each other. To improve reliability, marginally operating systems may be allowed to continue to function. At the same time, devices are not automatically beneficial, and they, like all technology, are associated with risk. Recent examples include ultrasound equipment that often does not comply with electrical safety guidelines, leakage of insulin pumps, defective artificial cardiac valves, and reactions of the body to materials used in implants.

21.1.3 LEGAL ASPECTS OF SAFETY—REITERATED

Limiting legal liability is one of the goals of system safety. Ideally, tort law complements safety regulation by deterring the production of harmful products, along with its primary purpose of compensating injured individuals. The impact of product liability judgments and the increase in insurance rates as a consequence have been highly controversial. While data are difficult to acquire, it is clear that new theories of liability have expanded the number of potential lawsuits and that there has been a trend toward larger compensatory and punitive damage awards. Medical devices have been the focus of several mass tort actions, including thousands of cases brought against the producers of tampons for toxic shock injuries and against A.H. Robbins, the manufacturer of the Dalkon Shield. There is no question that product liability has increased the costs of doing business in some sectors of the medical device industry.

The three most common theories of liability for which a manufacturer may be held liable for personal injury caused by its product are

- Negligence
- Strict liability
- Breach of warranty

These are referred to as common-law causes of action, which are distinct from causes of action based on federal or state statutory law. Although within the last decade federal legislative action that would create a uniform federal product liability law has been proposed and debated, no such law exists today. Thus, such litigation is governed by the laws of each state.

The basic idea of negligence law is that one should have to pay for injuries that he or she causes when acting below the standard of care of a reasonable, prudent person participating in the activity in question. This standard of conduct relates to a belief that centers on potential victims: that people have a right to be protected from unreasonable risks of harm. A fundamental aspect of the negligence standard of care resides in the concept of foreseeability. In one of the most famous torts opinions, in what today would be called a products liability case, it was written that there is a duty "to use ordinary skill and care to avoid...danger" when "one person is by circumstances placed in such a position with regard to another...that every one of ordinary sense who did think would at once recognize" the risk of danger "if he did not use ordinary care and skill." It is interesting that in this nineteenth century opinion, long before the coinage of the term "products liability," this formulation emerged from a case dealing with a ship painter's allegation that the failure of a rope on a scaffold caused him to fall.

Under the theory of negligence, a manufacturer that does not exercise reasonable care or fails to meet a reasonable standard of care in the manufacture, handling, or distribution of a product may be liable for any damages caused. For example, if it can be established that not having a reliability program constitutes a failure to meet an industry-wide practice that is found to be an applicable standard of care, then the manufacturer may be subject to liability for negligence. This is also mandated by the wording of the mandate for the Food and Drug Administration (FDA), section 21 *Code of Federal Regulations* (*CFR*).

Unlike the negligence suit, in which the focus is on the defendant's conduct, in a strict liability suit, the focus is on the product itself. The formulation of strict liability states that one who sells any product in a defective condition unreasonably dangerous to the user or consumer or to his property is subject to liability for physical harm thereby caused to the ultimate user or consumer or to his property if the seller is engaged in the business of selling such a product, and it is expected to and does reach the user or consumer without substantial change to the condition in which it is sold. Therefore, the critical focus in a strict liability case is on whether the product is defective and unreasonably dangerous. A common standard applied in medical device cases to reach that determination is the risk/benefit analysis, that is, whether the benefits of the device outweigh the risks attendant with its use.

Strict Liability in tort had its modern origins in warranty and in the tort doctrine of *res ipsa loquitur* (the thing speaks for itself). The rationales for imposing strict liability are based on the fact that the manufacturer is in the best position to reduce the risk. The loss may be overwhelming to the injured person, but it can be effectively insured against by the manufacturer and distributed among the public as a cost of doing business. The manufacturer, even if not negligent, is responsible for the product being on the market.

There are three types of breaches of warranty that may be alleged:

- Breach of the implied warranty of merchantability
- Breach of the implied warranty of fitness for a particular purpose
- Breach of an express warranty

An express warranty is one which is stated explicitly, either orally, in a contract of sale, or in the labeling. For instance, let us assume that a clinical engineer requests and receives a written or oral statement that the medical device manufacturer followed a certain reliability protocol or that the medical device and its software will perform in a specific fashion. If the device causes an injury because it was not developed according to the stated reliability protocol or because it did not function as warranted, the manufacturer faces liability under the express warranty theory.

Sometimes, a warranty is not stated explicitly. By introducing a product into commercial distribution, the manufacturer implicitly warrants that the product is reasonably fit for the purposes that similar products are intended to serve. For instance, if all pacemaker manufacturers have reliability programs to ensure proper functioning of their products, and a new pacemaker manufacturer begins commercial distribution, there may be an implied warranty that the new manufacturer has a similar program in place.

These warranty causes of action do not offer any advantages for the injured plaintiff that cannot be obtained by resort to negligence and strict liability claims and, in fact, pose greater hurdles to recovery. Thus, although a breach of warranty claim is often pled in the plaintiff's complaint, it is seldom relied on at trial as the basis for recovery.

As a general proposition, a plaintiff is entitled to plead and prove as many counts or causes of action as they wish. The plaintiff is usually entitled to recover all foreseeable damages in a products liability suit. Such damages may cover the areas of emotional distress, punitive damages, and joint and several liability.

There is a division of authority as to whether recovery for emotional distress alone is allowable, where there is no accompanying physical injury. Courts have held that recovery for emotional distress without physical injury is permissible where the defendant's conduct is intentional or outrageous. A distinction is drawn between recovery for fear of future injury, and recovery for the risk of the injury itself. Some courts will not allow recovery for the risk of future injury, even where the chance that the risk will result in greater injury is greater than 50%. Others allow for recovery if the risk is more probable than not.

Perhaps no subject in tort law has generated more heated controversy in recent years that the recoverability of punitive damages in tort, including products liability. The evidence indicates that only a small fraction of cases result in punitive damages and many of these are business torts, rather than personal injury cases. A few cases have received disproportionate attention, however, and the specter of potentially large punitive recoveries has probably contributed significantly to substantial increase in products liability insurance premiums, as well as to the enactment at the state level of various restrictions on punitive recoveries. The statutory restrictions vary widely, from raising the burden of proof to clear and convincing evidence to requiring actual malice, to placing a cap on the amount of recovery, to requiring bifurcation of trial of the compensatory and punitive aspects of a case, to requiring managerial involvement in the misconduct, to requiring part of the recovery to be paid to the state, and other variations.

Another area in which extensive efforts have been made to modify the common law by statute is with regard to joint liability—whereby one tortfeasor (person who commits a civil wrongdoing) is held liable for all damages suffered by a claimant, even though other tortfeasors may also have contributed to the injury. If the damages are readily divisible, the tortfeasor would normally be liable only for his share. But liability for the full amount of damages is usually imposed when the damages are practically indivisible, as is often the case when there are multiple tortfeasors.

As can be seen, safety, as evidenced through products liability, will have an effect on a manufacturer, both in terms of finances and in reputation. The topic of products liability is discussed in greater detail in Chapter 15.

21.1.4 System Safety

Every system, no matter how complex it is, should be fail-safe, that is, it should be designed to fail into a safe and harmless state. Only a few simple functions should be required to enter or preserve the safe states by terminating or preventing potentially hazardous conditions. These functions, usually called interlocks, lockouts, or shutdown systems, should be designed to work properly despite the failure of other functions. Many regulations and guidelines specify that these safety functions should not be performed by the same computer system that provides normal operating functions and should perhaps not be performed by computers at all.

A very important part of the design process is identifying the safe states. A radiation therapy machine is in a safe state when the beam is turned "off" and all motions are stopped. An automatic drug infusing device is in a safe state when the infusion is stopped or, depending on the drug, when the infusion rate is at some constant, low value. Unlike some applications, like aviation, which require backup computers with considerable functionality, in medicine it is usually sufficient to provide a simple safety system that disconnects the computer, achieves the safe state, and turns on an alarm when faults are discovered. It can then be left to a human operator to correct the problem.

21.1.5 HARDWARE SAFETY

Computer hardware is less robust than electromechanical hardware. Modern solid-state electronics, including discrete logic modules and microprocessors are far more vulnerable to environmental stresses than are relays, for example. Extremes of heat and cold, modest electronic over voltages, even static electric charge carried on an operator's clothing can temporarily disrupt or permanently damage solid-state circuitry. Extremely brief electronic transients or "noise spikes" which can only be detected by special test equipment, may cause mystifying and irreproducible systems behavior. Control programs stored in what is supposed to be permanent, read-only memory may fade away as components age. Electrical interferences from unexpected sources can induce serious hazards.

Because solid-state electronics are more delicate than electromechanical devices, it is usually not possible to simply replace electromechanical controls with functionally equivalent solid-state equipment. New equipment often fails because it proves to be vulnerable to electrical interference and other environmental disturbances that the older equipment could easily tolerate. It is usually necessary to provide the computer with a more protected electrical environment, constructed according to good packaging, grounding. and shielding practices. In addition, special signal conditioning and isolation circuitry is often required. Techniques sufficient for personal computers and other consumer electronics are not always adequate for more demanding process control environments.

Electromechanical components usually fail one at a time. Consequently, many functions may continue to work even after one or more components have failed. Therefore, each component in a device must be analyzed for potential failures and safety concerns. There are several methods which aid in the analysis of components. Fault tree analysis is a methodology where potential failures are traced back to the components causing them. Failure mode analysis looks at each component and determines the effect of a failure of that component on the system.

Once the component has been analyzed, there are techniques that can be employed to reduce the potential for the failure of that component. Such techniques include component derating, increasing safety margins, and providing better load protection. The methodologies and techniques for analyzing and assuring component safety are discussed in detail in Chapter 10.

21.1.6 SOFTWARE SAFETY

Software is not, in itself, unsafe. Only the physical systems that it may control can do damage. Safety considerations hardly arise for programs that perform conventional data processing or scientific computation. In these applications, the computer only displays results on paper or on video screens. It is presumed that the users will review these results, bringing their informed judgment to bear before acting upon them. It is only when computers are used to directly control systems that are themselves potentially unsafe, that safety issues arise.

It is usual for all software functions to fail simultaneously. In computer-controlled systems, many different functions are usually performed by a single processor. Distributed or multi-processor control systems having more than one computer usually have more functions than they have processors. Different functions are controlled by different parts of the control program, which are run in rapid sequences on a single processor. This control program replaces, in effect, a large number of relays or other discrete components. Concentrating so much complexity into software provides much of the economy and flexibility of computer-controlled systems, but also makes then more vulnerable to errors.

Most programming languages provide some way to divide a program text into sections variously called subroutines, procedures, functions, modules, tasks, or processes. These sections are sometimes called software components, but this analogy is misleading. Even when the text of two program sections appears to be completely independent, they are in fact much more tightly coupled than is usual for electronic components. They share a vital resource—the processor itself.

Certain kinds of software errors can cause the process to interrupt its normal sequence of operations and enter an abnormal state that prevents it from doing useful work. Such an event is colorfully termed a crash. Crashes can be caused by many programming errors that are easy to commit: attempting to divide by zero, attempting to compute a number larger or smaller than can be accommodated by the processor hardware, attempting to read or write into a memory location that is not populated by a memory chip, attempting to use an array element whose subscript is larger than the size of the array, and so forth. Crashes can also be induced by hardware failures, such as intermittent faults or electrical interference.

The behavior of a crashed program is completely unpredictable. It may halt in some apparently random state, or it may continue on, generating random output. Another kind of global program failure occurs when a particular program section seizes control of the processor and will not release it. These occurrences in which the computer appears to be stuck or hung result from common programming errors, such as infinite loops and deadlocks. All functions in a crashed or hung program stop working, not just the one containing the error. System in which several processors share common memory can be vulnerable because errors in one processor's program may cause it to corrupt instructions or data needed by other processors.

In data processing and scientific computing, program crashes do not contribute to accidents because they do not release energy directly. In these environments, programs run under control of a supervisory program called an operating system, which can usually recover control from a crashed or stuck program. If the operating system itself crashes, the operator can shutdown and restart the computer, which often clears the problem. Process control systems, on the other hand, often have no operating system or include customized program sections that perform some of the functions of the operating system. There may be no opportunity for the operator to intervene in any useful way. Consequences of failure can be very serious. A runaway program could drive a radiation therapy machine gantry into a patient. A hung program could fail to terminate a radiation exposure and deliver an overdose.

The expanded use of software in medical devices has offered the promise of increased product functionality and more efficient manufacturing. The type of failures caused by poor design practices, however, can result in costs that can easily exceed projected benefits. One way to prevent this is to study past software safety-related failures and develop methods to prevent those failures.

Fortunately, there are enough sources of information available detailing past software safety failures to construct history of such occurrences. FDA's "Device Recalls: A Study of Quality Problems" documents 85 preproduction quality problems caused by software design. It also relates 93 episodes in which a change in some aspect of the device or its manufacturing process led ultimately to a recall. Another FDA publication titled "Evaluation of Software Related Recalls for the Period FY83-FY89" identifies 116 problems in software quality that resulted in medical device recalls.

21.1.7 VERIFICATION AND VALIDATION OF SAFETY

A proof of safety involves a choice or combination of (1) showing that a fault cannot occur, that is, the device cannot get into an unsafe state or (2) showing that if a fault occurs, it is not dangerous. It has been argued that verification systems that prove the correspondence of devices to concrete specifications are only fragments of verification systems. Verification systems must capture the semantics of the hardware, the software code, and the system behavior.

Another verification methodology for safety involves the use of fault tree analysis. Once the design is completed, fault tree analysis procedures can be used to work backward from critical faults determined by the top levels of the fault tree through the device to verify whether the device can cause the top level event or mishap.

Since the goal of safety verification is to prove that something will not happen, it is helpful to use proof by contradiction. That is, it is assumed the device has produced an unsafe action and it is shown that this could not happen since it leads to a logical contradiction. Although a proof of correctness should theoretically be able to show that a device is safe, it is often impractical to accomplish this because of the sheer magnitude of the proof effort involved and because of the difficulty of completely specifying correct behavior.

21.1.8 EFFECTIVE SAFETY PROGRAM

Any effective safety program requires procedures and expertise in formal hazard identification and analysis techniques. In addition, several expected-hazard mitigation controls should be implemented in any medical device system. These controls include checking the status of hardware on start-up, monitoring hardware equipment during runtime, checking data ranges to reduce the likelihood of operator entry errors, defining system fail-safe states in the case of failures, implementing securing controls, and conducting formal low-level testing and review of safety-critical functions. Applying hazard mitigation controls that adhere to good engineering practices is essential for developing an effective safety program.

A truly effective safety program includes implementation of internal hazard analysis procedures, a firm grasp of regulatory and other standards, and an awareness of the current industry practices regarding safety controls. Such programs consume considerable time and resources, but failing to make the investment increases the risk of product recalls for medical device manufacturers.

Safety analysis begins when the project is conceived and continues throughout the product development life cycle. Due to the variety of medical devices with many degrees of complexity, the following should be included in a safety analysis program:

- Safety review personnel must have a thorough understanding of the operation of the device. Personnel should review pertinent documentation, such as drawings, test reports, and manuals before the analysis.
- Make a representative device available for the review. It will be subject to disassembly.
- Use a checklist for the analysis especially prepared for the particular device.
- Address all areas of concern immediately. Safety release is not granted until the device has no apparent areas of concern.
- Safety releases the device via a release letter only after all areas of concern are addressed.
- Retain the checklist and release letter as part of the product file.

Specifically prepare a comprehensive checklist for the device under analysis. Areas to be addressed in the checklist include, but are not limited to

- Voltages
- Operating frequencies

Characteristic	Comments
Operating voltages	
Operating frequencies	
Leakage currents	
Dielectric withstand	
Environmental specifications	
Grounding impedance	
Power cord and plug	
Electrical insulation	
Abnormal operations	
Physical stability	
Corrosion protection	
Circuit breakers/fuses	
Color coding	
Ergonomic specifications	
Standards conformance	
Alarms and warnings	
Mechanical design integrity	
Cleaning solutions	

FIGURE 21.1 Safety analysis checklist. (From Fries, R.C., *Reliable Design of Medical Devices*, Marcel Dekker, New York, 1997.)

- Leakage currents
- Dielectric withstand
- Environmental specifications
- Grounding impedance
- Power cord and plug
- Electrical insulation
- Abnormal operations
- Physical stability
- Corrosion protection
- Circuit breakers and fuses
- Color coding
- Ergonomic specifications
- Standards conformance
- Alarms, warnings, and indicators
- Mechanical design integrity

The checklist should be signed by the analyst(s) after completion of the analysis. Figure 21.1 shows an example of one page of such a checklist.

21.2   ACCIDENT RECONSTRUCTION AND FORENSICS

Biomedical engineers, due to their generally broad-based education, may sometimes be called upon to analyze accidents. Analysis of medical device accidents will first be discussed, followed by a brief discussion on biomechanics and accident (physical injury due to car, etc. impact) investigation. Both of these have implications for improved designs of devices and processes that biomedical engineers may be involved in.

21.2.1 MEDICAL DEVICE ACCIDENT INVESTIGATIONS

Medical device accident investigation follows a fairly typical chain of events, of which most are in common for accidents in general. The overall process for a medical device accident investigation takes roughly the following outline:

- An incident occurs, someone is injured, and a cause for action is established.
- You are contacted by the wronged person, by his/her lawyer, or by one of the parties or their representative needing an investigation.
- After an initial familiarization with the problem, you may opt to work on the problem, or opt out.
- You need to collect data. This means you must inspect the equipment and scene (if any), photograph or sketch the environment as necessary, gather evidence, and read whatever written documentation exists at this point. This may include operative notes, nurses' notes, some preliminary testimony, machine charts, etc.
- You need to research the device or process in question. This means that you will use manufacturers and users device experience (MAUDE) if necessary. You will need to access the operators' manual for the device, as necessary. You will need to investigate maintenance manuals, if necessary. You may need to run simulations on the device, if necessary. You likely will need to use the Web, other than just MAUDE for keyword searches. You may need to obtain agreement for the use of specialists, such as personnel who perform calibration or maintenance on the devices as necessary. You may need to do some basic research and mockups of the device as necessary.
- As a result of your work, you will need to estimate causes and their likelihood. If you can demonstrate the error, so much the better.
- Branching point is reached here. A report (oral or written, this should be preagreed to) should be submitted to the person who contacted and contracted you. You must be prepared to continue investigation, await further court action, or be released from further work. The latter is generally the case when you find for someone other than those who hired you.
- You must be prepared to answer questioning from opposition lawyers if necessary. This can take the form of both oral and written testimony as to the current status of the investigation.
- Most of the time, a final formal report designating the fault will end your work. On a small number of occasions, expect to go to court, get sworn in, and testify regarding your work.

Several brief cases below will serve to illustrate the range of efforts that may come of a medical device accident investigation. Forensics and consulting in the context of licensure will be discussed again in Section 22.6.

21.2.1.1 Enteral Feeding Tube Complication

An elderly male patient was sent home from a nursing facility with an enteral feeding pump (direct to stomach tube feeding), a supply of feeding compound, and a supply of enteral feeding pump tubes. On the first use of the pump, the patient wound up with too high a flow of food such that food filled the stomach and entered the lungs. He expired due to pneumonia induced by the flow within a few days.

Lawsuits were filed against the skilled nursing facility, the makers of the enteral feeding tube, the makers of the enteral feeding apparatus (pump mechanism), the physician involved, etc. After a very brief overview of the material, the manufacturer of the enteral feeding apparatus suggested a panel meeting of all involved parties under rule 26, in order to attempt to place blame and suggest a method of discovery if necessary. A panel was convened comprised of representatives of the nursing facility, a biomedical engineer from academia (author King), a representative from the company that manufactured the pump, and the opposition lawyers. Within 5 min the determination was made that the pump had been sent home with the wrong manufacturers' pump tubing installed. The tubing that was in place allowed for gravity feed of the feeding fluid independent of the pump speed. Thus, a direct cause of the accident was found in a timely manner.

21.2.1.2 Pressure Limited Respiration System

A pressure limited pump was used to ventilate a very young child who had a very small plastic airway directly in place in the throat (tracheotomy tube). The child was found asphyxiated after the airway had withdrawn from the child. The unit, though the nursing service has presumably properly set the upper and lower pressure controls, was not alarming.

The unit was obtained and taken to a clinical engineering service to be tested. All testing was videotaped. Without the tracheotomy tube in place, the device alarmed. It was determined that the extremely small airway element enabled sufficient backpressure to the system that the recommended pressure settings were meaningless. With the tracheotomy tube in place, the unit did not alarm, and continued ventilating "nothing."

It was surmised that the child managed to move and dislodge the tracheotomy tube while being ventilated. Without the lower pressure alarm being set with this disconnection pressure, the unit would never alarm. No adequate information could be found in the operator's manual to account for such a situation. A report indicating these facts was submitted, the nursing service and the ventilator manufacturer settled out of court.

It is worth noting that ventilators that rely only on pressure alarms are thus not necessarily safe in disconnect and pinched tubing settings. A far better alarm system for a patient ventilator will use both pressure alarm settings and CO_2 waveform detection. If the CO_2 waveform is not evident— the alarms must be sounded—the patient is either disconnected or not breathing.

21.2.1.3 Intramedullary Nail Accident

A veterinarian was using a commercially available chuck system which held a double pointed intramedullary rod for insertion into the broken femur of a fairly large dog. The chuck (similar to those used on drills) used a friction system to hold the rod in place. There was a short section of tubing at the back end of the chuck to keep the nail straight during insertion.

The veterinarian was overly hasty in his attempt at insertion of the rod, instead of using a hammer at the end of the tubing on the insertion device, he used his hand. The rod went through his hand, putting him out of business as a vet for several months. He sued for pain and suffering and lost income.

In the ensuing court case the lack of a backstop for the intermedullary rod was pointed out as a design flaw. However, it was presented that there were clear instructions with the insertion device as to its use with a hammer or other safe impact facilitation devices. He further had been instructed in veterinarian school as to the proper use of the device. He lost his case.

He was further instructed to pay court costs.

21.2.1.4 To Assist in a Suit, or Not?

Circa 1993 author King was asked to investigate and to be an expert witness for a case involving multiple failed operations. Specifically, the case involved a man who had had three successive failed penile implants. The implants were of a pump-type mechanism for the purposes of male "enhancement" when natural mechanisms fail.

After a very brief investigation, the case was refused. At the time, the specific operation done was only successful 75% of the time. Each time the patient had signed an operative consent form which had indicated that he had a 25% chance of failure. After three operations he was still at a risk of failure at the 1.56% level (1/64). He had successively and knowingly signed on to a surgery with a high risk of failure and thus did not have a case in the eyes of this author.

21.2.1.5 Blood Oxygenator Malfunction

A patient was on a heart–lung machine during open heart surgery. A portion of the device was a holding/oxygenation system which was designed with an ultrasonic high-level blood detection system and two low (low, lower) level ultrasonic detection systems. The system was designed to alarm on the high and low level of blood in the device. The lowest level of blood alarm system was designed to shutoff the pump system entirely.

The system additionally had a sensor on the return line to the patient; this sensor was an infrared detection system. This sensor would only operate if there were no color in the return line (e.g., straight ringers), it would not alarm if it "saw" foamy blood or normal blood.

The first set of ultrasonic alarms could be turned off during the start of any procedure such that the alarms would not bother the surgeon and other staff. Likewise, they could be turned off when the surgeon needed special low flow, etc. conditions.

According to hand kept records, there was one case where a patient received an air embolus during on-pump surgery, no alarms sounded. The patient died; a lawsuit ensued.

The records were inspected for probable cause, as was the device and preliminary testimony. Two design flaws stood out. First, the alarms on the level of the blood could be turned off even if the surgical conditions requiring low flow (and increased vigilance on the part of the pump operator). It is likely that the pump operator failed to turn the alarms back on when the pump was returned to full flow. For the flow rates that were common during the majority of the operation, it would have required only 1.4 s of time to completely empty the blood reservoir. Having done so, foam and air would have been passing by the final alarm device—the infrared detector before the patient. The second design flaw was the use of the infrared system, which could be fooled by foam. A state of the art device would have used a final ultrasonic blood detection system at this point as they are virtually foolproof. This case was settled out of court.

21.2.1.6 Failure to Monitor

A patient suffering from heart palpations and dizziness was admitted to an overcrowded hospital environment. The patient was placed on a gurney and screened of from public view. The patient was connected to a computerized vital signs monitoring system that was connected to the local area network and a central monitoring network.

The patient was found dead the next morning due to cardiac arrest. The next of kin began a lawsuit. An inspection of records showed that the monitor had a function called "admit patient," this function was never initiated. Thus, the patient was never monitored. This case was settled out of court. The monitoring system has been redesigned.

21.2.1.7 Failure to Perform

A ventilator was in use on a small child during transport in a private vehicle. The ventilator was being powered via a cord inserted into the cigarette light socket. The air-conditioning system on the car was in use. The caretakers noted after a time (>30 min) that the child was not being ventilated and was turning blue. No alarms were sounding. Emergency care was immediately attempted; an initial return to normal ventilation was noted. The child died within a few days however, due to this injury.

An inspection of the data log for this device indicated that it was rebooting for a period of ~20 min during the time of the incident. It was surmised that this was due to a poor connection with the car via the cigarette lighter, or poor power filtering of spikes from the car power system (and thus interference with the ventilator computer system).

This case was settled out of court. Of interest is the fact that the company recalled their supplied (car) power cords about this same time. This recall was followed by a recall of most of their devices for replacement of the power supply board on their systems.

21.2.2 BIOMECHANICS AND TRAFFIC ACCIDENT INVESTIGATIONS

A very basic understanding of biomechanics is necessary before any undertaking in which a bioengineer may be involved in design or accident investigation. Some of the concepts that need to be understood are the following:

1. Data collection: data for analyses involving traffic accidents involve data collected from reported and analyzed accidents. Large data sets are collected by individual states, some of which is recollected and analyzed by the National Highway Transportation Safety Administration (NHTSA). The agency also specifically maintains data on fatal accidents, including information on vehicle type, rollover, ejection, alcohol use, etc. A smaller data set includes data for cases specifically investigated by the agency, data that include medical information as well as more specific conclusions as to the cause of the accident and many other details. Other related data sets have been obtained from cadaver studies, anthropometrical dummy studies, animal studies, and mathematical modeling analyses. Much of the data obtained and related issues may be found on the NHSTA Web site (http://www.nhtsa.dot.gov/) and in the *Proceedings of the Annual STAPP Car Crash Conferences*.

2. Injury estimation: In studies of human survival following trauma an early scheme involved the development of an abbreviated injury scale (AIS), this scale ranges from 0 (minor sprain) to 6 (unsurvivable injury). This scale is developed for each of the six body regions of interest in survivability, the head, face, chest, abdomen, extremities (including pelvis), and external. The highest squared scores from the three most injured areas are added together to generate a new score, the Injury Severity Score (ISS). With the exception that any "6" AIS rating automatically yields the maximum ISS score of 75, this score relates linearly to rates of mortality, morbidity, and length of hospital stay.*

3. Impact analyses: Often, an engineer must estimate the relative speeds of the vehicles and personnel involved. This means that the engineer must, from the data involved in the accident report, crush patterns on the vehicles involved, vehicle data sheets, weather conditions reported, etc., estimate the relative speeds, angles of impact, and probable outcome of an accident. For example, working backward from skid length data, one can find that a vehicles initial velocity before the skid is directly related to the square root of twice the product of skid length, skid friction coefficient, and the value of gravity, g. The skid friction coefficient is a function of the type of surface (e.g., pavement vs. dirt road), the weather conditions (dry, wet, or icy), and the type of braking system the vehicle has (two or four wheel, antilock, etc.). If a subject has been thrown or ejected from a vehicle, simple trajectory analysis can be done to determine the initial velocity, if sufficient information exists. If there is little body damage on two vehicles, a combination of conservation of momentum analysis and elastic collision analysis might apply, along with skid analysis. Alternatively, damage analysis combined with inelastic collision analysis must be used.

The biomedical engineer doing design or doing forensic analysis after the fact in on matters involving vehicular accidents must understand the above, and be able to apply background material learned in a biomechanics class to real-life problems. A few examples follow:

---

* See www.trauma.org for more information.

- Occupant restraint systems may be designed to absorb energy during an impact. Consider the alternatives for air bags, especially for situations with low body weight passengers.
- During a motorcycle–truck accident, the helmet of the motorcyclist came off, resulting in death of the motorcyclist due to blunt head trauma. Where was the error in the design of the helmet system?
- Current seat belts are a trade-off between convenience and safety. Determine the "ideal" design.

One case example will enforce the above. A husband and wife were in a private vehicle, stopped near the midline of a road, waiting to turn into a driveway. Both claimed to have been belted in. They were rear ended by another vehicle which was traveling at excessive speed. The wife (passenger) wound up in the inside back of the vehicle, severely injured. The driver (husband) had minor injuries.

A lawsuit ensued against the manufacturer of the vehicle that the husband/wife team occupied. An investigation into the design of the vehicle turned up the fact that the wife's hand could have struck the seatbelt release button (due to the rear impact), thus releasing her to fly about the cabin due to this release. This case was settled out of court.

EXERCISES

1. Visit the new car assessment pages at the NHTSA Web site (http://www.nhtsa.dot.gov/NCAP/Info.html), copy the frequently asked questions list, and comment on five of the particular items.
2. Visit the Stapp Conference Web site (www.stapp.org), determine the history of the conferences.
3. Do a MAUDE search for deaths caused by Enteral Feeders Print out and discuss at least one case.
4. Do a MAUDE search for deaths caused in 1 week of the year. Comment on your results.
5. One of the authors of this book owns a 1996 Chrysler Voyager Van and a 1995 Volvo 950. Visit the NHSTA site to determine which is the safer car.
6. Find data for the chance of survival for a patient with a major liver laceration and a closed tibial fracture as a result of a vehicular injury.
7. A lawyer asks you to testify about an injury that was received during a low-speed (10 mph or less) two-vehicle collision. Specifically he asks that you testify that no data exist that can prove the correct speeds of the vehicles and the likelihood of injury. Is this correct?
8. A 3 year-old girl sustained neck injuries on a child roller coaster at a theme park. What would you do to prove or disprove this claim? By the way, the father has a videotape of the injury occurring, and the girl seemed to have a "long neck." This particular ride had been in use for 10 years.
9. A child sustained a severe cut on his nose due to him falling off of a motorbike. The helmet he was wearing caused the cut. What was the design flaw here, and who was at fault?
10. Find and report on the use of the Apgar score. Compare this to the AIS score in this section.
11. How might tissue engineering change the field of trauma care?
12. Brain poses a special case when studying injury patterns. Research the term contrecoup, report on its significance.

BIBLIOGRAPHY

Boardman, T. A., and DiPasquale, T. Product liability implications of regulatory compliance or noncompliance, in *The Medical Device Industry—Science, Technology, and Regulation in a Competitive Environment.* New York: Marcel Dekker, 1990.

Boumil, M. M. and Clifford E. E., *The Law of Medical Liability in a Nutshell*. St. Paul, MN: West Publishing Company, 1995.

Department of Defense, MIL-STD-882: System Safety Program Requirements. Washington DC: U.S. Government Printing Office, 1984.

Food and Drug Administration, *Device Recalls: A Study of Quality Problems*. Rockville, MD: FDA, 1990.

Food and Drug Administration, *Evaluation of Software Related Recalls for the Period FY83-FY89*. Rockville, MD: Food and Drug Administration, Center for Devices and Radiological Health, 1990.

Fries, R. C., *Reliability Assurance for Medical Devices, Equipment and Software*. Buffalo Grove, IL: Interpharm Press, 1991.

Fries, R. C., *Reliable Design of Medical Devices*. New York: Marcel Dekker, 1997.

Goel, A. L., Software reliability models: Assumptions, limitations, and applicability, *IEEE Transactions on Software Engineering*, SE-11(12), December 1985, 1411–1423.

Jacky, J., Safety of computer-controlled devices, in *Developing Safe, Effective and Reliable Medical Software*. Arlington, VA: Association for the Advancement of Medical Instrumentation, 1991.

Jahanian, F. and Mok, A. K., Safety analysis of timing properties in real time systems, *IEEE Transactions on Software Engineering*, SE-12(9), September 1986.

Jorgens, J. III, The purpose of software quality assurance: A means to an end, in *Developing Safe, Effective and Reliable Medical Software*. Arlington, VA: Association for the Advancement of Medical Instrumentation, 1991.

Levenson, N. G., Software safety: Why, what and how, *Computing Surveys*, 18(2), June 1986.

Levenson, N. G., *Safeware*. Reading, MA: Addison-Wesley, 1995.

Musa, J. D., Anthony Iannonio and Kazuhira Okumoto, *Software Reliability: Measurement, Prediction, Application*. New York: McGraw-Hill, 1987.

Nahum, A. M. and Melvin, J. W. *Accidental Injury, Biomechanics and Prevention*, 2nd ed. New York: Springer, 2002.

Olivier, D. P., Software safety: Historical problems and proposed solutions, *Medical Device and Diagnostic Industry*, 17(7), July 1995.

Phillips, J. J., *Products Liability in a Nutshell*. St. Paul, MN: West Publishing and Company, 1993.

Shapo, M. S., *Products Liability and the Search for Justice*. Durham, NC: Carolina Academic Press, 1993.

22 Professional Issues

A man's ethical behavior should be based effectually on sympathy, education, and social ties; no religious basis is necessary. Man would indeed be in a poor way if he had to be restrained by fear of punishment and hope of reward after death.

Albert Einstein

This chapter discusses several professional issues relating to professionalism in biomedical engineering. Specifically it will cover some of the alphabet soup of professional societies that many biomedical engineers are members of and need to be familiar with as they are also standards setting groups. Next it covers licensing of engineers and the ramifications for practicing biomedical engineers, especially those working in the area of forensics. Lastly, it briefly discusses issues relating to continuing education for both licensed and unlicensed engineers.

22.1 BME-RELATED PROFESSIONAL SOCIETIES

Society memberships, properly chosen, can be an invaluable aid in professional pursuits. Memberships should allow for one to meet others with related professional interests, assist in professional advancement through relevant newsletters and professional magazines, and should allow for the sharing of knowledge and the acquisition of new knowledge through regularly scheduled reasonably convenient national meetings. Most will also have a Web presence and a means for distribution of job opportunities. The results of these meetings should be archived and be a part of the membership benefits of the organization. Some of the groups also provide standards setting functions; the ability to sit on such committees should be a function of experience with the group and its goals. These groups will be relisted as necessary in the following section.

22.1.1 BIOMEDICAL ENGINEERING SOCIETIES

Many campuses have a small number of societies that relate directly to biomedical engineering; the choice of societies can widen dramatically upon graduation and a first or later job. Some of the major societies are as follows:

1. AAMI (Association for the Advancement of Medical Instrumentation): This society is aimed at designers, managers, users, and regulators of medical technologies. As such it is heavily hospital user and medical industry oriented, with a large clinical engineering emphasis. For product and process design engineers, it is a comprehensive and useful organization in which to be a member (see www.aami.org for more information).
2. ACM (Association for Computing Machinery): This society has a special interest group—Special Interest Group-BIOlogy (SIGBIO), which emphasizes medical informatics and other topics such as multimedia and molecular databases. This group sponsors several workshops and conferences each year (see www.acm.org, search for sigbio).
3. AMIA (American Medical Informatics Association): This society is devoted to developing and using information technologies to improve health care and is composed of individual, institutional, and corporate members. This group holds one major and one minor congress each year devoted to the application of informatics to problems in heath care, and

collaborates with the international medical informatics association (see www.amia.org and www.imia.org for additional information).

4. BMES (Biomedical Engineering Society): This society aims to promote the increase of biomedical engineering knowledge and its utilization. This group is heavily academic (students to professors) oriented, provides one national meeting, one newsletter and one annals. Many campuses have a student chapter (see www.bmes.org for additional information).

5. IBE (Institute of Biological Engineering): This society aims to encourage interest and promote inquiry into biological engineering in its broadest manner, with potential application to the improvement of the human condition. This group is very broad in nature and includes many participants from agricultural engineering. Many of their yearly conferences are held in conduction with other groups that have some overlap in interests, such as the BMES (see www.ibeweb.org for additional information).

6. IEEE-EMBS (Institute of Electrical and Electronics Engineers-Engineering in Medicine and Biology Society): This is a multinational society that represents many working in the electronics and related industries; one of its 36 societies is the EMBS. The membership in this group exceeds 10,000, with about 25% of this membership outside the United States. The group publishes *Transactions on Biomedical Engineering, Transactions on Nanoscience, Transactions on Neural Systems and Rehabilitation Engineering, Transactions on Information Technology*, as well as a bimonthly magazine, the *Engineering in Medicine and Biology Magazine*. It collaborates on three other publications, one on medical imaging, one on neural networks, and one on machine intelligence. This society also sponsors one international conference each year. This full-service group represents the largest number of biomedical engineers of any organization (see www.ieee.org, search for the EMBS group).

7. RESNA (Rehabilitation Engineering and Assistive Technology Society of North America): This society is an interdisciplinary association of people with a common interest in technology and disability. As might be expected from the title, this group is composed of a broad range of professionals interested in various aspects of assistive care and technology. This group holds an annual meeting; selected student design teams are invited (see www.resna.org).

8. SPIE (Society of Photo-Optical Instrumentation Engineers): This is an international society specializing in photo-optical systems, many of which have biomedical applications (see www.spie.org).

Several of the major classical discipline-oriented groups have focus groups relating to biomedical engineering. The American Society for Mechanical Engineers (ASME) has a bioengineering division as one of its many subdivisions; this group holds a small conference each year. Most biomed-related papers are part of the yearly ASME meeting or publication (see www.asme.org/bed/). The American Society of Civil Engineers and the American Institute of Chemical Engineers do not have specific subdivisions relating to biomedical engineering, but they do publish papers relevant to various aspects of the field (see www.asce.org and www.aiche.org). The American Society for Engineering Education sponsors a biomedical engineering division; this group sponsors a number of sessions at the yearly conferences (see www.asee.org). The site www.biomat.net is a resource for those working with biomaterials.

22.2 STANDARDS SETTING GROUPS

To establish minimal standards for biomedical devices and some processes many groups have established written standards in areas within their areas of expertise. These standards are then typically available for purchase; documentation that standards have been met then becomes a part of the continuing certification that a process or product meets specifications.

TABLE 22.1
Representative U.S. Standards Setting Organizations

Agency	Web Site
American Heart Association	www.aha.org
American Dental Association	www.ada.org
American Medical Association	www.ama-assn.org
American Society for Quality Control	www.asqc.org
American Society for Testing of Materials	www.astm.org
American Society of Mechanical Engineers	www.asme.org
Association for the Advancement of Medical Instrumentation	www.aami.org
Federal Communications Commission	www.fcc.gov
Institute of Electrical and Electronic Engineers	www.ieee.org
Joint Commission on Accreditation of Healthcare Organizations	www.jcaho.org
National Council on Radiation Protection and Measurements	www.ncrp.com
National Electrical Manufacturers Association	www.nema.org
National Fire Protection Association	www.nfpa.org
National Safety Council	www.nsc.org
Occupational Safety and Health Administration	www.osha.gov
Underwriters Laboratory	www.ul.com

In the United States, standards setting is done by a mixture of professional societies, nongovernmental agencies, and governmental agencies. For example, AAMI sets standards in the areas of biomedical equipment, dialysis equipment, and sterilization. The American National Standards Institute (ANSI) (www.ansi.org), an independent organization, coordinates U.S. voluntary standards and is the U.S. representative to the International Organization for Standardization (ISO). ANSI has a small number of standards that are uniquely theirs; they colist with many of the other standards as being in agreement with those standards. The major governmental organization involved in standards is the Occupational Safety and Health Organization (OSHA); the majority of the standards here relate to health and safety of workers in the workplace. Specific standards apply to the health industry. A partial listing of U.S. Standards setting agencies and groups may be seen in Table 22.1.

Many nations have the majority of their standards setting functions imbedded in a governmental sponsored standards body. For a partial listing of such sites and representative standards, a good starting point is the publication *The Guide to Biomedical Standards* (Aspen Publishers, Gaithersburg, Maryland, ISBN 0-8342-1692-2, 1999).

The most influential international standards organization is the ISO (http://www.worldyellow pages.com/iso/) which is a worldwide federation of national standards bodies from some 110 countries, one from each country (ANSI in the Unites States). The mission of the ISO is the development of consensus standards to facilitate the international exchange of goods and services, and to developing cooperation in the spheres of intellectual, scientific, technological, and economic activity. ISO's work results in international agreements that are published as international standards. If successful, these standards will supplant the potentially 110 or more individual country standards as time progresses.

22.3 PROFESSIONAL ENGINEERING LICENSURE

An extremely important decision in an engineer's career is that of applying for professional licensure. All states in the United States have statutes that establish the registration requirements for architects, engineers, landscape architects and interior designers, and describe the size and scope of projects for which a registrant is needed. To improve the level of professional conduct and to establish a standard of care, the licensing board also enacts Rules of Professional Conduct. A typical state licensure board holds its purpose one of safeguarding life, health, and property, and the promotion of the public welfare through the establishment of standards and regulating the practice of engineering within the state. It does this through general requirements regarding educational attainment, participation in practice, examination and licensure, continuing education requirements, and the publication and enforcement of codes of conduct for the practice of engineering.

The implications of licensure are increased earnings, better employment possibilities, and a legal status for private practice opportunities, such as consulting and expert witnessing. According to the National Council of Examiners for Engineering and Surveying, licensed engineers enjoy salaries 15%–25% higher than nonlicensed engineers. State regulations specify the conditions under which a licensed engineer must be supervisory, certain projects cannot be undertaken without this supervision, which includes certification with a signature and stamp. If called upon to testify in court regarding areas of your expertise, for example, a medical device accident investigation, the professional engineering license and your experience as evidenced by your vita is generally enough to convince a judge that you are a credible witness. Professional licensure is a two-step process involving engineering internship and examination and registration as a professional engineer (PE).

22.3.1 ENGINEERING INTERNSHIP

To become an engineering intern (also known as an engineer-in-training) the following conditions must (typically) be satisfied:

1. Graduation (or a senior in good standing) from a minimum 4 year undergraduate engineering curriculum accredited by the Accreditation Board for Engineering and Technology (ABET) or substantially equivalent; or individuals who have an undergraduate degree determined to be substantially equivalent to an ABET-accredited degree
2. Passage of the Fundamentals of Engineering Examination (a full day general comprehensive examination, passage is set at 70%)

The exam is generally given twice a year. It is generated by the National Council of Examiners for Engineering and Surveying and administered by state-delegated examiners (see www.ncees. org/exams for additional information). Under discussion in 2007 is the potential requirement that the examinee must have a master's degree as an entry, rather than the bachelor's degree.

Fees for the exam are reasonable (~$50). Pass rates vary by state, dependent in part on whether or not the exam is mandatory for graduation from college. A pass rate of 60% or better is common. For many states, once this barrier is passed, the exam does not need to be retaken.

22.3.2 REGISTRATION AS A PROFESSIONAL ENGINEER

The following requirements must typically be satisfied for professional engineering licensure:

1. Graduation from a minimum 4 year undergraduate engineering curriculum (or equivalent, as above) accredited by the ABET or substantially equivalent
2. Four years of progressive engineering experience satisfactory to the board (often certified via plans developed, etc.)
3. Certification as an engineer intern or 12 years of progressive engineering experience satisfactory to the board
4. Passage of the Principles and Practice of Engineering Examination in one of the 18 areas tested (mechanical, electrical, etc. a major day-long exam)

Requirement 3 may take the form of an oral examination and the documentation of experience as an engineering intern. Fees for this exam are reasonable. Pass rates vary considerably by state and by discipline. Tennessee had an overall pass rate just above 50% in 2006. Licenses are state dependent; thus you must make application to practice as a practicing engineer in another state, and pay any relevant license and privilege fees.

Once a person has passed the above registration process, the license must be maintained by

1. Yearly license renewal fee payment
2. Yearly or other privilege tax payment (if mandated)
3. Proof of continuing education efforts (if requested)
4. Abiding by the rules of conduct as set forward by the state

22.4 RULES OF PROFESSIONAL CONDUCT

The following are general guidelines regarding the rules of professional conduct for the practice of engineering:

1. Registrants must recognize that the welfare of the public is paramount. If it is felt that the decisions made by one's employer (or client, etc.) are counter to this it is the registrant's responsibility to report the decision to the appropriate authorities and to refuse to carry out the decision.
2. Registrants must perform service only in areas of personal competence. This service will typically be noted by the affixing of his or her signature and seal to documents prepared in this way. The affixing of this seal or signature to other documents can lead to dismissal or fines. Similar punishments will ensue due to violation of any regulations and acts of incompetence due to malpractice or disability.
3. Professional reports and expert testimony made by the registrant must be objective and truthful. If the registrant is speaking on behalf of another party, that fact must be clearly enunciated.
4. Registrants must avoid conflicts of interest; if any arise it must be disclosed to the employer or client. Compensation must be above board and only for services performed (no acceptance of bribes, perks, kickbacks, etc.).
5. Registrants must be honest in all matters regarding their professional qualifications. Registrants must not offer any gift of any kind for the awarding of a contract.

State licensing boards have the power to fine and suspend engineers violating the rules above, or assisting others to violate the rules. Suspension typically can also occur if the registrant is convicted of a felony, or has had his or her license suspended in another state (for cause). Suspension or fines can occur for nonpayment of privilege tax (license fee), practice without a license, improper use of seal (validation of designs not done by the engineer), etc.

22.5 CODES OF ETHICS

Most major societies prominently post and endorse a code of ethics. In general, these amount to reiterations and refinements of the above stated five rules of professional conflict.

The *IEEE Code of Ethics* has 10 points; the IEEE makes explicit the additional ethical rules of nondiscrimination, rules against slander, and suggests the role of mentor for associations with coworkers (see http://www.ieee.org/about/whatis/code.html for details). The National Society for Professional Engineers (see http://www.nspe.org/ethics/eh1-code.asp) reiterates the above five rules as six fundamental canons. They then refine and expand each of these terms in a rules of practice section, which is then followed by an interesting section on professional obligations.

This section suggests such topics as participation in public affairs for the common good, and publication in the lay press, along with sections that further refine the above rules of practice section. This site further has links to case studies and the engineers' creed.

Several online Web sites offer links to codes of ethics and case studies. One of the larger relating to engineering is http://www.onlineethics.org/. Should the need arise, this center offers assistance in solving ethical questions.

22.6   FORENSICS AND CONSULTING

At some point in many engineer's career, they may acquire sufficient knowledge in an area that they can become forensic engineers or consulting engineers. Both can be very interesting and highly remunerative careers.

Forensic engineers typically research, to assist in the determination of fault, the cause of an accident. Finding fault or placing blame allows one to proceed with litigation, if necessary or justified. In the field of biomedical engineering, cases can run the gamut from determination of the potential for injury in a low speed auto accident, to the determination as to who is at fault for a death due to an air embolus. The first case would require that the engineer be well versed in biomechanics and accident reconstruction and the databases maintained on automobile accident injuries. The second case would require an in-depth look at all the instrumentation used in the case, the personnel involved, all records kept, etc.

A typical case involving a medical device accident involves an initial telephone contact between a lawyer (or sometimes a relative of the injured party) and the engineer (or firm). Paper or e-mail contacts are not generally used, as this material is discoverable. An initial familiarization with the accident being investigated is strongly recommended before accepting the case, to determine if one has the credentials and the desire to pursue a given case. This may involve a review of the operative notes from a case or other documentation involving the injury or death. An hour or two of study should allow one to accept or decline a case. If the engineer agrees to investigate, details such as timing (when might this go to court, how fast a response is needed, etc.) and payment schedule (rates per hour, contingencies, expense payments, etc.) need to be agreed to. Other details, such as the need for access to records and devices need to be taken care of as soon as feasible.

A personal philosophy regarding acceptance of cases should be developed. For example, the question "can my work ameliorate some of the harm done to the client?" might be a useful guideline. "Is this going to be an interesting case?" might be another.

There is no typical investigation. A broken device may be investigated and documented with data taken from the Food and Drug Administration manufacturers and users device experience database system (medical device error reporting system). The clinical engineering services group in a related hospital may be queried about similar incidents. The device in question may be linked up to a patient simulator to determine error conditions in an assumed scenario. Determination of the fault is often a function of the imagination and resourcefulness of the investigating engineer.

In a significant fraction of cases, the engineer will find fault with the conduct of the clients of the lawyer who retained the engineer, at which point the engineer is typically relieved of further duty and employment on the case. If the data obtained is sufficient for the case of the employer, negotiations will often become the job of the lawyer, with an out-of-court settlement generally a goal. A step in this process may (or may not) involve the generation of a document under Rule 26 (a federal statute), in which your opinion regarding a case may requested in a document that introduces your case, then outlines your qualifications, fees, materials reviewed, and summarizes your opinion at this time. Another step in this process might also involve a request by the opposition lawyer for a copy of all documents you have generated for a case; it thus behooves one to keep clean notes without bias.

Rarely do cases make it to trial. When and if they do, it behooves the engineer to have the credentials of licensing and adequate proof of experience with the device or process in question.

A good command of the facts in question and the ability to accept questioning under stressful circumstances is also of value, to put it mildly.

A typical hourly rate for a forensic engineer is about 1/1000 of the engineers' annual gross salary, or more. Daily rates generally are capped at 1/100 of the annual salary to not overcharge for time spent waiting for a court appearance, traveling, etc. Other reasonable fees (mileage, meals, and hotel) are charged as applicable. Additional charges (for technician help, etc.) need to be negotiated in advance. With fees in this range, the engineer's fees will be near, but typically slightly lower, than the lawyer's fees.

Consulting practices generally involve the use of an expert in a particular area as an assistant in the solution of a closed-end problem. Thus an expert in optics might be hired by an anesthesia machine development company to assist in the correction of a system to measure CO_2 in expired air. An expert in bioinformatics may be asked to advice on the development of a new database system. A biotechnologist may be called in to help advice on a new pharmaceutical generation system.

Consultants can be paid on an hourly basis, with rates negotiated by the consultants. These are often higher than those of the forensic engineer, due to the specialization of the task(s). Often consultants are kept on a retainer basis, and their expertise requested on an as-needed basis.

22.7   CONTINUING EDUCATION

To maintain licensure, many states require that licensed engineers obtain a minimum of relevant continuing education hours per year (e.g., 24 h per year in Tennessee). These may be via attendance (sometimes with presentation) at technical or professional meetings, via seminars (corporate or correspondence), or via attendance in college or university courses. The courses must be relevant to the practice of engineering as the licensee practices it.

Membership in at least one relevant society and attendance at one 3 day society meeting per year would meet this minimum requirement, and is strongly recommended. Additional attendance at trade shows and related seminars (e.g., Medical Design and Manufacturing show, held three times a year) is highly recommended as a means of staying current. Staying current is necessary in this competitive world.

EXERCISES

1. Perform a Web search using the terms "engineer" and "code of ethics." Briefly document the number and variety of sources you find.
2. Perform a Web search using the term "forensic engineering." Summarize your data. Discuss one firm or case of interest to you.
3. As a forensic engineer, you are called in on a case to determine how air entered a patient undergoing heart catheterization. What sources would you use to determine the cause of death?
4. As a forensics engineer, you have been let go by the client that hired you, as your results were not conducive to them winning their case. The opposition lawyers ask to hire you. What is your answer and why?
5. You have submitted a written report detrimental to the company that hired you as a consultant. Their lawyer asks that all further discussions with them be oral rather than written. Why?
6. Search the Web for the details on licensure in your home state. Compare to those mentioned in the text (Tennessee).

23 Design Case Studies

When possible make the decisions now, even if action is in the future. A reviewed decision usually is better than one reached at the last moment.

William B. Given, Jr.

The goal of this chapter is to review a mixture of design case studies with the aim of illuminating the design processes elaborated on in the previous chapters. Each example is assumed to be at the level of a senior biomedical engineering student, with the required course or experience in their background. It is not meant to be complete in its coverage; this chapter is too small to be that comprehensive.

23.1 MULTIDETECTOR BRAIN SCANNING SYSTEM DEVELOPMENT

23.1.1 BACKGROUND

Nuclear medicine is a branch of radiology whereby radioactive elements are injected into a study subject. The elements are typically such that they are concentrated in the body in known processes; this information is mapped using radiation detectors to diagnose normal or abnormal function. For example, technetium 99-m is a gamma emitter with chemical characteristics similar to calcium. Thus, it will concentrate in areas with a high metabolism, such as in tumors. In the early days of nuclear medicine, these distributions were mapped with single detector systems, which were translated in a rectilinear grid.

The detection system consisted of a collimator (typically lead) section, shielding, a scintillation crystal, and a photodetector system. Shielding allowed the system to have directional sensitivity, cutting holes in the collimator allowed the system to have a sensitivity which is depth (distance from collimator) dependent. Figure 23.1 shows a typical point source response for a focusing collimator with a 3 in. depth of focus. The lead shielding is on the left (crosshatched), the detector crystal and electronics would be on the far left.

The depth response of the collimator assists in determining how deep a system can sense radiation in the body. A straight bore collimator shows a fairly strict adherence to an inverse square of the distance sensitivity; the collimator shows an enhanced response to radiation (hence tumors) at a depth of about 3 in.

23.1.2 PROBLEM STATEMENT

The design problem to be addressed is this: given one or more focusing collimators (and related electronics), develop a brain scanning system that shows increased sensitivity compared to a rectilinear scan using a single detector, and which gives a cross-sectional image of the brain more suited for surgical planning than rectilinear scans.

23.1.3 SOLUTION 1: MULTICRYSTAL TOMOGRAPHIC SCANNER

An initial solution attempt (James et al., 1971) is seen in Figure 23.2. Eight detectors were placed on pivot points on an annular support structure. Each of the eight detectors was slaved to a common drive point through a series of slotted guide bars. The patient's head was to be placed in the

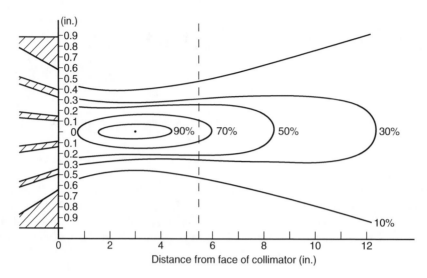

FIGURE 23.1 Point source response map for a converging focus collimator.

apparatus; the level at which the scanners were placed determined the section to be scanned. The slave point was then scanned in a rectilinear raster and data collected and displayed on a monitor. The data obtained from all eight detectors were simply stored together at the current raster data point in computer memory.

This early system was used for a few patient studies, and did prove to be a useful construct. However, additional sensitivity was desired. As configured, the drive point (number 6 in Figure 23.2) needed to traverse a roughly 9×7 sq in. rectangle to scan an adult head; this forced the diameter of the annulus to be roughly 20 in., which severely decreased the response of the detectors to information from points far away during the scanning procedure.

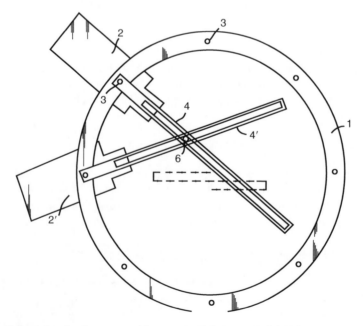

FIGURE 23.2 Schematic of early tomographic scanner, U.S. Patent 3,591,806.

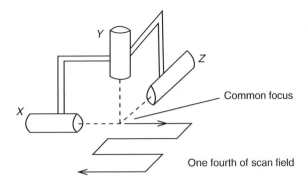

FIGURE 23.3 Mutually orthogonal schematic arrangement.

23.1.4 SOLUTION 2: MUTUALLY ORTHOGONAL MULTICRYSTAL TOMOGRAPHIC SCANNER

Figure 23.3 shows the next iteration of the attempts to obtain a good tomographic image of the human head. Based upon the work above, it was reasoned that the detector system could be replaced by four sets of three mutually orthogonal detector systems (12 in total, compared to the 8 above). The overall construct could be done such that each of the four sets of detectors need only scan one fourth of the patient in the raster scan, the data could then be properly placed in the correct x–y location in memory (David, 1977). The common scanning point for each of the triads was to be the focal point of the collimators used.

One enhancement of this apparatus is the fact that one is no longer constrained to a horizontal cross-sectional scan; the device can scan in any plane desired as long as it does not contact the patients' head. Further, the data obtained by the detectors could be signal processed to enhance the final image by looking at the individual rates and making some inference about the data actually contained in the source. For example, if one detector has a high count rate and the other two do not, the detector with a high count rate is likely getting data from off-focal point sources and the data to be stored should be related to the minimum of the count rates, rather than the sum.

This system too was tested on patients, and did prove to be a useful improvement over the original system. It never made it to production as a useful clinical tool as computerized axial tomography soon appeared as a clinical tool. Other developments, such as the Anger camera, also supplanted this technology due to higher count rates and better spatial sensitivity.

The basic concept in both these instruments is quite simple: additional data are always useful; this was achieved by combining in space homogenous objects destined for contiguous operations (TRIZ principle 5).

Data collection in scanning devices may be enhanced by proper geometry. Such a straightforward conclusion has led to the original work being cited in 12 subsequent patents that were granted.

23.2 TESTING OF ANESTHETISTS

23.2.1 BACKGROUND

The major job of an anesthetist is the preservation of a patient's well-being during surgical procedures. This implies that the patient will not react during the insults of surgery, that is, the patient will not hear or otherwise respond or remember anything that occurs during a procedure, and that the patient is medicated, ventilated, and transfused, to maintain a somewhat status quo physiological state. Anesthetists include CRNAs, who are graduates from nurse anesthetist training programs and medical doctors who have just begun to those who have completed a residency specializing in anesthesiology.

Training for anesthetists begins with didactic classroom and laboratory work in nursing or medical school. The hands-on portion is typically a gradual affair; new residents (post MD) typically spend 3 years under the supervision of practicing anesthesiologists with extensive experience in the field, gradually taking more and more responsibility for the care of the patient.

Training aids have been developed to assist in the education of these personnel; these aids range from simple plastic models of the throat to full-scale human patient simulators.* Of interest to this chapter is the human patient simulator, specifically the METI unit (see Figure 23.4).

The METI simulator consists of a full-scale plastic manikin and associated sensors, actuators, and computer system. The manikin has an airway and lungs; gas exchanges simulate normal and abnormal human responses. The simulator additionally has heartbeat and breath sounds that are audible at the surface, pulses at the wrists and neck, and a hand that responds to a neuromuscular stimulator. The computer system, through various transducers and actuators, simulates responses to drugs (administered through a flow sensor with bar code reader) and gas mixture administration (anesthesia machine, bagging, or room air) and controls gas exchange and other physiological responses. The patient may be programmed to be one of a multitude of patients (standard man [STAN], truck driver, etc.); the system will respond to the drugs and gasses administered in a mathematically modeled manner. The system is powered by both electrical and pneumatic methods, an anesthesia machine is necessary to administer agents and to display simulated vital signs approximating what might be seen in surgery. Data are archived on a regular basis (every 5 s) by the simulator. Various protocols or simulations may be run under the control of the simulation computer or by anyone adept at using the provided interfaces.[†]

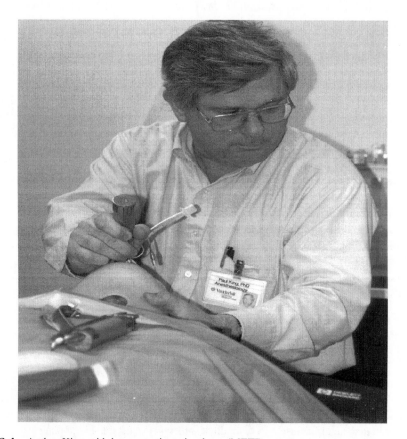

FIGURE 23.4   Author King with human patient simulator (METI).

---

* Medical Education Technology (Sarasota, Florida) and The Eagle Patient Simulator (Palo Alto, California) were two U.S. manufacturers of simulators.
[†] *METI Simulation Manual*, circa 1995.

23.2.2 PROBLEM STATEMENT

Given an anesthesiology department, a METI simulator, and the requisite personnel, devise a means to test the competency of residents and others involved in the provision of anesthesia care. Scores are to be numeric, and will hopefully indicate the level of training and therefore presumed competency of the person being tested.

23.2.3 PROBLEM SOLUTION

The above problem statement actually implies two elements: (1) the development of a test or testing method and (2) the development of a method of quantification of the test results.

A review of the literature for testing methods yielded that most were aimed at stressing the examinee. Many included scenarios whereby the examinee was put in charge of a patient in midcase. Something would then go wrong and the examinee was expected to react properly.

The design decision was made to generate a standardized scenario that would be independent of the need for multiple actors and the potential for variability between examiners. Rather the computerized protocol would be consistent between exams, variability would only be introduced into the process through the simulated patients' response to medications given by the examinee as the case progressed. A moderately stressful procedure (abdominal surgery) was generated as a testing protocol, testing took roughly 30 min per candidate (King et al., 1996).

The second design decision was to let the data speak for itself, rather than to use a mixture of subjective (examiner) and objective data. As recorded blood pressures, heart rate, and pulse oximetry data are normally recorded parameters in the operating room, only these data were used to quantify the tightness of control by the examiners. For a normotensive human, for example, with a preoperative blood pressure of 120/70, a $\pm20\%$ variation in systolic blood pressure will put the blood pressures from clinically hypertensive to hypotensive. Hypertensive events in surgery have been linked to postoperative cardiac and kidney problems, hypotensive episodes can lead to oxygenation problems. This 20% bound was applied to blood pressure and heart rate, any deviation below or above the preoperative value plus or minus this value was considered out of range. For pulse oximetry data, a tighter range of $\pm5\%$ was used, as variation in this parameter is more critical.

This protocol has been tested on a novice, a second year, and a postgraduate anesthetist. There was a clear demarcation of their abilities to hold patient parameters within the above-defined ranges (King et al., 1996).

23.3 APNEA DETECTION SYSTEM

23.3.1 BACKGROUND

In the United States, the current emphasis on reduction of labor costs in hospitalized care has excluded many patients with risk factors for respiratory depression from being cared for in traditional respiratory monitoring suites in critical care units. Many types of hospitalized patients are at risk for respiratory failure; those who are receiving postoperative opioids (morphine) are most at risk. Such patients are those who have undergone major joint surgery and need opioids for pain relief. They are often placed postsurgically in hospital environments where surveillance by hospital personnel is periodic, rather than continuous.

23.3.2 PROBLEM STATEMENT

A means of monitoring patients for potential respiratory depression needs to be found, such that the patient in a step-down unit may be better monitored than just with simple periodic visits by hospital personnel.

23.3.3 SOLUTION (PARTIAL)

A partial solution included the following items:

- CO_2 monitoring was selected as the indicator of choice for respiratory depression. Too low a CO_2 level as detected by a commercial unit, or no waveform (apnea) as detected by the unit was selected as the measurement system of choice. Sampling was initially achieved with a single capillary tube placed near the patient's nostrils. The units used had an alarm level that could be set manually; the alarm signal could be accessed for use in other devices.
- Chance conversation with a person installing an autodialing motion detection alarm system in the author's offices led to the acquisition of an autodialing system that was connected to the capnometer. The system would then dial the charge a nurses' beeper when the system alarmed. The beeper was unique to the patient being monitored.
- Basic system had to be modified with an on/off switch that could be activated when the patient was talking (seen as a high respiration rate) or eating.
- Special cannula had to be obtained to sample air from both nostrils and from near the mouth. This was necessary for patients who mouth breathed due to snoring or stuffy noses.

Twenty-two patients were studied with this system; these patients were selected as being at risk due to recent surgery and the prescription of opioids, either through epidural injection or patient-controlled analgesia. Alarms were generated on 21 patients during the period of the study. Several of the alarms were due to a displaced cannula (seven patients), a few were due to talking or mouth breathing (three patients), which led to the above change in cannula type and the on/off switch, several were due to legitimate concerns (apnea, occlusions, eight patients).

Due to a variety of reasons, this study only made it to the feasibility stage, and was formally presented at only one meeting (Smith et al. 1999). The complete solution, perhaps utilizing the technologies mentioned here, remains to be determined.

23.4 CANCER CLINIC CHARTING

23.4.1 BACKGROUND

Many hospital clinics service a mixture of well to quite ill patients. Such clinics are multiuse, serving as a screening clinic for the majority of the clients and as a triage and referral clinic for others. One such clinic that has been the subject of a design study is the Breast Diagnostic Center at Vanderbilt.* The clinic patient pathway was in need of study to determine areas for improvement in services and in patient perceptions of the process.

23.4.2 PROBLEM STATEMENT

While charting the pathway of patients through a screening clinic (breast cancer screening) a means had to be found to display not only the process but also the patient perception of the process.

23.4.3 PROBLEM SOLUTION

The student involved in the process painstakingly tracked patients through the clinic. The final flowchart for the process was very comprehensive, six pages in Micrografx FlowCharter. As several patients went through the clinic, the student additionally interviewed the patients as to their perceptions of the process. The patient concerns were overlaid on the clinic flowchart; the mood of the patient with diagnosed cancer was expressed in a thermometer form also on the chart. The overall combined process and patient perception flowchart is extremely informative, as a glance at Figure 23.5 should indicate.

---

* Design study by Michelle Kandcer, supervised by Dr. Doris Quinn, available at http://vubme.vuse.vanderbilt.edu/kandcer/.

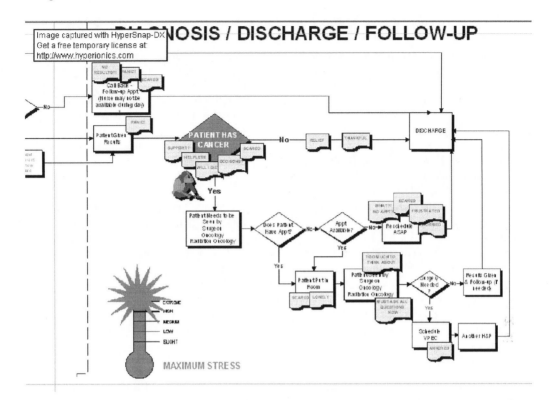

FIGURE 23.5   Cancer clinic diagnosis/discharge/follow-up section.

The total chart was used to identify points of stress for patients during their clinic visit, and will be used in the redesign of the clinic operation. (For other useful ways of envisioning information, the reader is referred to the texts by Edward R. Tufte and also to http://129.59.92.139/srdesign/ 1998/kandcer/final.flo)

23.5 EKG ANALYSIS TECHNIQUES

23.5.1 BACKGROUND

Paul H. King was approached by a physician, who stated the following:

> we have obtained (name deleted) equipment to record telemetered ECGs in live unrestrained mice, and have gotten what seem to be reasonable recordings out of the animals. The data are pretty noisy and others who have the same system tell me that they often need to resort to signal averaging. We are looking at drug effects in wild-type and animals in which cardiac ion channel proteins have been knocked out. This would be a great project for an overambitious student, and one bit a while back but never got back to me. Resources would also be available to support collaboration at a more senior level, e.g., a percentage of someone's effort. It will also become an institutional imperative to be able to do this sort of experiment in mice. It even occurs to me that you may have software "lying around" since I think (name deleted) has the same equipment that he is using for rats. Can you point me in the right direction?

23.5.2 PROBLEM STATEMENT

As is normal with the above open-ended statement, development of a problem statement was an evolutionary process. The initial request from the physician consisted of the following (edited):

> ...give you a single large file from each experiment, and have a machine derive values from 10–20 averaged ECGs at multiple points in time as well as RR plots as a function of time. I look forward to

hearing your verdict on how easy or difficult this may be. [Data sets were to consist of 2–3 hours of single lead ECGs sampled at 1000 samples per second.]

My ideal would be to have (technician) do the implant and drug administration parts of the experiments, and to record everything (best in a single large file) and give it to you for an RR time series (every beat), interval measurements at specified times (e.g., 5 min, 10 min, etc, averaged over some number of beats like 5–20), and perhaps a look at particular parts of the record that happened to be interesting from a rhythm point of view (identifiable from the RR plot).... What are the prospects of getting some system in place in the next month or so, so we could generate data for the ... deadline

... is a pediatric cardiology/electrophysiology fellow who is very interested in joining this ECG project. He will be in contact with you about looking at data and acting as a go between for you and (technician)—and looking at the data with a cardiologist's eye to make sure we are seeing what we want to see and aren't losing data by reducing it too much ...

I hope that by the end of the summer we will have a relatively sophisticated add-on to the current system that will allow interval analysis (possibly automated or at least semi-automated), RR analyses, and perhaps RR analysis in the frequency domain. We also need to think a bit about arrhythmia analysis (first cut = how many beats don't fit the sinus template, especially if they are preceded by different RRs), since some of the mice that are coming may be arrhythmia prone ...

From an engineering point of view, the above statements required some serious decision-making about the software platform to be utilized, the speed of the software, and the platform to be used for analysis. The data files generated were huge; the time to analyze them was of concern.

23.5.3 PROBLEM SOLUTION

Small sets of data were initially analyzed using Microsoft Excel and Visual Basic subroutines to get a feel for the size of the problem and the techniques needed for analysis as per the physician's request. This was a stopgap measure to get some results for analysis of archived data. It was too time-consuming for a full analytical procedure on the entire data set.

The following list of alternatives was considered:

- Microsoft Access and Microsoft Excel and Visual Basic
- IDL (interactive data language) and Visual C
- PV-WAVE and C
- MATLAB and Visual C++ and Excel and Visual Basic

The following sources were consulted for advice on alternatives:

- Others using software on the list (and doing similar work) at Vanderbilt
- Manufacturer of the data collection system
- Others in industry (referrals) using the above software packages
- Web site information from the above companies
- Documentation provided with the above software packages

The following criteria were applied to a final determination of software choice:

- Completeness of documentation
- Ease of use (analysis, display, archiving)
- Ability to handle data files up to 13 million data points in length
- User base at this location
- Ease of transferal of updating and maintenance tasks to others, such as graduate or undergraduate students
- Cost of one or more licenses as necessary

- License particulars
- Speed of program in the environment (PC based)
- Prior application and documentation in a similar situation (e.g., recommendation from data capture device company, etc.)

A formal procedure, such as using a quality function deployment (QFD) diagram for this analysis, was not done; it is left as an exercise for the reader as applied to their environment. MATLAB was chosen for the task at hand.

23.6 EKG ANALYSIS MODULE

23.6.1 BACKGROUND

Few people consider this, but planning a course from scratch is a design process (Waks, 1995). Modifying a course fits nicely into a plan-do-study-act type of cycle, as input from constituencies is evaluated (graduates, employers, etc.) How would you design a course?

23.6.2 PROBLEM STATEMENT

Given the mandate to do so, design an introductory freshman module (one-credit hour) to allow students potentially interested in your department to sample what they might experience in their remaining 3 years should they elect your department's major.

23.6.3 ONE SOLUTION

A module titled Electrocardiogram Capture and Analysis was designed for this purpose. Based in part on the above experiences, the specific goals of this course include the introduction of the student to

- Data analysis techniques in electrocardiography
- Medical and engineering nomenclature
- Engineering and engineering applied to medicine
- Technologies involved in cardiology and electrocardiography
- Societal ramifications of heart-related research

Specific topics covered via lectures were

- Cardiac anatomy and normal cardiac rhythm
- Abnormalities of the heart
- History of cardiology, from stethoscope to galvanometer to chart form
- Basics of EKG analysis from the chart
- Data capture techniques and A/D conversion
- Basic rhythm analysis using Excel
- Introduction to analysis using MATLAB (in parallel with a three hour common freshman course teaching MATLAB)
- Electrical pacing, advanced diagnostic procedures
- Defibrillators (external and implantable)
- Transmitter systems
- Holter monitors, databases
- Visit to a human patient simulator lab
- Visit to a clinical research facility
- Basic medical nomenclature, etc.

Each topic and lecture was aimed at bridging between engineering, science, and medicine, demonstrating how the principles to be stressed later in the curriculum would apply.

23.7 CHOOSING THE CORRECT PLASTIC MATERIAL

23.7.1 Background

Medical devices are cleaned with many different types of cleaning agents that vary widely in pH. Based on its pH value, a chemical may cause crazing and cracking when placed on various types of plastics. The chemical reaction may also affect the natural tensile strength of the material.

23.7.2 Problem Statement

The purpose of the test is to provide rough insight into whether plastic materials subjected to stress are compatible with common cleaning substances. The materials being considered are Valox, Zytel, Cycoloy, and Thermocomp.

23.7.3 Problem Solution

Protocol

1. Three ASTM Type-I tensile test samples (dogbones) will be used for each of nine commonly used cleaning substances:[*,†]
   (a) KleenAseptic®
   (b) Sporicidin
   (c) Cidex Plus
   (d) Aldiced
   (e) Virex 256
   (f) Wescodyn
   (g) Acetone
   (h) Bleach
   (i) Isopropyl alcohol
   Six samples will be used as controls.
2. Test half the control samples in the tensile test fixture at the commencement of the test.
   (a) Test each sample to failure
   (b) Use a speed of 5 mm/min (0.2 in./min) ±25%.
   (c) Measure the tensile strength, percent elongation, and modulus of elasticity.
3. Samples, including half the control samples, will be placed in fixtures built by the Louisville facility. The fixtures force each sample to bow, as depicted in Figures 23.1 and 23.2 in the GE Plastics publication. A strain of 1% will be used for this test. A base length of 8.19 in. will be used for the Cycoloy and Valox plastics. A base length of 8.28 in. will be used for Zytel, and 7.6875 in. will be used for Thermocomp.
4. Small length of cheesecloth will be saturated with each substance and wrapped around each sample at the center of each bow. Plastic film will be wrapped around the cheese-cloth to prevent evaporation of the cleaning substance. The multiple samples used for each cleaning substance will be distributed among the fixtures; the three samples for a given cleaning substance will not all be located on the same fixture. The samples will be exposed in this way to the cleaning substances for 7 days.
5. At the end of the exposure period, remove each specimen and wipe clean.

---

[*] *General Electric*, A Simplified Environmental Stress Cracking, Chemical Resistance Test, from *Design Tips*, Pittsfield, MA: GE Plastics publication (currently owned by SABIC Plastics) 10-89TSS, Number 16.
[†] ASTM, Standard Test Method for Tensile Properties of Plastics, ASTM Designation D 638–698.

TABLE 23.1
Cycoloy Results

Chemical Applied	Sample Averages	
	Tensile Strength (psi)	Young's Modulus (psi)
Control, nonstressed	8,202	148,911
Control, stressed	7,862	140,766
Acetone	N/A (all failed)	N/A (all failed)
Alcide LD	6,277	147,854
Alcohol	4,483 (two samples)	143,564
Bleach	6,238	143,484
Cidex Plus	3,120	127,530
Kleen-Aseptic	N/A (all failed)	N/A (all failed)
Sporicidin	2,099 (one sample)	138,267 (one sample)
Virex 256	N/A (all failed)	N/A (all failed)
Wescodyne	6,589	148,944

Notes: Acetone samples were blanched and cracked after test. All Kleen-Aseptic, Virex 256, and acetone samples failed. Two Sporicidin and one alcohol samples failed.

6. Perform the following:
   (a) Examine each sample for indications of crazing or embrittlement.
   (b) Bend the bars with the chemically exposed area in tension (at outermost point of the bend).
   (c) Record any visible effects.
7. Test the samples in the tensile test fixture.
   (a) Test each sample to failure
   (b) Use a speed of 5 mm/min (0.2 in./min) $\pm 25\%$.
   (c) Measure the tensile strength, percent elongation, and modulus of elasticity

Results

   Clearly, Cycoloy would be an inappropriate material for use with the cleaning chemicals, as several of the chemicals fractured the samples outright. By inspection, the Zytel samples appear to have been affected by the chemicals as well. With the Thermocomp and Valox, any differences are less distinct. Tables 23.1 through 23.4 indicate the test results:

TABLE 23.2
Zytel Result

Control, nonstressed	10,795	152,378
Control, stressed	9,482	111,868
Second control, nonstressed	11,059	138,446
Acetone	9,087	99,053
Alcide LD	8,039	85,013
Alcohol	8,565	86,903
Bleach	8,675	96,596
Cidex Plus	8,279	87,297
Kleen-Aseptic	8,156	92,696
Sporicidin	8,036	78,989
Virex 256	8,020	77,531
Wescodyne	8,011	87,061

TABLE 23.3
Thermocomp Results

Chemical Applied	Sample Averages	
	Tensile Strength (psi)	Young's Modulus (psi)
Control, nonstressed	6,032	146,531
Control, stressed	6,108	131,205
Acetone	5,975	135,291
Alcide LD	6,001	132,608
Alcohol	6,024	131,036
Bleach	6,025	135,290
Cidex Plus	5,997	134,990
Kleen-Aseptic	5,997	131,733
Sporicidin	6,029	129,006
Virex 256	5,991	135,704
Wescodyne	6,002	136,874

Note: Large pieces of glass were found embedded in the resin.

23.8 SELECTING APPROPRIATE MATERIAL FOR AUTOCLAVING

23.8.1 BACKGROUND

When humidity is required for breathing assistance in infants, incubators contain reservoirs where demineralized water is kept until the humidifier distributes it to the patient area. As part of maintaining the overall system and keeping it free of pathogens, the reservoir must be autoclaved.

23.8.2 PROBLEM STATEMENT

Certain materials react well after being subjected to autoclave cycles. Other materials can show crazing and cracking following repeated autoclave cycles. This can be dependent upon the type of material used, the types of bends in the material made when it was formed, or a combination of both. The purpose of this project was to determine the proper plastic material to survive a minimum of 75 autoclave cycles.

TABLE 23.4
Valox Results

Chemical Applied	Sample Averages	
	Tensile Strength (psi)	Young's Modulus (psi)
Control, nonstressed	7,769	116,676
Control, stressed	7,755	104,042
Acetone	7,121	107,147
Alcide LD	7,520	105,698
Alcohol	7,911	110,644
Bleach	7,566	120,864
Cidex Plus	7,663	114,774
Kleen-Aseptic	7,691	107,404
Sporicidin (two samples)	7,564	113,090
Virex 256	7,520	117,873
Wescodyne	7,540	108,279

23.8.3 PROBLEM SOLUTION

Protocol

Parts made of injection-molded polysulfone and Radel-R will be tested. Two of each part (marked A and B) will be subjected to the test cycles. One of each part (marked C) will be used as a control and will not be subjected to the test cycles.

1. Check each part for color and structure.
2. Separate the top and bottom parts. Subject them to the following wash cycle:
   (a) Machine wash cycle: 0.5 h
   (b) Machine dry cycle: 0.5 h
   (c) Wash temperature: 49°C
   (d) Water: soft
   (e) Detergent: Alcojet
3. Place the top and bottom parts together (both parts marked A and both parts marked B). Subject the parts to five autoclave cycles consisting of the following parameters:
   (a) Conditioning time: ~3.5 min
   (b) Sterilizing time: 20 min at 134°C (273°F) and 32 psi
   (c) Exhaust time: ~21 min
   (d) Total cycle time: ~45 min
4. Remove the parts from the autoclave. Examine each part for discoloration and/or structure changes. Record all observations, listed by the number of autoclave cycles that have been completed.
5. Repeat steps 2 through 4, until a minimum of 75 autoclave cycles have been achieved or significant changes in the color and structure of the parts are observed.

Results

The polysulfone material showed crazing and small cracks following 25 autoclave cycles. The crazing and cracking increased as the number of autoclave cycles was increased (Figure 23.6). The Radel-R material survived the 75 autoclave cycles with no crazing or cracking.

FIGURE 23.6   Crazing of polysulfone after 40 autoclave cycles.

23.9 CHOOSING THE CORRECT CLEANING MATERIAL

23.9.1 BACKGROUND

Due to the concern over AIDS issues, many hospitals have begun to use very harsh cleaning agents to clean plastic parts. Because the pH of these materials is very basic, the chemicals react with the plastics, causing crazing and cracking. One example of this type of cleaning agent is Cidex Plus. Recently, Canada has outlawed the use of Cidex Plus to clean medical plastics. An alternative is Cidex OPA.

23.9.2 PROBLEM STATEMENT

The purpose of this test is to determine the chemical reaction of Cidex OPA to plastic flow sensors made of cycoloy material. Parts will be treated with Cidex Plus as a benchmark. Previous work has indicated cycoloy crazes and cracks when treated with Cidex Plus. Three sensors will be treated with Cidex OPA, three with Cidex Plus, and three will receive no treatment as a control.

23.9.3 PROBLEM SOLUTION

Protocol

1. Check each part for color and structure.
2. Place one set of sensors in a container of Cidex OPA, the other in a container of Cidex Plus for 30 min.
3. Remove the traps, rinse them thoroughly, and let them air dry for 60 min.
4. Visually inspect the traps for color or structure changes. Compare to the control group. Record any observations, including the number of soak cycles completed.
5. Repeat steps 2 through 4, completing four soak and air dry cycles per day.
6. Repeat steps 2 through 5 until the material shows any color or structural changes. Record the number of cycles completed.

Results

Crazing and cracking was noted on the cycoloy material when Cidex Plus was used. No crazing and cracking was noted when cleaning with Cidex OPA.

EXERCISES

1. Rectilinear scanner systems in Section 23.1 have been replaced by systems using one or more gamma cameras and back projection algorithms. Diagram the two systems and discuss how the gamma camera systems improve image collection efficiency.
2. Section 23.2 discusses testing of anesthetist competency. Discuss the objections you might have if you were the examinee. As the examiner, what ethical questions might come up if an examinee fails a test? How would you approach such a question if the examinee were a first year resident? A seasoned physician in practice for several years?
3. Apnea detection system in Section 23.3 was not continued as the principals involved dispersed due to various conditions. Go to the literature and determine (or estimate) the number of patients lost each year due to episodes of apnea, estimate the device market available if you were to develop an inexpensive monitor of respiration. Given current technology, suggest a design for your device.
4. One of the consequences of the use of certain illegal drugs is apnea. Perform a literature search to determine the causes of this effect, the drugs that cause it, and current suggested ways of prevention of death in drug users due to apnea. Suggest two or more ways to decrease the number of deaths, and discuss the ethics of your choices.

5. Cancer clinic charting system detailed in Section 23.4 is not unique. Based upon you experience or that of one of your acquaintances, outline (flowchart) a clinic visit and the emotions the visit caused.

6. EKG analysis technique outlined in Section 23.5 could have been presented in a preferences or evaluation chart form (see Section 2.6.) Take the four proposed problem solution techniques and generate an evaluation chart.

7. EKG analysis module in Section 23.6 could be extended to a number of other bioelectric signals. Select one and outline the course content for a one-credit hour module for freshmen engineering students.

8. Ethylene oxide has been used for sterilization of medical instruments. Research the several harmful effects of this gas.

REFERENCES

David, R. P. III, The design, construction, and preliminary testing of a mutually orthogonal coincident focal point tomographic brain scanner. M.S. thesis, Vanderbilt University, Tennessee, December 1977.

James, P. et al., US Patent 3,591,806, July 6, 1971.

King, P. H. et al., Simulator training in anesthesiology: An answer? *Biomedical Instrumentation and Technology*, July/August 1996, pp. 341–345.

Smith, B. E. et al., Automated end-tidal CO_2 monitoring in the postoperative patient, "contact author King for more information," available at http://shr.hamamed.ac.jp/iscaic18/AbstractSymposium/Smith-CO2.htm, 1999.

Waks, S. *Curriculum Design, From an Art Towards a Science*. Hamburg, Germany: Tempus Publications, 1995.

24 Future Design Issues

When it comes to the future, there are three kinds of people: those who let it happen, those who make it happen, and those who wonder what happened.

Richardson, John M. Jr.

"May you live in interesting times" is a saying meant to be a curse, implying that the recipient of the curse would be overwhelmed in an environment that—in contrast to the current—is interesting. The quotation is variously attributed to ancient Chinese literature (unproven), John Kennedy in 1966 (proven), or early science fiction literature (1950, proven). These are interesting times in the field of design in general and in the proliferation of topics for design projects in biomedical and biological engineering. This chapter provides a review of the topical areas involved in biomedical engineering and several of the important design and research issues that are studied here. The remainder of the chapter will highlight several of the funding sources for this work. It is left as an exercise for the reader to actively pursue one or more of the topical areas and the sources of funding for a successful career in this field.

24.1 GENERAL AREAS OF ENDEAVOR IN BIOMEDICAL ENGINEERING

*The Biomedical Engineering Handbook** gives an excellent overview of the field of biomedical engineering, describing and classifying the endeavor as being composed of some 15 areas of concentration. These areas form the basis of this discussion.

The study of bioelectric phenomena comprises one major subset of biomedical engineering. Basic to these studies are works done in the areas of cardiology, electromyography, and electroencephalography. Work will continue in this area to elucidate basic cardiac malfunctions, such as conduction blocks. Improved defibrillation devices continue to be a major effort. Improved control of artificial limbs, with a goal of sensory feedback, is desired. Improved brain–machine interfaces are a goal of several working with the EEG.

Biomaterials research will continue to expand, with efforts continuing in both hard and soft tissue replacements and implants. The use of implant materials as a scaffold for the growth of structures such as bones will be a major area of endeavor. The continued development of techniques to grow items such as heart valves (in vivo and ex vivo) will be a major endeavor.

Biomedical sensors and medical instrumentation will continue to be a mainstay of the field. Of concern is the development of smaller, lighter, and more sophisticated systems that will be of greater value in the daily monitoring and control of health issues, such as arise with the older people. Less invasive technologies, such as systems that use saliva rather than blood samples for diagnostic purposes, will be developed.

Biomedical imaging modalities will continue to be developed, with a goal of improved detection with decreased cost (radiation and financial) a major concern. The combination of imaging systems with therapeutic systems will a general thrust in the field. Systems such as tunable monochromatic x-ray systems will hopefully see its application in the fields of holographic imaging and x-ray triggered therapeutic dose release in tumors.

---

* Bronzino, J., Ed., *The Biomedical Engineering Handbook*, Boca Raton, FL: CRC Press, 1995.

Tissue and cell engineering research will hopefully get a boost from new methods of stem cell generation. New methods for growing cells and cell systems will lead to continued progress toward partial organ replacements. Gene therapy techniques will continue to be developed using these systems and in conjunction with biologists. Pharmaceutical research will also be enhanced by this research.

Continued development in prosthetics and rehabilitation engineering will in part be driven by the increases in the number of older people and by loss of limbs due to war and other injuries. Much development work will be, in the area of prosthetics, aimed at improved control of systems and the use of sensory feedback. At-home monitoring of patients health and rehabilitation efforts are stressed to improve overall patient care with a lower cost.

Clinical engineering departments' futures will depend heavily on the size of the hospital that is being served and the foresight of the hospital managers. In large medical centers, the overall system is well served by a clinical engineering service that does not only service work, but also is involved in the informatics and research endeavors of the enterprise. Teaming up with a biomedical engineering department locally can be a win–win collaboration for both groups. For small hospitals and clinics, outsourcing of the service responsibilities is likely the best decision.

Other areas of research and design endeavors involve tissue engineering, biotechnology, human factors and human performance engineering, medical informatics, physiological modeling, artificial intelligence, signal analysis, and ethics. Each of these generally relates in part to the material just presented, and is worthy areas of study too.

24.2   RESEARCH AND DESIGN SUPPORT

The material just discussed was a light overview of research and design endeavors in the field of biomedical engineering. The material that follows will be a partial overview of the driving force for these efforts, the financial and goal-oriented nature of various governmental and private organizations. The list is not meant to be comprehensive, but a guide to the novice regarding some of the support structures for work in biomedical engineering design and research.

24.2.1   NSF DESIGN, MANUFACTURING, AND INDUSTRIAL INNOVATION DIVISION

The division of the National Science Foundation (NSF) that is most responsible for assisting in the development of the process of design in both academia and industry is the division of design, manufacturing, and industrial Innovation. Two of the arms of this group, the SBIR/STTR (Small Business Innovation Research/Small Business Technology Transfer, in the Industrial Innovation Program) and the Engineering Design Program (in the Engineering Decision Systems Program), are of major importance to this section, which are discussed in Sections 24.2.1.1 and 24.2.1.2.

24.2.1.1   SBIR/STTR

The SBIR and STTR programs are designed to assist small businesses in the generation of new and innovative products, devices, processes, or services to facilitate the competitiveness of the industrial sector of the nation. The NSF sets the standards for acceptance of proposals, generally looking for end results that will translate into new jobs or other social benefits. To that end, it guides potential grantees to work in areas it feels is of importance. Thus, it generates listings of areas it deems worthy of funding. The listing below (1–20) was posted on the NSF Web site solicitation for year 2002 proposals (some editing by the authors was done to define or expand terms used.*)

---

* http://www.eng.nsf.gov/sbirspecs/BT/bt.htm.

1. Genomics (the study of genes and their interaction and influence on biological pathways and physiology): New capabilities enabling the rapid and massive sequencing of entire genomes of organisms, from microbes to humans, are transforming biological research. Exciting opportunities for commercialization activity have been created, with more yet to be proposed.

2. Proteomics (the study of protein structure, function, and interaction): The full complement of proteins expressed by complete genomes is now susceptible to analysis, prediction, and modification of structure, function, and interactions, giving rise to new commercial opportunities.

3. Bioinformatics (the science and art of converting data to knowledge): Computer power and new mathematical methods are required to harness the vast and expanding data sets that are being explosively generated through genomics and proteomics, creating bioinformatics business opportunities.

4. Biochips: These are biologically based microarray and microfluidic devices used for analysis and synthesis. How can they be made at lower cost? How can their applications be expanded?

5. Combinatorial biotechnology: Proposals are welcome on potential commercial applications of combinatorial biosynthesis, combinatorial biocatalysis, and biologically oriented combinatorial chemistry.

6. Computational biotechnology: Research with commercial objectives is needed for the development and implementation of algorithms and software for
   (a) Characterization of the relationship of DNA and protein sequence to biological function
   (b) Design of small molecules with biological activity
   (c) Analysis of complex dynamic biological systems
   (d) Multiscale ecological modeling

7. Environmental biotechnology (including bioremediation): How can the power of biology be applied to improve and protect the environment?

8. Ecological engineering and biocomplexity in the environment: Research with commercialization potential is sought for the design and management of ecosystems based on ecological principles and incorporating the self-organizing capacity of natural systems. Specific areas include ecosystem rehabilitation, habitat construction or enhancement, and flood prevention or mitigation.
   The term "biocomplexity" refers to phenomena that arise as a result of dynamic interactions that occur within living systems, including human beings, and between these systems and the physical environment, both natural and artificial. Biocomplexity encompasses ecological engineering as well as other areas. For further discussion, see <www.nsf.gov/home/crssprgm/be>.

9. Agricultural and food biotechnology: How can biotechnology be applied to crops and food products? How can it enhance food safety? Biological control of pests is included in this subtopic.

10. Marine biotechnology and aquaculture: How can biotechnology be used to enhance the search for valuable products from the sea and to improve their production?

11. Industrial bioproducts: Bioproducts such as industrial enzymes, biopolymers, neutraceuticals, and bioreagents are opening up new opportunities for small businesses.

12. Biosensors: What new biosensors can be developed for commercial applications?

13. Bioprocessing and bioconversion: Proposals are welcome on new commercial applications for involving bioreactors, bioseparations and purification, and biotechnology for a sustainable environment, for example, biomining and bioleaching serve as alternatives to smelting.

14. Biomedical engineering/research to aid persons with disabilities: Bioengineering research with commercial objectives is sought to help improve health care and reduce its costs. Proposals are welcome in such areas as

(a) Deriving information from cells, tissues, organs, and organ systems; extracting useful information from complex biomedical signals to derive new approaches to the design of structures and materials for eventual medical use

(b) Devising new means for characterizing, restoring, and substituting normal functions in humans, such as advanced prosthetics, hearing, speech, vision technologies, and other assistive technologies

(c) Novel and improved medical imaging technologies such as in vivo molecular and cellular imaging and probes

(d) Biomedical photonics, such as optical coherence tomography, and two-photon imaging/microscopy/spectroscopy

(e) Home care technologies such as mobility enhancement, manipulation ability, cognitive function, and remote patient monitoring

15. Tissue engineering: Tissue engineering technologies have opened commercial opportunities for developing polymer/cell structures and systems for biomedical applications.

16. Metabolic engineering: How can the metabolic pathways in organisms be altered in a targeted and purposeful manner to enable or improve the generation of useful products?

17. Biomaterials: Proposals are sought on developing new materials for bioengineering applications.

18. Pharmaceutical drug delivery: What systems, devices, or materials can be developed to enable or improve pharmaceutical dose applications or regimens?

19. Biotechnology at the nanoscale: Research is encouraged on fabrication at the nanoscale involving biomolecules and biosystems for potential commercial applications.

20. Newly emerging developments in biotechnology: Proposals are welcome in creative new biotechnology areas as they emerge.

24.2.1.2 Engineering Design Program

The following are some relevant issues addressed by the Engineering Design Program at NSF, as published on the Web (minor modifications again by the authors):*

Rapid generation of design alternatives: Designs are chosen from among a set of alternatives. Alternatives are generated by engineers using tools such as computer-aided design (CAD). We need tools that can take natural language descriptions and other natural forms of inputs and quickly derive candidate designs.

Easy evaluation of candidate designs: Evaluation of candidate designs typically requires various forms of analysis, such as finite element analysis, thermal analysis, hydrodynamic analysis, and so on. Many computer codes exist and new codes are under development for such analyses. But it can be very difficult to interface these codes to work together or even to use CAD representations as the inputs to these analyses. We need methods to facilitate quick, accurate, and complete evaluation of candidate designs.

Rigorous evaluation of design decisions: A view of engineering design that is providing the basis for many of the significant advances in the field today is that design is a decision-making process. In accordance with this view, design decisions are subject to rigorous analysis using well-established principles of decision theory. We need theory and tools for the rigorous evaluation and comparison of design alternatives. This theory could build, for example, on the von Neumann–Morgenstern axioms of utility theory.

Optimization of designs: In virtually all cases, a designer is confronted with an infinity of possible design alternatives. Selection of an optimal design can be extraordinarily difficult, and generally impossible. First, it is often not even possible to create a finite taxonomy of design alternatives.

---

* http://www.eng.nsf.gov/dmii/Message/EDS/ED/ed.htm.

Methods are needed to assist designers in creating and categorizing alternatives. Second, the range of alternatives is usually too great even to give consideration to all classes of alternatives. A method for discarding classes of alternatives early in the design process, under substantial uncertainty and risk, is needed. Third, consideration of a class of design alternatives demands that the class be modeled. Better methods for creating system models and methods for reuse of models are needed. Fourth, design intensely involves decision-making under uncertainty and risk. Convenient methods of modeling system performance including uncertainty and risk are needed, and these methods must be compatible with the goal of system optimization. Fifth, virtually all products, processes, or systems require huge numbers of variables to describe. Thus, their optimization runs into problems of dimensionality. We need better approaches to the issue of dimensionally in design optimization.

Design information systems: A great deal of data may be generated during the design process. This may include the design itself, documentation of the rationale for the design, listing of all requirements for the design, and listing of the verification activities for each requirement. We need to find ways of capturing these data and maintaining them in an accessible database. There is also a need for several engineers working together to simultaneously access a design database and to make changes in the database that are instantly accessible by others on the team. At the same time, there is a need for conveying these data to designers at remote locations while providing a high level of security on proprietary designs.

Collaborative design: More than ever before, engineering design has become a collaborative process that may involve teams and individuals working remotely. Particularly as these teams may be comprised of engineers representing a wide range of different organizations, their objectives may not precisely overlap, and hence the design becomes in the mathematical sense a cooperative game. This raises many issues: How can we manage design teams to assure rational design? How can the results provided by team members be effectively integrated to obtain a desired result? What are the best protocols for the transfer of design data? What forms of communication between team members work well?

Design education: An emerging view of engineering design holds that design intensely involves decision-making under conditions of uncertainty and risk. But current engineering curricula rarely include any principles of decision theory. Value or utility theory, central to all decision-making, is largely neglected and almost always treated incorrectly in the engineering community. And probability theory, which comprises the basic mathematics needed for the assessment of uncertainty and risk, is taught in only about half the engineering curricula. We need new pedagogy for design education. We need practical examples, particularly of real design cases. And we need much better approaches to the integration of design education across the engineering curriculum.

Two specific examples of the application of the above list of needs are worth mentioning. The first is the open workshop on decision-based design,* which is primarily an interactive Web site sponsored for the purpose of exchange of information about the general process of decision-based design. Recognizing that indeed design is a decision-making activity, it seeks exchange of information about that activity, with a parallel interest in the development of definitions and taxonomies of design. It also intends to help develop laws and axioms relating to design, perhaps to parallel the work done by Suh at MIT on axiomatic methods.[†]

The second major development that the NSF has helped sponsor is the publication of *Advanced Engineering Environments: Achieving the Vision: Phase 1* (1999)[‡] and *Design in the New Millennium: Advanced Engineering Environments: Phase 2*, National Academies Press,[§] (2000). The

---

* http://dbd.eng.buffalo.edu/.
[†] Suh, N. P., *Axiomatic Design: Advances and Applications*, New York: Oxford University Press, 2001.
[‡] http://www.nap.edu/catalog/9597.html.
[§] http://www.nap.edu/books/0309071259/html/.

messages sent by these publications include the development of design environments, three-dimensional imaging and interaction with design computer systems, increased dispersal of design personnel, increased and justifiable use of design software packages, and decreased time-to-market. Computational assistance will become pervasive in the design process.

24.2.2 National Institute of Biomedical Imaging and Bioengineering

The National Institute of Biomedical Imaging and Bioengineering is a newly formed institute (2001) within the National Institutes of Health (NIH).* The goal of the institute is to support basic and applied research and research training that improves health by promoting fundamental discoveries, design and development, translation, and assessment of technological capabilities in biomedical imaging and bioengineering. Research projects should be enabled by areas of engineering, the physical sciences, mathematics, and the computational sciences and should result in discoveries that can be translated into applications for specific diseases, disorders, or biological processes. Integrated, multidisciplinary, and collaborative approaches to addressing biomedical research are encouraged. Proposals can be based on hypothesis-, design-, needs-, development-, or problem-driven research. Thus, this new institute is fully supportive of design activities, as opposed to hypothesis-driven research as is the case in much of the rest of the NIH.

Some of the current design activities sponsored include

1. Development of probes for microimaging the nervous system
2. Development of novel technologies for in vivo imaging
3. Development of new and improved instruments or devices for research
4. Development of new methodologies for biomedical research
5. Development of software to be used in biomedical research
6. Development of rapid, accurate diagnostics for natural and bioengineered microbes and toxins (botulism, anthrax, plague, etc.)

Many of these are in conjunction with other braches of the NIH, and are only brief listings of all activities related to design.

24.2.3 National Institute of Science and Technology

The Advanced Technology Program (ATP)[†] at the National Institute of Science and Technology (NIST) is aimed at supporting high-risk, high-payoff proposals from all technology areas. For example, projects covering cutting-edge developments in the areas of tools for DNA diagnostics, photonics, manufacturing, and component-based software have all been supported by the ATP. The program's focus areas include chemistry and the life science, electronics and photonics technology, information technology and applications, and economic assessments. Current information on this and related topics may be found on the NIH bioengineering consortium Web site.[‡]

24.2.4 DARPA

The DARPA (Defense Advanced Research Projects Agency) mission is to develop imaginative, innovative, and often high-risk research ideas offering a significant technological impact that will go well beyond the normal evolutionary developmental approaches; and, to pursue these ideas from the demonstration of technical feasibility through the development of prototype systems.[§] Some of the more recent solicitations made by this agency include development work in the areas of

---

* http://www.nibib.nih.gov/research/investigators.htm#NIBIBfunding.
[†] http://www.becon.nih.gov/becon_news.htm#20011220.
[‡] http://www.becon.nih.gov/becon.htm.
[§] http://www.arpa.mil/.

- Biooptic synthetic systems
- Biomagnetic interfacing concepts
- Biological input/output (BIO) systems
- Brain–machine interfaces
- Biomolecular motors
- Evidence Extraction and Link Discovery Program
- BIO-surveillance system
- Effective, affordable, reusable speech-to-text
- Augmented cognition
- Speech in noisy environments
- Microelectronics
- Microelectromechanical systems
- Optoelectronics and photonics technology

As may be expected, all of these have defense ramifications as well as potential applications in the general field of human and animal health and welfare.

24.2.5 OTHER SUPPORT AGENCIES

Many other support agencies exist. One of the major drivers is the Alfred Mann Foundation, which has endowed several universities with funds (multimillion) to develop specific design and development efforts to speed research efforts from the laboratory to the marketplace. The Walter H. Coulter Foundation is also a major contributor to areas of translational research. The Robert Wood Johnson Foundation funds research in the areas of public health, childhood obesity, and health insurance coverage. The American Cancer Society supports cancer diagnosis, cure, and prevention research.

24.3   MISCELLANEOUS AND OTHER AREAS OF FUTURE DESIGN ACTIVITY

Potential developments in the field of design in biomedical engineering encompass many more areas than have been elaborated above. A sampler of topics include*

- Neural computing, interfacing between the living and the inanimate
- Biotechnology in general, design of organs specifically
- Genetic modification to relieve disease or genetic disorders
- Cloning
- Improved vision correction systems
- Biometrics and biometric technology for site protection
- Improved biomedical optics systems, automated cancer laser surgery
- Improved pharmacogenomics
- Improved mass analysis techniques (microarrays)
- Improved and miniaturized detection systems (biochips)
- Stem cell applications
- Agriculture and food technology interaction with BME
- Advanced medical informatics
- Improved biomaterials
- Improved design software and visualization tools
- Robotic surgery, including nanoscale
- Computational biotechnology

As long as there is money and interest, there will be design work to be done.

---

* Thanks to the class of 2002 at Vanderbilt for the majority of these topics.

24.4   CONCLUSIONS

The future of design in biomedical engineering is indeed promising with plenty of challenges and opportunities. Increasing technical sophistication is being accompanied by increasing breadth of research areas, such that the distinction between biomedical engineering and biological engineering is beginning to blur. This blurring of boundaries will give way to enormous numbers of new opportunities now and in the future.

EXERCISES

1. For any of the items 1–20 listed above in Section 24.2.1.1 perform a Web search and report out an expanded definition of the terms used. Hypothesize or find an expected benefit from research and development in this field. List several groups performing this activity.
2. Obtain a copy of *Design in the New Millennium* (refer to note 6). Read the summary and write a brief review of this chapter.
3. Investigate and report on the current list of NIBIB-sponsored projects as reported on the Web. Report on one subject of interest to you.
4. Investigate and report on the current list of NIST-sponsored projects as reported on the Web. Report on one subject of interest to you.
5. Investigate and report on the current list of DARPA-sponsored projects as reported on the Web. Report on one subject of interest to you.
6. Investigate and report on the current list of miscellaneous projects. Report on one subject of interest to you.

Appendix 1: Chi Square Table

v/γ	0.975	0.950	0.900	0.050	0.100	0.050	0.025
1	0.001	0.004	0.016	0.455	2.706	3.841	5.024
2	0.051	0.103	0.211	1.386	4.605	5.991	7.738
3	0.216	0.352	0.584	2.366	6.251	7.815	9.438
4	0.484	0.711	1.064	3.357	7.779	9.488	11.143
5	0.831	1.145	1.610	4.351	9.236	11.070	12.832
6	1.237	1.635	2.204	5.348	10.645	12.592	14.449
7	1.690	2.167	2.833	6.346	12.017	14.067	16.013
8	2.180	2.733	3.490	7.344	13.362	15.507	17.535
9	2.700	3.325	4.168	8.343	14.684	16.919	19.023
10	3.247	3.940	4.865	9.342	15.987	18.307	20.483
11	3.816	4.575	5.578	10.341	17.275	19.675	21.920
12	4.404	5.226	6.304	11.340	18.549	21.026	23.337
13	5.009	5.892	7.042	12.340	19.812	22.362	24.736
14	5.629	6.571	7.790	13.339	21.064	23.685	26.119
15	6.262	7.261	8.547	14.339	22.307	24.996	27.488
16	6.908	7.962	9.312	15.338	23.542	26.296	28.845
17	7.564	8.672	10.085	16.338	24.769	27.587	30.191
18	8.231	9.390	10.865	17.338	25.989	28.869	31.526
19	8.907	10.117	11.651	18.338	27.204	30.144	32.852
20	9.591	10.851	12.443	19.337	28.412	31.410	34.170
21	10.283	11.591	13.240	20.337	29.615	32.671	35.479
22	10.982	12.338	14.041	21.337	30.813	33.924	36.781
23	11.688	13.091	14.848	22.337	32.007	35.172	38.076
24	12.401	13.848	15.659	23.337	33.196	36.415	39.364
25	13.120	14.611	16.473	24.337	34.382	37.652	40.646
26	13.844	15.379	17.292	25.336	35.563	38.885	41.923
27	14.573	16.151	18.114	26.336	36.741	40.113	43.194
28	15.308	16.928	18.939	27.336	37.916	41.337	44.461
29	16.047	17.708	19.768	28.336	39.087	42.557	45.722
30	16.791	18.493	20.599	29.336	40.256	43.773	46.979

Appendix 2: Percent Rank Tables

	Sample Size = 1						
Order Number	**2.5**	**5.0**	**10.0**	**50.0**	**90.0**	**95.0**	**97.5**
1	2.50	5.00	10.00	50.00	90.00	95.00	97.50

	Sample Size = 2						
Order Number	**2.5**	**5.0**	**10.0**	**50.0**	**90.0**	**95.0**	**97.5**
1	1.258	2.532	5.132	29.289	68.377	77.639	84.189
2	15.811	22.361	31.623	71.711	94.868	97.468	98.742

	Sample Size = 3						
Order Number	**2.5**	**5.0**	**10.0**	**50.0**	**90.0**	**95.0**	**97.5**
1	0.840	1.695	3.451	20.630	53.584	63.160	70.760
2	9.430	13.535	19.580	50.000	80.420	86.465	90.570
3	29.240	36.840	46.416	79.370	96.549	98.305	99.160

	Sample Size = 4						
Order Number	**2.5**	**5.0**	**10.0**	**50.0**	**90.0**	**95.0**	**97.5**
1	0.631	1.274	2.600	15.910	43.766	52.713	60.236
2	6.759	9.761	14.256	38.573	67.954	75.140	80.588
3	19.412	24.860	32.046	61.427	85.744	90.239	93.241
4	39.764	47.287	56.234	84.090	97.400	98.726	99.369

	Sample Size = 5						
Order Number	**2.5**	**5.0**	**10.0**	**50.0**	**90.0**	**95.0**	**97.5**
1	0.505	1.021	2.085	12.945	36.904	45.072	52.182
2	5.274	7.644	11.223	31.381	58.389	65.741	71.642
3	14.663	18.926	24.644	50.000	75.336	81.074	85.337
4	28.358	34.259	41.611	68.619	88.777	92.356	94.726
5	47.818	54.928	63.096	87.055	97.915	98.979	99.495

	Sample Size = 6						
Order Number	**2.5**	**5.0**	**10.0**	**50.0**	**90.0**	**95.0**	**97.5**
1	0.421	0.851	1.741	19.910	31.871	39.304	45.926
2	4.327	6.285	9.260	26.445	51.032	58.180	64.123
3	11.812	15.316	20.091	42.141	66.681	72.866	77.722
4	22.278	27.134	33.319	57.859	79.909	84.684	88.188
5	35.877	41.820	48.968	73.555	90.740	93.715	95.673
6	54.074	60.696	68.129	89.090	98.259	99.149	99.579

(*continued*)

Appendix 2: Percent Rank Tables (continued)

			Sample Size = 7			
Order Number	2.5	5.0	10.0	50.0	90.0	95.0	97.5
1	0.361	0.730	1.494	9.428	28.031	34.816	40.962
2	3.669	5.338	7.882	22.489	45.256	52.070	57.872
3	9.899	12.876	16.964	36.412	59.618	65.874	70.958
4	18.405	22.532	27.860	50.000	72.140	77.468	81.595
5	29.042	34.126	40.382	63.588	83.036	87.124	90.101
6	42.128	47.930	54.744	77.151	92.118	94.662	96.331
7	59.038	65.184	71.969	90.752	98.506	99.270	99.639

			Sample Size = 8			
Order Number	2.5	5.0	10.0	50.0	90.0	95.0	97.5
1	0.316	0.639	1.308	8.300	25.011	31.234	36.942
2	3.185	4.639	6.863	20.113	40.625	47.068	52.651
3	8.523	11.111	14.685	32.052	53.822	59.969	65.086
4	15.701	19.290	23.966	44.016	65.538	71.076	75.514
5	24.486	28.924	43.462	55.984	76.034	80.710	84.299
6	34.914	40.031	46.178	67.948	85.315	88.889	91.477
7	47.349	52.932	59.375	79.887	93.137	95.361	96.815
8	63.058	68.766	74.989	91.700	98.692	99.361	99.684

			Sample Size = 9			
Order Number	2.5	5.0	10.0	50.0	90.0	95.0	97.5
1	0.281	0.568	1.164	7.413	22.574	28.313	33.627
2	2.814	4.102	6.077	17.962	36.836	42.914	48.250
3	7.485	9.775	12.950	28.624	49.008	54.964	60.009
4	13.700	16.875	21.040	39.308	59.942	65.506	70.070
5	21.201	25.137	30.097	50.000	69.903	74.863	78.799
6	29.930	34.494	40.058	60.692	78.960	83.125	86.300
7	39.991	45.036	50.992	71.376	87.050	90.225	92.515
8	51.750	57.086	63.164	82.038	93.923	95.898	97.186
9	66.373	71.687	77.426	92.587	98.836	99.432	99.719

			Sample Size = 10			
Order Number	2.5	5.0	10.0	50.0	90.0	95.0	97.5
1	0.253	0.512	1.048	6.697	20.567	25.887	30.850
2	2.521	3.677	5.453	16.226	33.685	39.416	44.502
3	6.674	8.726	11.583	25.857	44.960	50.690	55.610
4	12.155	15.003	18.756	35.510	55.173	60.662	65.245
5	18.709	22.244	26.732	45.169	64.578	69.646	73.762
6	26.238	30.354	35.422	54.831	73.268	77.756	81.291
7	34.755	39.338	44.827	64.490	81.244	84.997	87.845
8	44.390	49.310	55.040	74.143	88.417	91.274	93.326
9	55.498	60.584	66.315	83.774	94.547	96.323	97.479
10	69.150	74.113	79.433	93.303	98.952	99.488	99.747

			Sample Size = 11			
Order Number	2.5	5.0	10.0	50.0	90.0	95.0	97.5
1	0.230	0.465	0.953	6.107	18.887	23.840	28.491
2	2.283	3.332	4.945	14.796	31.024	36.436	41.278
3	6.022	7.882	10.477	23.579	41.516	47.009	41.776

Appendix 2: Percent Rank Tables (continued)

Order Number	Sample Size = 11						
	2.5	5.0	10.0	50.0	90.0	95.0	97.5
4	10.926	13.508	16.923	32.380	51.076	56.437	60.974
5	16.749	19.958	24.053	41.189	59.947	65.019	69.210
6	23.379	27.125	31.772	50.000	68.228	72.875	76.621
7	30.790	34.981	40.053	58.811	75.947	80.042	83.251
8	39.026	43.563	48.924	67.620	83.077	86.492	89.074
9	48.224	52.991	58.484	76.421	89.523	92.118	93.978
10	58.722	63.564	68.976	85.204	95.055	96.668	97.717
11	71.509	76.160	81.113	93.893	99.047	99.535	99.770

Order Number	Sample Size = 12						
	2.5	5.0	10.0	50.0	90.0	95.0	97.5
1	0.211	0.427	0.874	5.613	17.460	22.092	26.465
2	2.086	3.046	4.524	13.598	28.750	33.868	38.480
3	5.486	7.187	9.565	21.669	38.552	43.811	48.414
4	9.925	12.285	15.419	29.758	47.527	52.733	57.186
5	15.165	18.102	21.868	37.583	55.900	60.914	65.112
6	21.094	24.530	28.817	45.951	63.772	68.476	72.333
7	27.667	31.524	36.228	54.049	71.183	75.470	78.906
8	34.888	39.086	44.100	62.147	78.132	81.898	84.835
9	42.814	47.267	52.473	70.242	84.581	87.715	90.075
10	51.586	56.189	61.448	78.331	90.435	92.813	94.514
11	61.520	66.132	71.250	86.402	95.476	96.954	97.914
12	73.535	77.908	82.540	94.387	99.126	99.573	99.789

Order Number	Sample Size = 13						
	2.5	5.0	10.0	50.0	90.0	95.0	97.5
1	0.195	0.394	0.807	5.192	16.232	20.582	24.705
2	1.921	2.805	4.169	12.579	26.784	31.634	36.030
3	5.038	6.605	8.800	20.045	35.978	41.010	45.447
4	9.092	11.267	14.161	27.528	44.426	49.465	53.813
5	13.858	16.566	20.050	35.016	52.343	57.262	61.426
6	19.223	22.396	26.373	52.508	59.824	64.520	68.422
7	25.135	28.705	33.086	50.000	66.914	71.295	74.865
8	31.578	35.480	40.176	57.492	73.627	77.604	80.777
9	38.574	42.738	47.657	64.984	79.950	83.434	86.142
10	46.187	50.535	55.574	72.472	85.839	88.733	90.908
11	54.553	58.990	64.022	79.955	91.200	93.395	94.962
12	63.970	68.366	73.216	87.421	95.831	97.195	98.079
13	75.295	79.418	83.768	94.808	99.193	99.606	99.805

Order Number	Sample Size = 14						
	2.5	5.0	10.0	50.0	90.0	95.0	97.5
1	0.181	0.366	0.750	4.830	15.166	19.264	23.164
2	1.779	2.600	3.866	11.702	25.067	29.673	33.868
3	4.658	6.110	8.148	18.647	33.721	38.539	42.813
4	8.389	10.405	13.094	25.608	41.698	46.566	50.798
5	12.760	15.272	18.513	32.575	49.197	54.001	58.104

(continued)

Appendix 2: Percent Rank Tables (continued)

Sample Size = 14

Order Number	2.5	5.0	10.0	50.0	90.0	95.0	97.5
6	17.661	20.607	24.316	39.544	56.311	60.959	64.862
7	23.036	26.358	30.455	46.515	63.087	67.497	71.139
8	28.861	32.503	36.913	53.485	69.545	73.642	76.964
9	35.138	39.041	43.689	60.456	75.684	79.393	82.339
10	41.896	45.999	50.803	67.425	81.487	84.728	87.240
11	49.202	53.434	58.302	74.392	86.906	89.595	91.611
12	57.187	61.461	66.279	81.353	91.852	93.890	95.342
13	66.132	70.327	74.933	88.298	96.134	97.400	98.221
14	76.836	80.736	84.834	95.170	99.250	99.634	99.819

Sample Size = 15

Order Number	2.5	5.0	10.0	50.0	90.0	95.0	97.5
1	0.169	0.341	0.700	4.516	14.230	18.104	21.802
2	1.658	2.423	3.604	10.940	23.557	27.940	31.948
3	4.331	5.685	7.586	17.432	31.279	36.344	40.460
4	7.787	9.666	12.177	23.939	39.279	43.978	48.089
5	11.824	14.166	17.197	30.452	46.397	51.075	55.100
6	16.336	19.086	22.559	36.967	53.171	57.744	61.620
7	21.627	24.373	28.218	43.483	59.647	64.043	67.713
8	26.586	29.999	34.152	50.000	65.848	70.001	73.414
9	32.287	35.957	40.353	56.517	71.782	75.627	78.733
10	38.380	42.256	46.829	63.033	77.441	80.914	83.664
11	44.900	48.925	53.603	69.548	82.803	85.834	88.176
12	51.911	56.022	60.721	76.061	87.823	90.334	92.213
13	59.540	63.656	68.271	82.568	92.414	94.315	95.669
14	68.052	72.060	76.443	89.060	96.396	97.577	98.342
15	78.198	81.896	85.770	95.484	99.300	99.659	99.831

Sample Size = 16

Order Number	2.5	5.0	10.0	50.0	90.0	95.0	97.5
1	0.158	0.320	0.656	4.240	13.404	17.075	20.591
2	1.551	2.268	3.375	10.270	22.217	26.396	30.232
3	4.047	5.315	7.097	16.365	29.956	34.383	38.348
4	7.266	9.025	11.380	22.474	37.122	41.657	45.646
5	11.017	13.211	16.056	28.589	43.892	48.440	52.377
6	15.198	17.777	21.041	34.705	50.351	54.835	58.662
7	19.753	22.669	26.292	40.823	56.544	60.899	64.565
8	24.651	27.860	31.783	46.941	62.496	66.663	70.122
9	29.878	33.337	37.504	53.059	68.217	72.140	75.349
10	35.435	39.101	43.456	59.177	73.708	77.331	80.247
11	41.338	45.165	49.649	65.295	78.959	82.223	84.802
12	47.623	51.560	56.108	71.411	83.944	86.789	88.983
13	54.354	58.343	62.878	77.526	88.620	90.975	92.734
14	61.652	65.617	70.044	83.635	92.903	94.685	95.953
15	69.768	73.604	77.783	89.730	96.625	97.732	98.449
16	79.409	82.925	86.596	95.760	99.344	99.680	99.842

Appendix 2: Percent Rank Tables (continued)

Order Number	2.5	5.0	10.0	50.0	90.0	95.0	97.5
				Sample Size = 17			
1	0.149	0.301	0.618	3.995	12.667	16.157	19.506
2	1.458	2.132	3.173	9.678	21.021	25.012	28.689
3	3.779	4.990	6.667	15.422	28.370	32.619	36.441
4	6.811	8.465	10.682	21.178	35.187	39.564	43.432
5	10.314	12.377	15.058	26.940	41.639	46.055	49.899
6	14.210	16.636	19.716	32.704	47.807	52.192	55.958
7	18.444	21.191	24.614	38.469	53.735	58.029	61.672
8	22.983	26.011	29.726	44.234	59.449	63.599	67.075
9	27.812	31.083	35.039	50.000	64.961	68.917	72.188
10	32.925	36.401	40.551	55.766	70.274	73.989	77.017
11	38.328	41.971	46.265	61.531	75.386	78.809	81.556
12	44.042	47.808	52.193	67.296	80.284	83.364	85.790
13	50.101	53.945	58.361	73.060	84.942	87.623	89.686
14	56.568	60.436	64.813	78.821	89.318	91.535	93.189
15	63.559	67.381	71.630	84.578	93.333	95.010	96.201
16	71.311	74.988	78.979	90.322	96.827	97.868	98.542
17	80.494	83.843	87.333	96.005	99.382	99.699	99.851

Order Number	2.5	5.0	10.0	50.0	90.0	95.0	97.5
				Sample Size = 18			
1	0.141	0.285	0.584	3.778	12.008	15.332	18.530
2	1.375	2.011	2.995	9.151	19.947	23.766	27.294
3	3.579	4.702	6.286	14.58	26.942	31.026	34.712
4	6.409	7.970	10.064	20.024	33.441	37.668	41.418
5	9.695	11.643	14.177	25.471	39.602	43.888	47.637
6	13.343	15.634	18.549	30.921	45.502	49.783	53.480
7	17.299	19.895	23.139	36.371	51.184	55.405	59.007
8	21.530	24.396	27.922	41.823	56.672	60.784	64.255
9	26.019	29.120	32.885	47.274	61.980	65.940	69.243
10	30.757	34.060	38.020	52.726	67.115	70.880	73.981
11	35.745	39.216	43.328	58.177	72.078	75.604	78.470
12	40.993	44.595	48.618	63.629	76.861	80.105	82.701
13	46.520	50.217	54.498	69.079	81.451	84.336	86.657
14	52.363	56.112	60.398	74.529	85.823	88.357	90.305
15	58.582	62.332	66.559	79.976	89.936	92.030	93.591
16	65.288	68.974	73.058	85.419	93.714	95.298	96.421
17	72.706	76.234	80.053	90.849	97.005	97.989	98.625
18	81.470	84.668	87.992	96.222	99.416	99.715	99.859

Order Number	2.5	5.0	10.0	50.0	90.0	95.0	97.5
				Sample Size = 19			
1	0.133	0.270	0.553	3.582	11.413	14.587	17.647
2	1.301	1.903	2.835	8.678	18.977	22.637	26.028
3	3.383	4.446	5.946	13.827	25.651	29.580	33.138
4	6.052	7.529	9.514	18.989	31.859	35.943	39.578
5	9.147	10.991	13.394	24.154	37.753	41.912	45.565
6	12.576	14.747	17.513	29.322	43.405	47.580	51.203
7	16.289	18.750	21.832	34.491	48.856	52.997	56.550
8	20.252	22.972	26.327	39.660	54.132	58.194	61.642

(*continued*)

Appendix 2: Percent Rank Tables (continued)

Order Number	\	\	\	Sample Size = 19	\	\	\
	2.5	5.0	10.0	50.0	90.0	95.0	97.5
9	24.447	27.395	30.983	44.830	59.246	63.188	66.500
10	28.864	32.009	35.793	50.000	64.207	67.991	71.136
11	33.500	36.812	40.754	55.170	69.017	72.605	75.553
12	38.358	41.806	45.868	60.340	73.673	77.028	79.748
13	43.450	47.003	51.144	65.509	78.168	81.250	83.711
14	48.797	54.420	56.595	70.678	82.487	85.253	87.424
15	54.435	58.088	62.247	75.846	86.606	89.009	90.853
16	60.422	64.057	68.141	81.011	90.486	92.471	93.948
17	66.682	70.420	74.349	86.173	94.054	95.554	96.617
18	73.972	77.363	81.023	91.322	97.165	98.097	98.699
19	82.353	85.413	88.587	96.418	99.447	99.730	99.867

Order Number	\	\	\	Sample Size = 20	\	\	\
	2.5	5.0	10.0	50.0	90.0	95.0	97.5
1	0.127	0.256	0.525	3.406	10.875	13.911	16.843
2	1.235	1.807	2.691	8.251	18.096	21.611	24.873
3	3.207	4.217	5.642	13.147	24.477	28.262	31.698
4	5.733	7.135	9.021	18.055	30.419	34.366	37.893
5	8.657	10.408	12.693	22.967	36.066	40.103	43.661
6	11.893	13.955	16.587	27.880	41.489	45.558	49.105
7	15.391	17.731	20.666	32.795	46.727	50.782	54.279
8	19.119	21.707	24.906	37.711	51.803	55.803	59.219
9	23.058	25.865	29.293	42.626	56.733	60.642	63.946
10	27.196	30.195	33.817	47.542	61.525	65.307	68.472
11	31.528	34.693	38.475	52.458	66.183	69.805	72.804
12	36.054	39.358	43.267	57.374	70.707	74.135	76.942
13	40.781	44.197	48.197	62.289	75.094	78.293	80.881
14	45.721	49.218	53.273	67.205	79.334	82.269	84.609
15	50.895	54.442	58.511	72.120	83.413	86.045	88.107
16	56.339	59.897	63.934	77.033	87.307	89.592	91.343
17	62.107	65.634	69.581	81.945	90.979	92.865	94.267
18	68.302	71.738	75.523	86.853	94.358	95.783	96.793
19	75.127	78.389	81.904	91.749	97.309	98.193	98.765
20	83.157	86.089	89.125	96.594	99.475	99.744	99.873

Order Number	\	\	\	Sample Size = 21	\	\	\
	2.5	5.0	10.0	50.0	90.0	95.0	97.5
1	0.120	0.244	0.500	3.247	10.385	13.295	16.110
2	1.175	1.719	2.562	7.864	17.294	20.673	23.816
3	3.049	4.010	5.367	12.531	23.405	27.055	30.377
4	5.446	6.781	8.577	17.209	29.102	32.921	36.342
5	8.218	9.884	12.062	21.891	34.522	38.441	41.907
6	11.281	13.245	15.755	26.574	39.733	43.698	47.166
7	14.588	16.818	19.619	31.258	44.771	48.739	52.175
8	18.107	20.575	23.632	35.943	49.661	53.594	56.968
9	21.820	24.499	27.779	40.629	54.416	58.280	61.565
10	25.713	28.580	32.051	45.314	59.046	62.810	65.979
11	29.781	32.811	36.443	50.000	63.557	67.189	70.219
12	34.021	37.190	40.954	54.686	67.949	71.420	74.287

Appendix 2: Percent Rank Tables (continued)

Order Number				Sample Size = 21			
	2.5	5.0	10.0	50.0	90.0	95.0	97.5
13	38.435	41.720	45.584	59.371	72.221	75.501	78.180
14	43.032	46.406	50.339	64.057	76.368	79.425	81.893
15	47.825	51.261	55.229	68.742	80.381	83.182	85.412
16	52.834	56.302	60.267	73.426	84.245	86.755	88.719
17	58.093	61.559	65.478	78.109	87.938	90.116	91.782
18	63.658	67.079	70.898	82.791	91.423	93.219	94.554
19	69.623	72.945	76.595	87.469	94.633	95.990	96.951
20	76.184	79.327	82.706	92.136	97.438	98.281	98.825
21	83.890	86.705	89.615	96.753	99.500	99.756	99.880

Order Number				Sample Size = 22			
	2.5	5.0	10.0	50.0	90.0	95.0	97.5
1	0.115	0.233	0.478	3.102	9.937	12.731	15.437
2	1.121	1.640	2.444	7.512	16.559	19.812	22.844
3	2.906	3.822	5.117	11.970	22.422	25.947	29.161
4	5.187	6.460	8.175	16.439	27.894	31.591	34.912
5	7.821	9.411	11.490	20.911	33.104	36.909	40.285
6	10.729	12.603	15.002	25.384	38.117	41.980	45.370
7	13.865	15.994	18.674	29.859	42.970	46.849	50.222
8	17.198	19.556	22.483	34.334	47.684	51.546	54.872
9	29.709	23.272	26.416	38.810	52.275	56.087	59.342
10	24.386	27.131	30.463	43.286	56.752	60.484	63.645
11	28.221	31.126	34.619	47.762	61.119	64.746	67.790
12	32.210	35.254	38.881	52.238	65.381	68.874	71.779
13	36.355	39.516	43.248	56.714	69.537	72.869	75.614
14	40.658	43.913	47.725	61.190	73.584	76.728	79.291
15	45.128	48.454	52.316	65.666	77.517	80.444	82.802
16	49.778	53.151	57.030	70.141	81.326	84.006	86.135
17	54.630	58.020	61.883	74.616	84.998	87.397	89.271
18	59.715	63.091	66.896	79.089	88.510	90.589	92.179
19	65.088	68.409	72.106	83.561	91.825	93.540	94.813
20	70.839	74.053	77.578	88.030	94.883	96.178	97.094
21	77.156	80.188	83.441	92.488	97.556	98.360	98.879
22	84.563	87.269	90.063	96.898	99.522	99.767	99.885

Order Number				Sample Size = 23			
	2.5	5.0	10.0	50.0	90.0	95.0	97.5
1	0.110	0.223	0.457	2.969	9.526	12.212	14.819
2	1.071	1.567	2.337	7.191	15.884	19.020	21.949
3	2.775	3.652	4.890	11.458	21.519	24.925	28.038
4	4.951	6.168	7.808	15.734	26.781	30.364	33.589
5	7.460	8.981	10.971	20.015	31.797	35.493	38.781
6	10.229	12.021	14.318	24.297	36.626	40.390	43.703
7	13.210	15.248	17.816	28.580	41.305	45.098	48.405
8	16.376	18.634	21.442	32.863	45.856	49.644	52.919
9	19.708	22.164	25.182	37.147	50.291	54.046	57.226
10	23.191	25.824	29.027	41.431	54.622	58.315	61.458
11	26.820	29.609	32.971	45.716	58.853	62.461	65.505
12	30.588	33.515	37.012	50.000	62.988	66.485	69.412

(continued)

Appendix 2: Percent Rank Tables (continued)

Sample Size = 23

Order Number	2.5	5.0	10.0	50.0	90.0	95.0	97.5
13	34.495	37.539	41.147	54.284	67.029	70.391	73.180
14	38.542	41.685	45.378	58.569	70.973	74.176	76.809
15	42.734	45.954	49.709	62.853	74.818	77.836	80.292
16	47.081	50.356	54.144	67.137	78.558	81.366	83.624
17	51.595	54.902	58.695	71.420	82.184	84.752	86.790
18	56.297	59.610	63.374	75.703	85.682	87.979	89.771
19	61.219	64.507	68.203	79.985	89.029	91.019	92.540
20	66.411	69.636	73.219	84.266	92.192	93.832	95.049
21	71.962	75.075	78.481	88.542	95.110	96.348	97.225
22	78.051	80.980	84.116	92.809	97.663	98.433	98.929
23	85.151	87.788	90.474	97.031	99.543	99.777	99.890

Sample Size = 24

Order Number	2.5	5.0	10.0	50.0	90.0	95.0	97.5
1	0.105	0.213	0.438	2.847	9.148	11.735	14.247
2	1.026	1.501	2.238	6.895	15.262	18.289	21.120
3	2.656	3.495	4.682	10.987	20.685	23.980	26.997
4	4.735	5.901	7.473	15.088	25.754	29.227	32.361
5	7.132	8.589	10.497	19.192	30.588	34.181	37.384
6	9.773	11.491	13.694	23.299	35.246	38.914	42.151
7	12.615	14.569	17.033	27.406	39.763	43.469	46.711
8	16.630	17.796	20.493	31.513	44.160	47.873	51.095
9	18.799	21.157	24.058	35.621	48.449	52.142	55.322
10	22.110	24.639	27.721	39.729	52.461	56.289	59.406
11	25.553	28.236	31.476	43.837	56.742	60.321	63.357
12	29.124	31.942	35.317	47.946	60.755	64.244	67.179
13	32.821	35.756	39.245	52.054	64.683	68.058	70.876
14	36.643	39.679	43.258	56.163	68.524	71.764	74.447
15	40.594	43.711	47.359	60.271	72.279	75.361	77.890
16	44.678	47.858	51.551	64.379	75.942	78.843	81.201
17	48.905	52.127	55.840	68.487	79.507	82.204	84.370
18	53.289	56.531	60.237	72.594	82.967	85.431	87.385
19	57.849	60.086	64.754	76.701	86.306	88.509	90.227
20	62.616	65.819	69.412	80.808	89.503	91.411	92.868
21	67.639	70.773	74.246	84.912	92.527	94.099	95.265
22	73.003	76.020	79.315	89.013	95.318	96.505	97.344
23	78.880	81.711	84.738	93.105	97.762	98.499	98.974
24	85.753	88.265	90.852	97.153	99.562	99.787	99.895

Sample Size = 25

Order Number	2.5	5.0	10.0	50.0	90.0	95.0	97.5
1	0.101	0.205	0.421	2.735	8.799	11.293	13.719
2	0.984	1.440	2.148	6.623	14.687	17.612	20.352
3	2.547	3.352	4.491	10.553	19.914	23.104	26.031
4	4.538	5.656	7.166	14.492	24.802	28.172	31.219
5	6.831	8.229	10.062	18.435	29.467	32.961	36.083
6	9.356	11.006	13.123	22.379	33.966	37.541	40.704
7	12.072	13.948	16.317	26.324	38.331	41.952	45.129
8	14.950	17.030	19.624	30.270	42.582	46.221	49.388

Appendix 2: Percent Rank Tables (continued)

Order Number	\multicolumn{7}{c}{Sample Size = 25}
	2.5	5.0	10.0	50.0	90.0	95.0	97.5
9	17.972	20.238	23.032	34.215	46.734	50.364	53.500
10	21.125	23.559	26.529	38.161	50.795	54.393	57.479
11	24.402	26.985	30.111	42.108	54.722	58.316	61.335
12	27.797	30.513	33.774	46.054	58.668	62.138	65.072
13	31.306	34.139	37.514	50.000	62.486	65.861	68.694
14	34.928	37.862	41.332	53.946	66.226	69.487	72.203
15	38.665	41.684	45.228	57.892	69.889	73.015	75.598
16	42.521	45.607	49.205	61.839	73.471	76.441	78.875
17	46.500	49.636	53.266	65.785	76.968	79.762	82.028
18	50.612	53.779	57.418	69.730	80.736	82.970	85.050
19	54.871	58.048	61.669	73.676	83.683	86.052	87.928
20	59.296	62.459	66.034	77.621	86.877	88.994	90.644
21	63.917	67.039	70.533	81.565	89.938	91.771	93.169
22	68.781	71.828	75.198	85.508	92.834	94.344	95.462
23	73.969	76.896	80.086	89.447	95.509	96.648	97.453
24	79.648	82.388	85.313	93.377	97.852	98.560	99.016
25	86.281	88.707	92.201	97.265	99.579	99.795	99.899

Appendix 3: 40 Inventive Principles, Engineering Parameters, and Conflict Matrix

40 INVENTIVE PRINCIPLES

1   Segmentation
2   Extraction
3   Local quality
4   Asymmetry
5   Combining
6   Universality
7   Nesting
8   Counterweight
9   Prior counteraction
10 Prior action
11 Cushion in advance
12 Equipotentiality
13 Inversion
14 Spheroidality
15 Dynamicity
16 Partial or overdone action
17 Moving to a new dimension
18 Mechanical vibration
19 Periodic action
20 Continuity of useful action
21 Rushing through
22 Convert harm into benefit
23 Feedback
24 Mediator
25 Self-service
26 Copying
27 An inexpensive short-life object instead of an expensive durable one
28 Replacement of a mechanical system
29 Use a pneumatic or hydraulic construction
30 Flexible film or thin membranes

31   Use of porous materials
32   Changing the color
33   Homogeneity
34   Rejecting and regenerating parts
35   Transformation of physical and chemical states of an object
36   Phase transition
37   Thermal expansion
38   Use strong oxidizers
39   Inert environment
40   Composite materials

INVENTIVE PRINCIPLES ORDERED BY FREQUENCY OF USE

35   Transformation of physical and chemical states of an object
10   Prior action
 1   Segmentation
28   Replacement of a mechanical system
 2   Extraction
15   Dynamicity
19   Periodic action
18   Mechanical vibration
32   Changing the color
13   Inversion
26   Copying
 3   Local quality
27   An inexpensive short-life object instead of an expensive durable one
29   Use a pneumatic or hydraulic construction
34   Rejecting and regenerating parts
16   Partial or overdone action
40   Composite materials
24   Mediator
17   Moving to a new dimension
 6   Universality
14   Spheroidality
22   Convert harm into benefit
39   Inert environment
 4   Asymmetry
30   Flexible film or thin membranes
37   Thermal expansion
36   Phase transition
25   Self-service
11   Cushion in advance
31   Use of porous materials
38   Use strong oxidizers
 8   Counterweight
 5   Combining
 7   Nesting

21   Rushing through
23   Feedback
12   Equipotentiality
33   Homogeneity
 9   Prior counteraction
20   Continuity of useful action

Contradiction Table

Feature To Improve	1 Weight of moving object	2 Weight of nonmoving object	3 Length of moving object	4 Length of nonmoving object	5 Area of moving object	6 Area of nonmoving object	7 Volume of moving object	8 Volume of nonmoving object	9 Speed	10 Force	11 Tension, pressure	12 Shape	13 Stability of object	14 Strength	15 Durability of moving object	16 Durability of nonmoving object	17 Temperature	18 Brightness
1 Weight of moving object			15, 8, 29, 34		29, 17, 38, 34		29, 2, 40, 28		2, 8, 15, 38	8, 10, 18, 37	10, 36, 37, 40	10, 14, 35, 40	1, 35, 19, 39	28, 27, 18, 40	5, 34, 31, 35		6, 20, 4, 38	19, 1, 32
2 Weight of nonmoving object				10, 1, 29, 35		35, 30, 13, 2		5, 35, 14, 2		8, 10, 19, 35	13, 29, 10, 18	13, 10, 29, 14	26, 39, 1, 40	28, 2, 10, 27		2, 27, 19, 6	28, 19, 32, 22	19, 32, 35
3 Length of moving object	8, 15, 29, 34				15, 17, 4		7, 17, 4, 35		13, 4, 8	17, 10, 4	1, 8, 35	1, 8, 10, 29	1, 8, 15, 34	8, 35, 29, 34	19		10, 15, 19	32
4 Length of nonmoving object		35, 28, 40, 29				17, 7, 10, 40		35, 8, 2, 14		28, 10	1, 14, 35	13, 14, 15, 7	39, 37, 35	15, 14, 28, 26		1, 40, 35	3, 35, 38, 18	3, 25
5 Area of moving object	2, 17, 29, 4		14, 15, 18, 4				7, 14, 17, 4		29, 30, 4, 34	19, 30, 35, 2	10, 15, 36, 28	5, 34, 29, 4	11, 2, 13, 39	3, 15, 40, 14	6, 3		2, 15, 16	15, 32, 19, 13
6 Area of nonmoving object		30, 2, 14, 18		26, 7, 9, 39						1, 18, 35, 36	10, 15, 36, 37		2, 38	40			2, 10, 19, 30	35, 39, 38
7 Volume of moving object	2, 26, 29, 40		1, 7, 4, 35		1, 7, 4, 17				29, 4, 38, 34	15, 35, 36, 37	6, 35, 36, 37	1, 15, 29, 4	28, 10, 1, 39	9, 14, 15, 7	6, 35, 4		34, 39, 10, 18	2, 13, 10
8 Volume of nonmoving object		35, 10, 19, 14	19, 14	35, 8, 2, 14						2, 18, 37	24, 35	7, 2, 35	34, 28, 35, 40	9, 14, 17, 15		35, 34, 38	35, 6, 4	
9 Speed	2, 28, 13, 38		13, 14, 8		29, 30, 34		7, 29, 34			13, 28, 15, 19	6, 18, 38, 40	35, 15, 18, 34	28, 33, 1, 18	8, 3, 26, 14	3, 19, 35, 5		28, 30, 36, 2	10, 13, 19
10 Force	8, 1, 37, 18	18, 13, 1, 28	17, 19, 6, 36	28, 10	19, 10, 15	1, 18, 36, 37	15, 9, 12, 37	2, 36, 18, 37	13, 28, 15, 12		18, 21, 11	10, 35, 40, 34	35, 10, 21	35, 10, 14, 27	19, 2		35, 10, 24	
11 Tension, pressure	10, 36, 37, 40	13, 29, 10, 18	35, 10, 36	35, 1, 14, 16	10, 15, 36, 25	10, 15, 35, 37	6, 35, 10	35, 24	6, 35, 36	36, 35, 21		35, 4, 15, 10	35, 33, 2, 40	9, 18, 3, 40	19, 3, 27		35, 39, 19, 2	
12 Shape	8, 10, 29, 40	15, 10, 26, 3	29, 34, 5, 4	13, 14, 10, 7	5, 34, 4, 10		14, 4, 15, 22	7, 2, 35	35, 15, 34, 18	35, 10, 37, 40	34, 15, 10, 14		33, 1, 18, 4	30, 14, 10, 40	14, 26, 9, 25		22, 14, 19, 32	13, 15, 32
13 Stability of object	21, 35, 2, 39	26, 39, 1, 40	13, 15, 1, 28	37	2, 11, 13	39	28, 10, 19, 39	34, 28, 35, 40	33, 15, 28, 18	10, 35, 21, 16	2, 35, 40	22, 1, 18, 4		17, 9, 15	13, 27, 10, 35	39, 3, 35, 23	35, 1, 32	32, 3, 27, 15
14 Strength	1, 8, 40, 15	40, 26, 27, 1	1, 15, 8, 35	15, 14, 28, 26	3, 34, 40, 29	9, 40, 28	10, 15, 14, 7	9, 14, 17, 15	8, 13, 26, 14	10, 18, 3, 14	10, 3, 18, 40	10, 30, 35, 40	13, 17, 35		27, 3, 26		30, 10, 40	35, 19
15 Durability of moving object	19, 5, 34, 31		2, 19, 9		3, 17, 19		10, 2, 19, 30		3, 35, 5	19, 2, 16	19, 3, 27	14, 26, 28, 25	13, 3, 35	27, 3, 10			19, 35, 39	2, 19, 4, 35
16 Durability of nonmoving object		6, 27, 19, 16		1, 10, 35					35, 34, 38				39, 3, 35, 23				19, 18, 36, 40	
17 Temperature	36, 22, 6, 38	22, 35, 32	15, 19, 9	15, 19, 9	3, 35, 39, 18	35, 38	34, 39, 40, 18	35, 6, 4	2, 28, 36, 30	35, 10, 3, 21	35, 39, 19, 2	14, 22, 19, 32	1, 35, 32	10, 30, 22, 40	19, 13, 39	19, 18, 36, 40		32, 30, 21, 16
18 Brightness	19, 1, 32	2, 35, 32			19, 32, 26										2, 19, 6		32, 35, 19	
19 Energy spent by moving object	12, 18, 28, 31		12, 28		15, 19, 25		35, 13, 18		8, 15, 35	16, 26, 21, 2	23, 14, 25	12, 2, 29	19, 13, 17, 24	5, 19, 9, 35	28, 35, 6, 18		19, 24, 3, 14	2, 15, 19
20 Energy spent by nonmoving object		19, 9, 6, 27								36, 37				27, 4, 29, 19	35			19, 2, 35, 32
21 Power	8, 36, 38, 31	19, 26, 17, 27	1, 10, 35, 37		19, 38	17, 32, 13, 38	35, 6, 38	30, 6, 25	15, 35, 2	26, 2, 36, 35	22, 10, 35	29, 14, 2, 40	35, 32, 15, 31	26, 10, 28	19, 35, 10, 38	16	2, 14, 17, 25	16, 6, 19
22 Waste of energy	15, 6, 19, 28	19, 6, 18, 9	7, 2, 6, 13	6, 38, 7	15, 26, 17, 30	17, 7, 30, 18	7, 18, 23	7	16, 35, 38	36, 38			14, 2, 39, 6	26			19, 38, 7	1, 13, 32, 15
23 Waste of substance	35, 6, 23, 40	35, 6, 22, 32	14, 29, 10, 398	10, 28, 24	35, 2, 10, 31	10, 18, 39, 31	1, 29, 30, 36	3, 39, 18, 31	10, 13, 28, 38	14, 15, 18, 40	3, 36, 37, 10	29, 35, 3, 5	2, 14, 30, 40	35, 28, 31, 40	28, 27, 3, 18	27, 16, 18, 38	21, 36, 39, 31	1, 6, 13
24 Loss of information	10, 24, 35	10, 35, 5	1, 26	26	30, 26	30, 16		2, 22	26, 32						10	10		19
25 Waste of time	10, 20, 37, 35	10, 20, 26, 5	15, 2, 29	30, 24, 14, 5	26, 4, 5, 16	10, 35, 17, 4	2, 5, 34, 10	35, 16, 32, 18		10, 37, 36, 5	37, 36, 4	4, 10, 34, 17	35, 3, 22, 5	29, 3, 28, 18	20, 10, 28, 18	28, 20, 10, 16	35, 29, 21, 18	1, 19, 26, 17
26 Amount of substance	35, 6, 18, 31	27, 26, 18, 35	29, 14, 35, 18		15, 14, 29	2, 18, 40, 4	15, 20, 29		35, 29, 34, 28	35, 14, 3	10, 36, 14, 3	35, 14	15,. 2, 17, 40	14, 35, 34, 10	3, 35, 10, 40	3, 35, 31	3, 17, 39	
27 Reliability	3, 8, 10, 40	3, 10, 8, 28	15, 9, 14, 4	15, 29, 28, 11	17, 10, 14, 16	32, 35, 40, 4	3, 10, 14, 24	2, 35, 24	21, 35, 11, 28	8, 28, 10, 3	10, 24, 35, 19	35, 1, 16, 11		11, 28,	2, 35, 3, 25	34, 27, 6, 40	3, 35, 10	11, 32, 13
28 Accuracy of measurement	32, 35, 26, 28	28, 35, 25, 26	28, 26, 5, 16	32, 28, 3, 16	26, 28, 32, 3	26, 28, 32, 3	32, 13, 6		28, 13, 32, 24	32, 2	6, 28, 32	6, 28, 32	32, 35, 13	28, 6, 32	28, 6, 32	10, 26, 24	6, 19, 28, 24	6, 1, 32
29 Accuracy of manufacturing	28, 32, 13, 18	28, 35, 27, 9	10, 28, 29, 37	2, 32, 10	28, 33, 29, 32	2, 29, 18, 36	32, 28, 2	25, 10, 35	10, 28, 32	28, 19, 34, 36	3, 35	32, 30, 40	30, 18	3, 27	3, 27, 40		19, 26	3, 32
30 Harmful factors acting on object	22, 21, 27, 39	2, 22, 13, 24	17, 1, 39, 4	1, 18	22, 1, 33, 28	27, 2, 39, 35	22, 23, 37, 35	34, 39, 19, 27	21, 22, 35, 28	13, 35, 39, 18	22, 2, 37	22, 1, 3, 35	35, 24, 30, 18	18, 35, 37, 1	22, 15, 33, 28	17, 1, 40, 33	22, 33, 35, 2	1, 19, 32, 13
31 Harmful side effects	19, 22, 15, 39	35, 22, 1, 39	17, 15, 16, 22		17, 2, 18, 39	22, 1, 40	17, 2, 40	30, 18, 35, 4	35, 28, 3, 23	35, 28, 1, 40	2, 33, 27, 18	35, 1	35, 40, 27, 39	15, 35, 22, 2	15, 22, 33, 31	21, 39, 16, 22	22, 35, 2, 24	19, 24, 39, 32
32 Manufacturability	28, 29, 15, 16	1, 27, 36, 13	1, 29, 13, 17	15, 17, 27	13, 1, 26, 12	16, 40	13, 29, 1, 40	35	35, 13, 8, 1	35, 12	35, 19, 1, 37	1, 28, 13, 27	11, 13, 1	1, 3, 10, 32	27, 1, 4	35, 16	27, 26, 18	28, 24, 27, 1
33 Convenience of use	25, 2, 13, 15	6, 13, 1, 25	1, 17, 13, 12		1, 17, 13, 16	18, 16, 15, 39	1, 16, 35, 15	4, 18, 39, 31	18, 13, 34	28, 13, 35	2, 32, 12	15, 34, 29, 28	32, 35, 30	32, 40, 3, 28	29, 3, 8, 25	1, 16, 25	26, 27, 13	13, 17, 1, 24
34 Repairability	2, 27, 35, 11	2, 27, 35, 11	1, 28, 10, 25	3, 18, 31	15, 13, 32	16, 25	25, 2, 35, 11	1	34, 9	1, 11, 10	13	1, 13, 2, 4	2, 35	11, 1, 2, 9	11, 29, 28, 27	1	4, 10	15, 1, 13
35 Adaptability	1, 6, 15, 8	19, 15, 29, 16	35, 1, 29, 2	1, 35, 16	35, 30, 29, 7	15, 16	15, 35, 29		35, 10, 14	15, 17, 20	35, 16	15, 37, 1, 8	35, 30, 14	35, 3, 32, 6	13, 1, 35	2, 16	27, 2, 3, 35	6, 22, 26, 1
36 Complexity of device	26, 30, 34, 36	2, 36, 35, 39	1, 19, 26, 24	26	14, 1, 13, 16	6, 36	34, 26, 6	1, 16	34, 10, 28	26, 16	19, 1, 35	29, 13, 28, 15	2, 22, 17, 19	2, 13, 28	10, 4, 28, 15		2, 17, 13	24, 17, 13
37 Complexity of control	27, 26, 28, 13	6, 13, 28, 1	16, 17, 26, 24	26	2, 13, 15, 17	2, 39, 30, 16	29, 1, 4, 16	2, 18, 26, 31	3, 4, 16, 35	36, 28, 40, 19	35, 36, 37, 32	27, 13, 1, 39	11, 22, 39, 30	27, 3, 15, 28	19, 29, 39, 25	25, 24, 6, 35	3, 27, 35, 16	2, 24, 26, 5
38 Level of automation	28, 26, 18, 35	28, 26, 35, 10	14, 13, 17, 28	23	17, 14, 13		35, 13, 16		28, 10	2, 35	13, 35	15, 32, 1, 13	18, 1	25, 13	6, 9		26, 2, 19	8, 32, 19
39 Productivity	35, 26, 24, 37	28, 27, 15, 3	18, 4, 28, 38	30, 7, 14, 26	10, 26, 34, 31	10, 35, 17, 7	2, 6, 34, 10	35, 37, 10, 2		28, 15, 10, 36	10, 37, 14	14, 10, 34, 40	35, 3, 22, 39	29, 28, 10, 18	35, 10, 2, 18	20, 10, 16, 38	35, 21, 28, 10	26, 17, 19, 1

Source: From Clarke, D.W. Sr., *TRIZ: Through the Eyes of an American TRIZ Specialist, A Study of Ideality, Contradictions, Resources,* Ideation International, 1997. By courtesy.

19	20	21	22	23	24	25	26	27	28	29	30	31	32	33	34	35	36	37	38	39
Energy spent by moving object	Energy spent by nonmoving object	Power	Waste of energy	Waste of substance	Loss of information	Waste of time	Amount of substance	Reliability	Accuracy of measurement	Accuracy of manufacturing	Harmful factors acting on object	Harmful side effects	Manufacturability	Convenience of use	Repairability	Adaptability	Complexity of device	Complexity of control	Level of automation	Productivity
35, 12, 34, 31		12, 36, 18, 31	6, 2, 34, 19	5, 35, 3, 31	10, 24, 35	10, 35, 20, 28	3, 26, 18, 31	3, 11, 1, 27	28, 27, 35, 26	28, 35, 26, 18	22, 21, 18, 27	22, 35, 31, 39	27, 28, 1, 36	35, 3, 2, 24	2, 27, 28, 11	29, 5, 15, 8	26, 30, 36, 34	28, 29, 26, 32	26, 35, 18, 19	35, 3, 24, 37
	18, 19, 28, 1	15, 19, 18, 22	18, 19, 28, 15	5, 8, 13, 30	10, 15, 35	10, 20, 35, 26	19, 6, 18, 26	10, 28, 8, 3	18, 26, 28	10, 1, 35, 17	2, 19, 22, 37	35, 22, 1, 39	28, 1, 9	6, 13, 1, 32	2, 27, 28, 11	19, 15, 29	1, 10, 26, 39	25, 28, 17, 15	2, 26, 35	1, 28, 15, 35
8, 35, 24		1, 35	7, 2, 35, 39	4, 29, 23, 10	1, 24	15, 2, 29	29, 35	10, 14, 29, 40	28, 32, 4	10, 28, 29, 37	1, 15, 17, 24	17, 15	1, 29, 17	15, 29, 35, 4	1, 28, 10	14, 15, 1, 16	1, 19, 26, 24	35, 1, 26, 24	17, 24, 26, 16	14, 4, 28, 29
		12, 8	6, 28	10, 28, 24, 35	24, 26	30, 29, 14		15, 29, 28	32, 28, 3	2, 32, 10	1, 18		15, 17, 27	2, 25	3	1, 35	1, 26	26		30, 14, 7, 26
19, 32		19, 10, 32, 18	15, 17, 30, 26	10, 35, 2, 39	30, 26	26, 4	29, 30, 6, 13	29, 9	26, 28, 32, 3	2, 32	22, 33, 28, 1	17, 2, 18, 39	13, 1, 26, 24	15, 17, 13, 16	15, 13, 10, 1	15, 30	14, 1, 13	2, 36, 26, 18	14, 30, 28, 23	10, 26, 34, 2
		17, 32	17, 7, 30	10, 14, 18, 39	30, 16	10, 35, 4, 18	2, 18, 40, 4	32, 35, 40, 4	26, 28, 32, 3	2, 29, 18, 36	27, 2, 39, 35	22, 1, 40	40, 16	16, 4	16	15, 16	1, 18, 36	2, 35, 30, 18	23	10, 15, 17, 7
35		35, 6, 13, 18	7, 15, 13, 16	36, 39, 34, 10	2, 22	2, 6, 34, 10	29, 30, 7	14, 1, 40, 11	25, 26, 28	25, 28, 2, 16	22, 21, 27, 35	17, 2, 40, 1	29, 1, 40	15, 13, 30, 12	10	15, 29	26, 1	29, 26, 4	35, 34, 16, 24	10, 6, 2, 34
		30, 6		10, 39, 35, 34		35, 16, 32, 18	35, 3	2, 35, 16		35, 10, 25	34, 39, 19, 27	30, 18, 35, 4	35		1		1, 31	2, 17, 26		35, 37, 10, 2
8, 15, 35, 38		19, 35, 38, 2	14, 20, 19, 35	10, 13, 28, 38	13, 26		18, 19, 29, 38	11, 35, 27, 28	28, 32, 1, 24	10, 28, 32, 25	1, 28, 35, 23	2, 24, 35, 21	35, 13, 8, 1	32, 28, 13, 12	34, 2, 28, 27	15, 10, 26	10, 28, 4, 34	3, 34, 27, 16	10, 18	
19, 17, 10	1, 16, 36, 37	19, 35, 18, 37	14, 15	8, 35, 40, 5		10, 37, 36	14, 29, 18, 36	3, 35, 13, 21	10, 23, 24	29, 37, 36	1, 35, 40, 18	13, 3, 36, 24	15, 37, 18, 1	1, 28, 3, 25	15, 1, 11	15, 17, 18, 20	6, 35, 10, 18	36, 37, 10, 19	2, 35	3, 28, 35, 37
14, 24, 10, 37		10, 35, 14	2, 36, 25	10, 36, 3, 37		37, 36, 4	10, 14, 36	10, 13, 19, 35	6, 28, 25	3, 35	22, 2, 37	2, 33, 27, 18	1, 35, 16	11	2	35	19, 1, 35	2, 36, 37	35, 24	10, 14, 35, 37
2, 6, 34, 14		4, 6, 2	2, 14	35, 29, 3, 5		14, 10, 34, 17	36, 22	10, 40, 16	28, 32, 1	32, 30, 40	22, 1, 2, 35	35, 1	1, 32, 17, 28	32, 15, 26	2, 13, 1	1, 15, 29	16, 29, 1, 28	15, 13, 39	15, 1, 32	17, 26, 34, 10
13, 19	27, 4, 29, 18	32, 35, 27, 31	14, 2, 39, 6	2, 14, 30, 40		35, 27	15, 32, 35		13	18	35, 24, 30, 18	35, 40, 27, 39	35, 19	32, 35, 30	2, 35, 10, 16	35, 30, 34, 2	2, 35, 22, 26	35, 22, 39, 23	1, 8, 35	23, 35, 40, 3
19, 35, 10	35	10, 26, 35, 28	35	35, 28, 31, 40		29, 3, 28, 10	29, 10, 27	11, 3	3, 27, 16	3, 27	18, 35, 37, 1	15, 35, 22, 2	11, 3, 10, 32	32, 40, 28, 2	27, 11, 3	15, 3, 32	2, 13, 28	27, 3, 15, 40	15	29, 35, 10, 14
28, 6, 35, 18		19, 10, 35, 38		28, 27, 3, 18	10	20, 10, 28, 18	3, 35, 10, 40	11, 2, 13	3	3, 27, 16, 40	22, 15, 33, 28	21, 39, 16, 22	27, 1, 4	12, 27	29, 10, 27	1, 35, 13	10, 4, 29, 15	19, 29, 39, 35	6, 10	35, 17, 14, 19
		16		27, 16, 18, 38	10	28, 20, 10, 16	3, 35, 31	34, 27, 6, 40	10, 26, 24		17, 1, 40, 33	22	35, 10	1	1	2	25, 34, 6, 35	1		10, 20, 16, 38
19, 15, 3, 17		2, 14, 17, 25	21, 17, 35, 38	21, 36, 29, 31		35, 28, 21, 18	3, 17, 30, 39	19, 35, 3, 10	32, 19, 24		22, 33, 35, 2	22, 35, 2, 24	26, 27	26, 27	4, 10, 16	2, 18, 27	2, 17, 16	3, 27, 35, 31	26, 2, 19, 16	15, 28, 35
32, 1, 19	32, 35, 1, 15	32	19, 16, 1, 6	13, 1	1, 6	19, 1, 26, 17	1, 19		11, 15, 32	3, 32	15, 19	35, 19, 32, 39	19, 35, 28, 26	28, 26, 19	15, 17, 13, 16	15, 1, 19	6, 32, 13	32, 15	2, 26, 10	2, 25, 16
		6, 19, 37, 18	12, 22, 15, 24	35, 24, 18, 5		35, 38, 19, 18	34, 23, 16, 18	19, 21, 11, 27	3, 1, 32		1, 35, 6, 27	2, 35, 6	28, 26, 30	19, 35	1, 15, 17, 28	15, 17, 13, 16	2, 29, 27, 28	35, 38	32, 2	12, 28, 35
				28, 27, 18, 31			3, 35, 31	10, 36, 23			10, 2, 22, 37	19, 22, 18	1, 4				19, 35, 16, 25			1, 6
16, 6, 19, 37			10, 35, 38	28, 27, 18, 38	10, 19	35, 20, 10, 6	4, 34, 19	19, 24, 26, 31	32, 15, 2	32, 2	19, 22, 31, 2	2, 35, 18	26, 10, 34	26, 35, 10	35, 2, 10, 34	19, 17, 34	20, 19, 30, 34	19, 35, 16	28, 2, 17	28, 35, 34
		3, 38		35, 27, 2, 37	19, 10	10, 18, 32, 7	7, 18, 25	11, 10, 35	32		21, 22, 35, 2	21, 35, 2, 22		35, 22, 1	2, 19		7, 23	35, 3, 15, 23	2	28, 10, 29, 35
35, 18, 24, 5	28, 27, 12, 31	28, 27, 18, 38	35, 27, 2, 31			15, 18, 35, 10	6, 3, 10, 24	10, 29, 39, 35	16, 34, 31, 28	35, 10, 24, 31	33, 22, 30, 40	10, 1, 34, 29	15, 34, 33	32, 28, 2, 24	2, 35, 34, 27	15, 10, 2	35, 10, 28, 24	35, 18, 10, 13	35, 10, 18	28, 35, 10, 23
		10, 19	19, 10			24, 26, 28, 32	24, 28, 35	10, 28, 23			22, 10, 1	10, 21, 22	32	27, 22				35, 33	35	13, 23, 15
35, 38, 19, 18	1	35, 20, 10, 6	10, 5, 18, 32	35, 18, 10, 39	24, 26, 28, 32		35, 38, 18, 16	10, 30, 4	24, 34, 28, 32	24, 26, 28, 18	35, 18, 34	35, 22, 18, 39	35, 28, 34, 4	4, 28, 10, 34	32, 1, 10	35, 28	6, 29	18, 28, 32, 10	24, 28, 35, 30	
34, 29, 16, 18	3, 35, 31	35	7, 18, 25	6, 3, 10, 24	24, 28, 35	35, 38, 18, 16		18, 3, 28, 40	13, 2, 28	33, 30	35, 33, 29, 31	3, 35, 40, 39	29, 1, 35, 27	35, 29, 25, 10	2, 32, 10, 25	15, 3, 29	3, 13, 27, 10	3, 27, 29, 18	8, 35	13, 29, 3, 27
21, 11, 27, 19		21, 11, 26, 31	10, 11, 35	10, 35, 29, 39	10, 28	10, 30, 4	21, 28, 40, 3		32, 3, 11, 23	11, 32, 1	27, 35, 2, 40	35, 2, 40, 26		27, 17, 40	1, 11	13, 35, 8, 24	13, 35, 1	27, 40, 28	11, 13, 27	1, 35, 29, 38
3, 6, 32		3, 6, 32	26, 32, 27	10, 16, 31, 28		24, 34, 28, 32	2, 6, 32	5, 11, 1, 23			28, 24, 22, 26	3, 33, 39, 10	6, 35, 25, 18	1, 13, 17, 34	1, 32, 13, 11	13, 35, 2	27, 35, 10, 34	26, 24, 32, 28	28, 2, 10, 34	10, 34, 28, 32
32, 2		32, 2	13, 32, 2	35, 31, 10, 24		32, 26, 28, 18	32, 30	11, 32, 1			26, 28, 10, 36	4, 17, 34, 26		1, 32, 35, 23	25, 10		26, 2, 18		26, 28, 18, 23	10, 18, 32, 39
1, 24, 6, 27	10, 2, 22, 37	19, 22, 31, 2	21, 22, 35, 2	33, 22, 19, 40	22, 10, 2	35, 18, 34	35, 33, 29, 31	27, 24, 2, 40	28, 33, 23, 26	26, 28, 10, 18			24, 35, 2	2, 25, 28, 39	35, 10, 2	35, 11, 22, 31	22, 19, 29, 40	22, 19, 29, 40	33, 3, 34	22, 35, 13, 24
2, 35, 6	19, 22, 18	2, 35, 18	21, 35, 2, 22	10, 1, 34	10, 21, 29	1, 22	3, 24, 39, 1	24, 2, 40, 39	3, 33, 26	4, 17, 34, 26							19, 1, 31	2, 21, 27, 1	2	22, 35, 18, 39
28, 26, 27, 1	1, 4	27, 1, 12, 24	19, 35	15, 34, 33	32, 24, 18, 16	35, 28, 34, 4	35, 23, 1, 24		1, 35, 12, 18		24, 2			2, 5, 13, 16	35, 1, 11, 9	2, 13, 15	27, 26, 1	6, 28, 11, 1	8, 28, 1	35, 1, 10, 28
1, 13, 24		35, 34, 2, 10	2, 19, 13	28, 32, 2, 24	4, 10, 27, 22	4, 28, 10, 34	12, 35	17, 27, 8, 40	25, 13, 2, 34	1, 32, 35, 23	2, 25, 28, 39		2, 5, 12		12, 26, 1, 32	15, 34, 1, 16	32, 26, 12, 17		1, 34, 12, 3	15, 1, 28
15, 1, 28, 16		15, 10, 32, 2	15, 1, 32, 19	2, 35, 34, 27		32, 1, 10, 25	2, 28, 10, 25	11, 10, 1, 16	10, 2, 13	25, 10	35, 10, 2, 16		1, 35, 11, 10	1, 12, 26, 15		7, 1, 4, 16	35, 1, 13, 11		34, 35, 7, 13	1, 32, 10
19, 35, 29, 13		19, 1, 29	18, 15, 1	15, 10, 2, 13		35, 28	3, 35, 15	35, 13, 8, 24	35, 5, 1, 10		35, 11, 32, 31		1, 13, 31	15, 34, 1, 16	1, 16, 7, 4		15, 29, 37, 28	1	27, 34, 35	35, 28, 6, 37
27, 2, 29, 28		20, 19, 30, 34	10, 35, 13, 2	35, 10, 28, 29		6, 29	13, 3, 27, 10	13, 35, 1	2, 26, 10, 34	26, 24, 32	22, 19, 29, 40	19, 1	27, 26, 1, 13	27, 9, 26, 24	1, 13	29, 15, 28, 37		15, 10, 37, 28	15, 1, 24	12, 17, 28
35, 38	19, 35, 16	19, 1, 16, 10	35, 3, 15, 19	1, 18, 10, 24	35, 33, 27, 22	18, 28, 32, 9	3, 27, 29, 18	27, 40, 28, 8	26, 24, 32, 28		22, 19, 29, 28	2, 21	5, 28, 11, 29	2, 5	12, 26	1, 15	15, 10, 37, 28		34, 21	35, 18
2, 32, 13		28, 2, 27	23, 28	35, 10, 18, 5	35, 33	24, 28, 35, 30	35, 13	11, 27, 32	28, 26, 10, 34	28, 26, 18, 23	2, 33	2	1, 26, 13	1, 12, 34, 3	1, 35, 13	27, 4, 1, 35	15, 24, 10	34, 27, 25		5, 12, 35, 26
35, 10, 38, 19	1	35, 20, 10	28, 10, 29, 35	28, 10, 35, 23	13, 15, 23		35, 38	1, 35, 10, 38	1, 10, 34, 28	18, 10, 32, 1	22, 35, 13, 24	35, 22, 18, 39	35, 28, 2, 24	1, 28, 7, 19	1, 32, 10, 25	1, 35, 28, 37	12, 17, 28, 24	35, 18, 27, 2	5, 12, 35, 26	

Appendix 4: Glossary

ABET: U.S.-based engineering program accrediting agency.

Accelerated testing: Testing at higher than normal stress levels to increase the failure rate and shorten the time to wear out.

Acceptable quality level: The maximum percent defective that, for the purpose of sampling inspection, can be considered satisfactory for a process average.

Acceptance: Sign-off by the purchaser.

Active redundancy: That redundancy wherein all redundant items are operating simultaneously.

Ambient: Used to denote surrounding, encompassing, or local conditions and is usually applied to environments.

Archiving: The process of establishing and maintaining copies of controlled items such that previous items, baselines, and configurations can be reestablished should there be a loss or corruption.

Assessment: Review and auditing of an organization's quality management system to determine that it meets the requirements of the standards, that it is implemented, and that it is effective.

Auditee: An organization to be audited.

Auditor: A person who has the qualifications to perform quality audits.

Axiomatic design: A vector-based approach to improved system design.

Baseline: A definition of configuration status declared at a point in the project life cycle.

Burn-in: The operation of items before their end application to stabilize their characteristics and identify early failures.

Calibration: The comparison of a measurement system or device of unverified accuracy to a measurement system or device of known and greater accuracy, to detect and correct any variation from required performance specifications of the measurement system or device.

Certification: The process which seeks to confirm that the appropriate minimum best practice requirements are included and that the quality management system is put into effect.

Certification body: An organization which sets itself up as a supplier of product or process certification against established specifications or standards.

Change notice: A document approved by the design activity that describes and authorizes the implementation of an engineering change to the product and its approved configuration documentation.

Checksum: The sum of every byte contained in an input/output record used for assuring the integrity of the programed entry.

Checklist: An aid for the auditor listing areas and topics to be covered by the auditors.

Client: A person or organization requesting an audit.

Code of Federal Regulations **(CFR):**   Federal statutes denoted *CFR*, Title 21 of this material relates the legal code pertaining to the FDA.

Compliance audit: An audit where the auditor must investigate the quality system, as put into practice, and the organization's results.

Concept map: A way of describing the elements of a process, using a series of concepts interconnected via labeled propositions (directed arrows) that describe the interrelationship of the concepts.

Conditioning: The exposure of sample units or specimens to a specific environment for a specified period of time to prepare them for subsequent inspection.

Confidence: The probability that may be attached to conclusions reached as a result of application of statistical techniques.

Confidence interval: The numerical range within which an unknown is estimated to be.

Confidence level: The probability that a given statement is correct.

Confidence limits: The extremes of a confidence interval within which the unknown has a designated probability of being included.

Configuration: A collection of items at specified versions for the fulfillment of a particular purpose.

Controlled document: Documents with a defined distribution such that all registered holders of control documents systematically receive any updates to those documents.

Corrective action: All action taken to improve the overall quality management system as a result of identifying deficiencies, inefficiencies, and noncompliances.

Creep: Continuous increase in deformation under constant or decreasing stress.

Critical item: An item within a configuration item which, because of special engineering or logistic considerations, requires an approved specification to establish technical or inventory control.

Cycle: An on/off application of power.

Debugging: A process to detect and remedy inadequacies.

Defect: Any nonconformance of a characteristic with specified requirements.

Degradation: A gradual deterioration in performance.

Delivery: Transfer of a product from the supplier to the purchaser.

Derating: Using an item in such a way that applied stresses are below rated values.

Design entity: An element of a design that is structurally and functionally distinct from other elements and that is separately named and referenced.

Design review: A formal, documented, comprehensive, and systematic examination of a design to evaluate the design requirements and the capability of the design to meet these requirements and to identify problems and propose solutions.

Design view: A subset of design entity attribute information that is specifically suited to the needs of a software project activity.

Deviation: A specific written authorization, granted before manufacture of an item, to depart from a particular requirement(s) of an item's current approved configuration documentation for a specific number of units or a specified period of time.

Device: Any functional system.

Discrete variable: A variable which can take only a finite number of values.

Document: Contain information which is subject to change.

Downtime: The total time during which the system is not in condition to perform its intended function.

Early failure period: An interval immediately following final assembly, during which the failure rate of certain items is relatively high.

Entity attribute: A named characteristic or property of a design entity that provides a statement of fact about the entity.

Environment: The aggregate of all conditions, which externally influence the performance of an item.

External audit: An audit performed by a customer or his representative at the facility of the supplier to assess the degree of compliance of the quality system with documented requirements.

Extrinsic audit: An audit carried out in a company by a third party organization or a regulatory authority, to assess its activities against specific requirements.

Fail-safe: The stated condition that the equipment will contain self-checking features which will cause a function to cease in case of failure, malfunction, or drifting out of tolerance.

Failure: The state of inability of an item to perform its required function.

Failure analysis: Subsequent to a failure, the logical, systematic examination of any item, its construction, application, and documentation to identify the failure mode and determine the failure mechanism.

Failure mode: The consequence of the mechanism through which the failure occurs.

Failure rate: The probability of failure per unit of time of the items still operating.

Fatigue: A weakening or deterioration of metal or other material, or of a member, occurring under load, specifically under repeated, cyclic, or continuous loading.

Fault: The immediate cause of a failure.

Fault isolation: The process of determining the location of a fault to the extent necessary to effect repair.

Feasibility study: The study of a proposed item or technique to determine the degree to which it is practicable, advisable, and adaptable for the intended purpose.

Firmware: The combination of a hardware device and computer instructions or computer data that reside as read-only software on the hardware device.

Form: The shape, size, dimensions, mass, weight, and other visual parameters which uniquely characterize an item.

Grade: An indicator, category, or rank relating to features or characteristics that cover different sets of needs for products or services intended for the same functional use.

Inherent failure: A failure basically caused by a physical condition or phenomenon internal to the failed item.

Inherent reliability: Reliability potential present in the design.

Inspection: The examination and testing of supplies and services to determine whether they conform to specified requirements.

Installation: Introduction of the product to the purchaser's organization.

Internal audit: An audit carried out within an organization by its own personnel to assess compliance of the quality system to documented requirements.

Item: Any entity whose development is to be tracked.

Maintainability: The measure of the ability of an item to be retained in or restored to a specified condition when maintenance is performed by personnel having specified skill levels, using prescribed procedures and resources, at each prescribed level of maintenance and repair.

Maintenance: The servicing, repair, and care of material or equipment to sustain or restore acceptable operating conditions.

Major noncompliance: Either the nonimplementation, within the quality system of a requirement of ISO 9001, or a breakdown of a key aspect of the system.

Minor noncompliance: A single and occasional instance of a failure to comply with the quality system.

Method: A prescribed way of doing things.

Metric: A value obtained by theoretical or empirical means to determine the norm for a particular operation.

Malfunction: Any occurrence of unsatisfactory performance.

Manufacturability: The measure of the design's ability to consistently satisfy product goals, while being profitable.

Manufacturers and users device experience: MAUDE database maintained by the FDA.

Mean time between failure: A basic measure of reliability for repairable items.

Mean time to failure: A basic measure of maintainability.

Mean time to repair: The sum of repair times divided by the total number of failures, during a particular interval of time, under stated conditions.

Minimum life: The time of occurrence of the first failure of a device.

Module: A replaceable combination of assemblies, subassemblies, and parts common to one mounting.

Noncompliance: Nonfulfillment of specified requirements.

Objective evidence: Qualitative or quantitative information, records, or statements of fact pertaining to the quality of an item or service or to the existence and the implementation of a quality system element, which is based on observation, measurement, or test, and which can be verified.

Observation: A record of an observed fact which may or may not be regarded as a noncompliance.

Parameter: A quantity to which the operator may assign arbitrary values, as distinguished from a variable, which can assume only those values that the form of the function makes possible.

Pareto chart: Generally a histogram of labeled problems, arranged in descending order. Occasionally a cumulative total chart is overlaid.

Parsing: The technique of marking system or subsystem requirements with specified attributes to sort the requirements according to one or more of the attributes.

Performance standards: Published instructions and requirements setting forth the procedures, methods, and techniques for measuring the designed performance of equipments or systems in terms of the main number of essential technical measurements required for a specified operational capacity.

Phase: A defined segment of work.

Population: The total collection of units being considered.

Precision: The degree to which repeated observations of a class of measurements conform to themselves.

Predicted: That which is expected at some future time, postulated on analysis of past experience and tests.

Preventive maintenance: All actions performed in an attempt to retain an item in specified condition by providing systematic inspection, detection, and prevention of incipient failures.

Probability: A measure of the likelihood of any particular event occurring.

Probability distribution: A mathematical model that represents the probabilities for all of the possible values a given discrete random variable may take.

Procedures: Documents that explain the responsibilities and authorities related to particular tasks, indicate the methods and tools to be used, and may include copies of, or reference to, software facilities or paper forms.

Product: Operating system or application software including associated documentation, specifications, user guides, etc.

Program: The program of events during an audit.

Prototype: A model suitable for use in complete evaluation of form, design, and performance.

Purchaser: The recipient of products or services delivered by the supplier.

Qualification: The entire process by which products are obtained from manufacturers or distributors, examined and tested, and then identified on a qualified products list.

Quality: The totality of features or characteristics of a product or service that bear on its ability to satisfy stated or implied needs.

Quality assurance: All those planned and systematic actions necessary to provide adequate confidence that a product or service will satisfy given requirements for quality.

Quality audit: A systematic and independent examination to determine whether quality activities and related results comply with planned arrangements and whether these arrangements are implemented effectively and are suitable to achieve objectives.

Quality control: The operational techniques and activities that are used to fulfill requirements for quality.

Quality function deployment: A customer-oriented graphical methodology for the determination of best approaches for product function and deployment planning.

Quality management: That aspect of the overall management function that determines and implements quality policy.
   A technique covering quality assurance and quality control aimed at ensuring defect-free products.

Quality policy: The overall intention and direction of an organization regarding quality as formally expressed by top management.

Management's declared targets and approach to the achievement of quality.

Quality system: The organizational structure, responsibilities, procedures, processes, and resources for implementing quality management.

Record: Provides objective evidence that the quality system has been effectively implemented.

A piece of evidence that is not subject to change.

Redundancy: Duplication or the use of more than one means of performing a function to prevent an overall failure in the event that all but one of the means fails.

Regression analysis: The fitting of a curve or equation to data to define the functional relationship between two or more correlated variables.

Reliability: The probability that a device will perform a required function, under specified conditions, for a specified period of time.

Reliability goal: The desired reliability for the device.

Reliability growth: The improvement a reliability parameter caused by the successful correction of deficiencies in item design or manufacture.

Repair: All actions performed as a result of failure, to restore an item to a specified condition.

Review: An evaluation of software elements or project status to ascertain discrepancies from planned results and to recommend improvement.

Review meeting: A meeting at which a work product or a set of work products are presented to project personnel, managers, users, customers, or other interested parties for comment or approval.

Revision: Any change to an original document which requires the revision level to be advanced.

Risk: The probability of making an incorrect decision.

Robust design: A design technique, originated by Taguchi, that seeks to improve processes by improving the fundamental operations of the device/process.

Safety factor: The margin of safety designed into the application of an item to insure that it will function properly.

Schedule: The dates on which the audit is planned to happen.

Screening: A process of inspecting items to remove those that are unsatisfactory or likely to exhibit early failure.

Service level agreement: Defines the service to be provided and the parameters within which the service provider is contracted to service.

Shelf life: The length of time an item can be stored under specified conditions and still meet specified requirements.

Simulation: A set of test conditions designed to duplicate field operating and usage environments as closely as possible.

Single point failure: The failure of an item which would result in failure of the system and is not compensated for by redundancy or alternative operational procedures.

Six sigma: A specialized business process improvement process.

Skunk works: A term originating in the defense industry, currently meant to indicate a self-sufficient design and development group.

Software: A combination of associated computer instructions and computer data definitions required to enable the computer hardware to perform computational or control functions.

Software design description: A representation of a software system created to facilitate analysis, planning, implementation, and decision-making.
  A blueprint or model of the software system.

Source code: The code in which a software program is prepared.

Specification: A document which describes the essential technical requirements for items, material, or services.

Standards: Documents that state very specific requirements in terms of appearance, formal and exact methods to be followed in all relevant cases.

Standard deviation: A statistical measure of dispersion in a distribution.

Standby redundancy: The redundancy wherein the alternative means of performing the function is not operating until it is activated upon failure of the primary means of performing the function.

Subcontractor: The organization which provides products or services to the supplier.

Supplier: The organization responsible for replication and issue of product.
The organization to which the requirements of the relevant parts of an ISO 9000 standard apply.

System: A group of equipments, including any required operator functions, which are integrated to perform a related operation.

System compatibility: The ability of the equipments within a system to work together to perform the intended mission of the system.

Testing: The process of executing hardware or software to find errors.
  A procedure or action taken to determine—under real or simulated conditions—the capabilities, limitations, characteristics, effectiveness, and reliability and suitability of a material, device, or method.

Tolerance: The total permissible deviation of a measurement from a designated value.

Tool: The mechanization of the method or procedure.

Total quality: A business philosophy involving everyone for continuously improving an organization's performance.

Traceability: The ability to track requirements from the original specification to code and test.

Trade-off: The lessening of some desirable factor(s) in exchange for an increase in one or more other factors to maximize a system's effectiveness.

Useful life period: The period of equipment life following the infant mortality period, during which the equipment failure rate remains constant.

Validation: The process of evaluating a product to ensure compliance with specified and implied requirements.

Variable: A quantity that may assume a number of values.

Variance: A statistical measure of the dispersion in a distribution.

Variant: An instance of an item created to satisfy a particular requirement.

Verification: The process of evaluating the products of a given phase to ensure correctness and consistency with respect to the products and standards provided as input to that phase.

Version: An instance of an item or variant created at a particular time.

Wearout: The process which results in an increase in the failure rate or probability of failure with increasing number of life units.

Wearout failure period: The period of equipment life following the normal failure period, during which the equipment failure rate increases above the normal rate.

Work instructions: Documents that describe how to perform specific tasks and are generally required only for complex tasks, which cannot be adequately described by a single sentence or paragraph with a procedure.

Worst-case analysis: A type of circuit analysis that determines the worst possible effect on the output parameters by changes in the values of circuit elements. The circuit elements are set at the values within their anticipated ranges, which produce the maximum detrimental output changes.

Index

A

Accident reconstruction and forensics
    biomechanics and traffic accident investigations,
        323–324
    medical device accident investigation (*see* Medical
        device(s), accident investigation)
Accreditation Board for Engineering and Technology
        (ABET), 183, 330, 377
Active Implantable Medical Devices Directive (AIMDD),
        245, 246, 255
Acute systemic toxicity, 179
Agency for Health care Research and Quality
        (AHRQ), 188
Alfred Mann Foundation, 357
Altshuller, Genrich, 20–21
American Institute of Chemical Engineers, 328
American Medical Informatics Association (AMIA),
        327–328
American National Standards Institute (ANSI),
        261, 329
American Society for Quality Control (ASQC),
        261, 329
American Society for Testing and Materials (ASTM),
        261, 329
American Society of Civil Engineers, 328
Americans with Disabilities Act (ADA), 303
Anesthetists testing, 337–339
Anthropometry, 144
Apnea detection system, 339–340
Apparatus claim, 279
Association for Computing Machinery (ACM), 327
Association for the Advancement of Medical
        Instrumentation (AAMI), 260, 292,
        327, 329
Autoclaving, material selection for, 345–347
Axiomatic design, 377
    goal of, 96
    process mapping, 97–98

B

Bhopal chemical plant disaster, 301
Bioelectric phenomena study, 351
Biological control tests, 177
Biomaterials uses, 169
Biomechanics, 222
Biomedical devices, standards for, 328–329
Biomedical engineering, 351–352
Biomedical Engineering Society (BMES), 328
Biomedical engineers, 1
    medical device accident investigation, 220, 221
    traffic accident investigation, 222
Biomedical imaging modality, 351
Biomedical processes, design of, 2–3
Biomedical sensor, 351

Biomedical system design, 2–3
Black box testing, 197
Blood oxygenator malfunction, 322
Brainstorming
    concept map for, 12
    design team, 11
Breach of warranty, 314–315; *see also* Liability,
        of manufacturer
British Standards Institute (BSI), 263, 264
Broadest claims, of patents, 278–279
Business proposal
    market need and market potential, 57
    product idea, 58
    project objectives, 56–57
    project planning, 59–60
    risk analysis and research plan, 58–59

C

Canadian Standards Association, 265
Cancer clinic charting, 340–341
Carcinogenicity studies, 180–181
CASE tools, 127
Cell engineering, 352
CE marking, 259–260
Center for Devices and Radiological Health
        (CDRH), 292
Chi square table, 359
Chronic toxicity, 180
CISPR, *see* International Special Committee on Radio
        Interference
Claims, for patent, *see* Patent, claims for
Class I devices, 227
Class II devices, 227
Class III devices, 227–228
Cleaning material, selection criteria for, 347–348
Clean room model, 115–116; *see also* Software
        development
Clinic flowchart
    blood pressure (hypertension)
        with decision point, 16
        linear, 14–15
Code of Federal Regulations (CFR), 55
Coding, 126–127
Collimator, 336
    depth response, 335
Comité Européen de Normalisation (CEN), 264
Comité Européen de Normalisation Electronique
        (CENELAC), 264
Component derating, 103–104
Computer-aided design (CAD), 354
Computer-aided software engineering tools,
        see CASE tools
Computer languages, *see* Programming languages
Concurrent design, 140
Conduction block, 351